UNDERSTANDING ENDOCRINOLOGY

UNDERSTANDING ENDOCRINOLOGY

By

Ashok Kumar

Dept. of Zoology
Bundelkhand University
Campus Department
Jhansi

DISCOVERY PUBLISHING HOUSE PVT. LTD.
NEW DELHI-110 002

First Published - 2010
Reprinted - 2016

ISBN: 978-81-8356-511-0

Understanding Endocrinology

Published by:
DISCOVERY PUBLISHING HOUSE PVT. LTD.
4383/4B, Ansari Road, Darya Ganj
New Delhi-110 002 (India)
Phone: +91-11-23279245, 43596064-65
Fax: +91-11-23253475
E-mail: discoverypublishinghouse@gmail.com
sales@discoverypublishinggroup.com
web: www.discoverypublishinggroup.com

Printed at:
Infinity Imaging Systems
Delhi

Preface

The present title "Understanding Endocrinology" has been written for those students interested in careers in diverse fields of biological sciences. It provides a structured approach to learning by covering all the important topics in a uniform, systematic format. The book has been comprehensively designed incorporating recent advances in this fast moving field. It also provides accessible information on endocrinology in compact form for undergraduate students in biology and related life sciences. It is intelligible to the educated layman, though it deals with some complex ideas. It is an adequate text for all the requirements of students in this area. In addition, busy lecturers who require a quick reference compendium will find it useful, particularly for tutional planning. Simple, yet hopefully clear figures and tables are provided throughout the book.

The over-riding goal of this book, and indeed of the whole *Understanding series*, is to present the essential information concering endocrinology in a compact, readily accessible form which leads itself to student learning and revision. The convergence of various approaches has generated a rich panorama of detail, the significance of which we are still attempting to unraval. The present text has been written as an introduction to this rapidly growing field.

To make the work more comprehensive and informative, the author has consulted many authoritative books, research journals, abstracts, monographs etc., so there can be no claim to originality except in the manner of treatment.

The author expresses his thanks to his friends and colleagues whose continue inspirations have initiated him to bring out this book.

The author expresses his gratitude to Mr. Wasan and staff of M/s Discovery Publishing House Pvt. Ltd. for their whole hearted co-operation in the publication of this book.

In the mean time, the author will remain sincerely responsible for any shortcomings of the book and be grateful to the readers for their suggestions and constructive criticism for the continuous betterment of the book. He takes this opportunity to appeal to the readers to send their suggestions straightaway to his Publisher.

Author

Contents

1

INTRODUCTION

The Phylum Chordata comprises the Subphylum Vertebrata and three other Subphyla which are collectively known as the Protochordata. The latter are of exceptional interest, because they combine certain features of vertebrate organization with other characteristics which provide something of a link with the Echinodermata, the group of invertebrates to which the vertebrates are now thought to be most closely related. Thus, we may reasonably expect that endocrinological investigations of the group will extend our understanding of the origin, organization and functioning of vertebrate hormonal systems.

The best-known protochordates are the Cephalochordata, represented by *Amphioxus*, and the Urochordata (Tunicata), including the sessile ascidians (sea-squirts) and various pelagic forms such as the salps. All of these are filter feeders, and they resemble each other in the possession of a large filtering pharynx, along the floor of which runs a groove or channel, the endostyle. This organ is provided with pairs of longitudinal glandular tracts, and it is generally held that it contributes to the mucus-like secretions with which the animals trap small particles of food material. An organ closely resembling it, and clearly homologous with it, is found in the ammocoete larva of the lamprey, the most primitive surviving vertebrate. It has long been known that at the metamorphosis of this larva a part of the endostyle is transformed into the thyroid gland, and it has been established by Gorbman and Creaser (1942), by Leloup and Berg (1954), and by Leloup (1955) that even in the larva the endostyle is trapping and binding iodine and is thereby forming monoiodotyrosine, diiodotyrosine, thyroxine, and probably triiodothyronine. We are thus led to enquire whether such

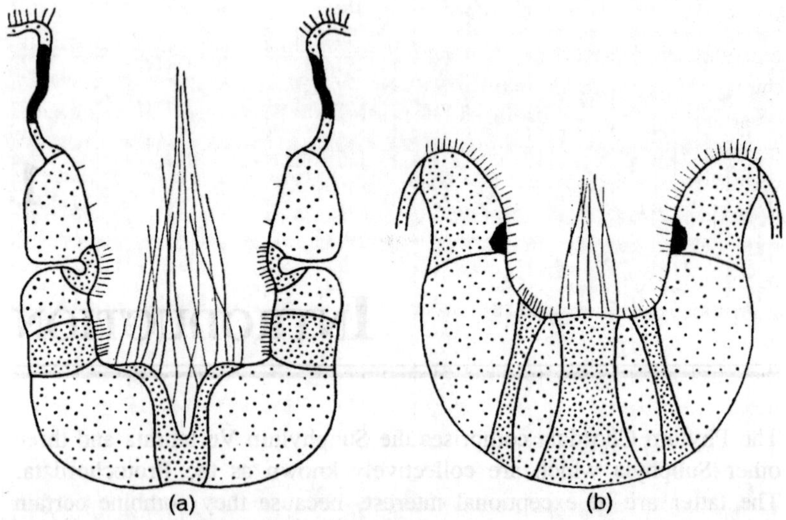

Fig. 1.1. Diagram of the endostyles: (a) Ciona; (b) Amphioxus.

thyroidal biosynthesis may not already be established in the endostyle of the protochordates; if it is, there is then opened up the possibility of learning something of the fundamental functions of the thyroidal hormones before they have been overlain by the extensive specializations of the vertebrates.

As regards *Amphioxus*, Thomas (1956) has recently shown that the endostyle of this animal does, in fact, contain organically bound iodine, demonstrable in autoradiographs, prepared from animals which have been immersed in sea water containing I^{131}. I have been investigating this situation further with a view to comparing his results with the observations which we at Nottingham were simultaneously and independently making on the Tunicata. He was inclined to ascribe some at least of the iodine binding in the endostyle to the more dorsal of the two pairs of glandular tracts. My autoradiographs, however, make it clear that the centre of iodination lies at, or near to, the surface of a group of cells lying immediately above these tracts; the latter do not themselves appear to be involved in iodination, and in this respect the situation resembles that already established for the endostyle of the ammocoete larva.

I have further been able to show that in *Amphioxus* the cells concerned produce a clearly defined secretion; this may reasonably be regarded as providing the molecular basis for the binding process and it is thus of obvious interest to compare it with thyroglobulin. Like the latter, it is strongly PAS-positive, but certain other tests reveal considerable differences. The significance of these is not easy to assess, owing to the lack of precise information as to the chemical composition of so-called muco-substances. It can be said, however, that epithelial mucins which are characteristic of alimentary tracts, and which we should reasonably expect to find in the pharynx of *Amphioxus*, have as a major component the carbohydrate polymers known as acid mucopolysaccharides or, as Kent and Whitehouse (1955) would prefer, as amino-polysaccharides, these being carbohydrates which contain the 2-aminosugars glucosamine and galactosamine. Secretions of this type react positively with mucicarmine and with alcian blue, and show the characteristic red colour of gamma metachromasia with toluidine blue.

The thyroglobulin of thyroid colloid, being a glycoprotein, is negative to these three tests, but the secretion of the iodine-binding cells of the *Amphioxus* endostyle is positive. This may well mean that in the latter animal we are witnessing an early stage in the evolution of thyroidal biosynthesis, one in which the process is associated with the presence of an alimentary muco-substance. I am not, of course, suggesting that the iodine is bound to a carbohydrate polymer. The position seems to be that aminopolysaccharides are often, and perhaps always, associated loosely with some protein; it is presumably to the latter that binding takes place, and the possibility that it is itself a thyroglobulin-like substance is not excluded by these tests. The point of interest is that iodine binding seems here to be evolving out of an essentially alimentary secretory process.

A further important conclusion which can be drawn from *Amphioxus* results from the clearly observable fact that cells other than those associated with iodine binding react positively to the tests for epithelial mucins, notably a group of cells lying at the base of the groove. No bound iodine has been found to be associated with these, however, and we can thus conclude that the binding is not a random result of any generalized property of mucins, but is a specialized property of one particular group of cells. From this it would seem to follow that the binding is a biochemically purposive act, giving rise to a secretion which is biochemically significant for the animal. This secretion, there is reason to believe, is mixed with the food material in the pharyngeal lumen, and can be absorbed through the wall of the alimentary canal.

In my published analysis of the binding of iodine in the endostyle of the ascidian *Ciona* it was suggested that the binding was actually taking place in an iodination centre in a clearly defined area of nonciliated epithelium which lies immediately above the three pairs of glandular tracts. The presence of iodine on the ciliated lip of the endostyle, and elsewhere on the pharyngeal epithelium, was explained as being the result of its adhesion to the ciliated cells which move the food-trapping mucus secretion over the pharyngeal wall, the bound iodine being incorporated into this secretion on its discharge from the endostyle. On this interpretation, the organization of the endostyle of the ascidians is fundamentally similar to that of *Amphioxus*, and subsequent work, as yet unpublished, has entirely confirmed this view both for *Ciona* and for other ascidian genera.

Good examples of the latter are provided by *Dendrodoa* and *Botryllus*, which, as members of the order Pleurogona, have a more specialized organization than *Ciona* and are regarded by the systematist as widely separated from it. In *Dendrodoa* a region corresponding to zone 7 is clearly defined, although it is less extended than in *Ciona*; its cells contain PAS-positive material, and in autoradiographs bound iodine is associated with them as well as the ciliated lip of the endostyle. *Botryllus* is particularly striking, for in this well-known colonial form the individual zooids are minute in size, ranging up to only 4 mm in length, as compared with the 25 mm of *Dendrodoa* or the 12 cm or more of *Ciona*. Despite this, the same regions are distinguishable in the endostyle, and clear autograph images indicate

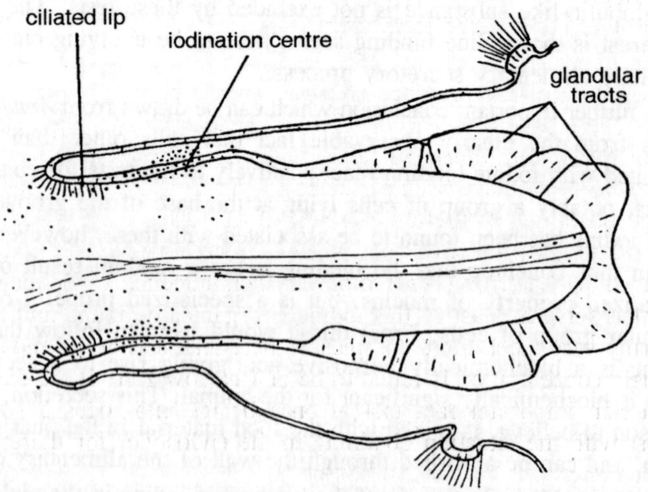

Fig. 1.2. Diagram of an autoradiograph of a transverse section of the endostyle of Botryllus.

an iodination centre in the upper part of zone 7, whereas elsewhere, on the ciliated lip and over the pharyngeal wall, the signs of bound iodine are irregular and may be quite inconspicuous.

At present the nature of the secretion of the iodine-binding cells is less easily definable in the ascidians than it is in *Amphioxus*, and individual genera may prove to vary in this respect. In *Ciona*, as I have pointed out elsewhere, it bears some resemblance to the secretion of the corresponding cells of the endostyle of the ammocoete larva. In *Dendrodoa* it is PAS-positive, but probably differs from the corresponding secretion in *Amphioxus* in being negative to alcian blue and to mucicarmine. In *Botryllus* PAS-positive droplets are sometimes distinguishable, but often there is no sign of secretion at all, probably, in part, because of the very small size of the cells and the difficulty of securing satisfactory fixation of them. This matter is still under investigation, but in the meantime it may be well to emphasize that the adult ascidians are probably not on the direct line of ascent to vertebrates, and are presumably much less closely related to the latter than is *Amphioxus*. Current theory derives the vertebrates, and also *Amphioxus* itself, by neoteny from the ascidian larval stage, although the further possibility that both *Amphioxus* and the *Tunicata* may have arisen independently from the third protochordate Subphylum, the Hemichordata, cannot be excluded. On either view it would seem that adult tunicates must have pursued their own independent specializations, which might in some cases he expected to parallel those of vertebrates and in others to diverge from them.

An important result of these studies on tunicates has been the demonstration of the presence of bound iodine in other parts of their bodies. There have been earlier reports of its occurrence in the stolonic septa of *Perophora annectens*, and in the canals connecting the zooids of *Botryllus*, but I have been particularly impressed by its accumulation in large quantities in the tunic or test of all forms I examined. This tunic is a tough protective coat which is secreted over the body and was shown by Cameron (1914) to be very rich in iodine. Autoradiographs show that some at least of this iodine is organically bound and that it is actually very much more conspicuous here than it is in the endostyle. The main concentration is found to be in a narrow zone at the extreme surface, an important fact that at once brings into focus a striking contrast with the situation obtaining in the endostyle, for there is no evidence that the binding of iodine in the tunic is related to any particular groups of cells. It is true that wandering cells derived from the mesenchyme are conspicuous throughout the tunic, but there is no constant accumulation of them in the surface layer, and there is quite certainly no epithelial layer developed there.

This matter also is still under investigation, but in the meantime certain properties of the tunic seem to merit attention. Its matrix is well known to contain a high proportion of cellulose, a condition unique in the animal kingdom. An early analysis of the composition of the tunic of *Ciona* showed it to be composed of 60.34 per cent cellulose, 27 per cent nitrogenous material, and 12.66 per cent inorganic material. According to recent work of Peres (1948a, b), the main mass of the tunic consists of cellulose associated with some glycoprotein, but the surface is formed of a distinct cuticle-like layer of pure protein. Here again there may well be variation from genus to genus; it appears, for example, that in *Clavelina* the cuticle is well-developed, particularly in the stalk region, whereas in *Ciona* it is much more inconspicuous. These facts certainly suggest, however, that we are dealing here with the phenomenon which has been shown to characterize a wide range of invertebrates, the association of iodine binding with the laying down of exoskeletal scleroproteins.

At this stage of the analysis it is clearly necessary to inquire as to the nature of the iodinated products which arise in the protochordates, for this has an obvious bearing on the question as to how far we are here dealing with an hormonal situation. As far as *Amphioxus* is concerned, the only positive evidence comes from the work of Sembrat (1953), who implanted 40 dried endostyles into one axolotl and 65 into another, and thereby produced metamorphic reduction of the fins and gills. Since control implants of muscle had at most only slight effects, this was regarded as giving some indication of the presence of thyroid hormone in the endostyle. My own material of *Amphioxus* has not been sufficiently plentiful to enable me as yet to study the problem in this animal, but I have applied radiochromatographic methods to *Ciona*, using Gleason's (1953) *tert*-amyl alcohol-2*N* ammonia solvent.

This work is not yet complete, but preliminary results have given satisfactory demonstrations of the presence of diiodotyrosine and thyroxine in extracts of the tunic, for using added carriers for identification, definite radioactive spots can be shown to coincide with spots of those particular substances, as revealed by the ceric sulphate-arsenious acid reaction. Triiodothyronine is probably also present, but it has not yet been possible to separate out diiodothyronine either from the extracts or from carrier mixtures. The activity of aqueous extracts of the tunic is weak, but treatment of the tissue with *N* NaOH at 50°C for 5 hours gives extracts of very much higher activity, a result which provides some support for the view, suggested above, that the bound iodine is associated with structural proteins. The radioactivity of aqueous extracts of the endostyle is very low, as is to be expected from the weak autoradiographs given by this organ, but the same radiochromatographic procedure has given good evidence for the presence in these extracts of diiodotyrosine, and it is highly probable that at least thyroxine is also present.

Perhaps at this point one might conveniently summarize the conclusions which seem to emerge from this review, always bearing in mind that these particular investigations are still in progress, and that the arguments should at this stage be regarded as primarily a reasoned directive for further research. As far as the endostyle is concerned, iodine binding is already established in it at the protochordate level of evolution, and it is established in a localized region of the organ in such a manner as to suggest that it is biochemically purposive and not a product of random iodine uptake. The cells concerned produce a secretion which, at least in *Amphioxus*, is iodinated at or near to the cell surface, as it apparently is in the cells of the thyroid gland of vertebrates. There is presumably little storage, and the iodinated secretion is discharged and becomes mixed with food cords, probably to be absorbed through the epithelium of the alimentary canal. It is highly probable that this iodine binding is indicative of an essentially thyroidal biosynthesis, and one may surmise that the replacement of the mucin-like secretion of *Amphioxus* by thyroglobulin may have provided a more efficient molecular basis for this purpose and for the storage of its products. It is interesting in this connection to note that Hooghwinkel et al. (1954) have interpreted thyroglobulin as a firm compound of mucopolysaccharide and protein, although as Gross (1957) has pointed out, their results seem to conform also to the more usual view that it is a glycoprotein. One may further surmise that this cytochemical evolution may well have been determined by the shortage of iodine which the chordates would have encountered when they migrated from marine to freshwater habitats, and this may also account for what seems to be a greater area of iodinating epithelium in the endostyle of the ammocoete larva as compared with that of the protochordates. Here, then, would seem to be an ecological adaptation, analogous to those which have been considered elsewhere in this symposium, but operating on a geological time scale.

As regards the tunic, the association of bound iodine with cuticular structures is, as I have already mentioned, by no means novel, for it has been recorded in annelids, arthropods, and molluscs. This association may well be a biological accident, with the degree of fixation and the yield of thyroxine being determined by the disposition of the tyrosine residues in scleroprotein molecules, and it has been plausibly argued that the thyroid gland might have evolved as a consequence of the ancestors of vertebrates becoming biochemically dependent on iodinated amino acids which were initially made available to them by such accidental means. It would be premature to develop this argument far in our present state of knowledge, but it is quite clear that conditions in the tunicates lend some support to it. Their tunic is often richly provided both with wandering mesenchyme cells and with blood vessels, so that it is by no means unlikely that some of its iodinated products might be released and transported through the body, and thus become available for utilization. Further, and perhaps more important, the tunic is not a passive tissue, but is constantly being secreted by the epidermis and lost from the surface, so that there would seem to be rather a good possibility of these ciliary feeding animals ingesting, and subsequently digesting, some of their own iodinated products, the more so in that they are often gregarious or colonial in habit. It is surely far from fantastic to suggest that in some such way tunicates might have become biochemically dependent upon these iodinated products at an early stage of their evolution; the iodinating properties of the endostyle might have arisen thereafter as an adaptation for the secretion of these substances in a form more readily available and in a position from which they could more easily be assimilated. At the moment, however, this can be no more than a suggestion, for it carries the implication that *Amphioxus* and the vertebrates evolved from tunicates, and we cannot be certain that this was, in fact, so. From this point of view it is therefore particularly important that we should learn more as to the significance of the association of bound iodine with the dermal glands of the Enteropneusta, for this rather puzzling group of Hemichordata appears to antedate the Tunicata and might have been ancestral to them.

I have deliberately devoted most of this review to a consideration of thyroidal biosynthesis, because it is the field in which I have been most directly concerned, and because it effectively illustrates the type of problem which the protochordates present to the endocrinologist. Other aspects, and particularly those related to reproduction, have been reviewed by Dodd (1955), who has drawn attention to the uncertainty of the evidence which has from time to time been held to establish the homology of the neural gland of the Tunicata with the pituitary of vertebrates. However much we may sympathize with the view that it is impossible to doubt this homology on morphological grounds, the fact is that we are still in no position to substantiate it on biochemical or physiological grounds, although we are equally unable to deny it. Dodd (1958) has now satisfied himself that no gonadotrophin, assayable by mouse or male toad methods, exists in the neural complex of breeding *Ciona*, nor is there any convincing evidence for the presence of vasopressin or of melanophore-expanding hormone. He certainly finds some support for the contention of earlier workers that an oxytocic substance is present, but the significance of this is quite obscure, for he has been able to show that its properties are very different from those of mammalian oxytocin, whereas a very similar substance can be extracted from various parts of the bodies of starfish and lugworms. Here, then, is a field in which recent work, in complete contrast to that on the endostyle, has failed to substantiate earlier views.

At the present time the clearest positive evidence as to the function of the neural gland of tunicates is that it has powers of phagocytosis and that it is capable of taking up finely divided particles which have entered the inhalent siphon in the incurrent stream of water. It seems well to emphasize once again in this connection that if vertebrates are derived from ascidians it is from their larvae and not from the adults, and that although the larvae possess an endostyle they certainly do not possess a neural gland, although the neotenous appendicularians are said to have a well-developed ciliated tubercle, this being the structure on which the duct of the gland opens in the adults of the other groups. It may, then, be more prudent to think of the neural gland not so much as a forerunner of the pituitary, but as an independent specialization of the Tunicata, although, of course, it might still be the expression of some genetic potentiality common to all the lower chordates.

Brambell and Cole (1939) have suggested that some such common potentiality for the production by invagination of an anterior pit-like organ may explain the development of the ciliated organ which they discovered at the base of the proboscis of the Enteropneusta. It might also account for the existence in *Amphioxus* of Hatschek's pit, and I should like to emphasize the potential importance of the latter, which we are now restudying at Nottingham, for there are excellent morphological and embryological grounds for homologizing it with the adenohypophysis, as Goodrich (1917) first demonstrated. Its function has often been said to be the secretion of mucus to aid in the trapping of food, but its cell structure is so elaborate that it is difficult to feel satisfied with such a simple interpretation. It is said to have no nerve supply, but some of its cells have long-drawn-out distal ends which project into the lumen of the pit, and one wonders whether they may not be sensitive to substances in the water current passing over them. Without wishing to indulge in too easy-going speculation, one cannot but wonder whether the evolutionary history of this organ may not have some analogy with the story of the endostyle outlined above. Just as a mucus-secreting pharyngeal organ, with powers of iodination, seems to have evolved into a glycoprotein-secreting endocrine gland, so perhaps might a mucus-secreting stomodaeal pit, sensitive to passing substances, have evolved into another glycoprotein-secreting gland, sensitive to materials reaching it in the blood. Here at least is an obvious field for further investigation.

Where, it may be asked, is one then to look for the homologue of the neurohypophysis? There is, in fact, in the floor of the cerebral vesicle of *Amphioxus* a group of peculiar cells referred to by earlier workers as an infundibular organ. Olsson and Wingstrand (1954) have shown that these are nerve cells containing Gomori-positive granules, although since they appear to secrete a Reissner's fibre they have something in common with the subcommissural organ of vertebrates, which is composed of ependymal cells secreting a Gomori-positive material. Clearly we need to know more as to the function of these cells, for Franz (1923) believed that they were light-sensitive. According to his interpretation, they respond to shadows cast upon them by the so-called "eye-spot" which is situated in front of them, in the anterior wall of the cerebral vesicle, and he believed them to be an essential element in the orientation of the animal to light and shade.

This particular problem has been reviewed in later publications by Olsson (1958), and further study will be needed before we can hope to resolve it or, indeed, any of the other problems which I have included in this review. I can only hope that my account of them will have served to demonstrate what I believe to be the great interest of the protochordates, and the possibilities which they offer us of enlarging our understanding of vertebral hormonal mechanisms, always provided that we approach these animals with due regard for their habitat and their specialized mode of life.

2

HORMONES OF ALIMENTARY CANAL

With the identification by Bayliss and Starling of the first blood-borne chemical messenger, secretin, produced by the duodenal mucosa, and the proposal for *gastrin*, from the antral stomach a few years later by Edkins, the existence of endocrine regulation of the mammalian digestive system was established and the discipline of endocrinology was born. Ironically the gastrointestinal (GI) hormones have been among the last to be chemically characterized, and it was only after isolation of the purified hormones in the 1960s that it became possible to confirm the diffusely distributed cellular types responsible for their secretion. Because of the lack of precision in observations, the complications of sorting out neutral and proposed endocrine factors, effects of known pharmacological agents and rapid proliferation of unsubstantiated factors, the entire area of GI endocrine research was easily over-shadowed when the nature of certain medically important hormones was discovered. Research with corticosteroids, reproductive hormones, thyroid hormones, insulin and epinephrine occupied the mainstream of endocrine research, whereas advances associated with GI factors were uncommon.

In recent years, there has been rapid expansion in GI peptide research, and there are now many peptides that have been isolated and chemically characterized. Yet we do not understand the physiological roles for many of these peptides. For others, the assignment of their physiological roles changes so frequently that it is bewildering. Even their names seem to be in a state of flux. Some GI peptides appear to function as classic hormones, others are paracrine secretions and still others may be neurotransmitters. Many GI peptides have been identified within the central nervous system and in certain endocrine glands. For simplicity, all of these regulatory substances are referred to here as GI peptides. Endocrine regulation of digestion must be considered an integral portion of the entire digestive process. Consequently it is necessary to review the entire digestive process to illustrate the regulatory actions of GI peptides. The following discussion emphasizes the human digestive system, including the mouth, pharynx, esophagus, stomach and small intestine. In addition to the roles played by these portions of the alimentary canal, digestion is aided by three essential exocrine glands: (1) the several pairs of salivary glands that secrete into the mouth, (2) the liver and (3) the exocrine pancreas; the last two secrete materials into the small intestine. The reader should keep in mind the many anatomical and functional differences that exist between the human digestive system and those of mammalian carnivores and herbivores with respect to the specific details of the following account.

Human Digestive System

The first detailed and systematic knowledge of human digestive processes came from the observations of William Beaumont on his patient Alexis St. Martin, a French Canadian who, while visiting Fort Mackinac, Michigan, was accidentally shot in the chest from close range by a shotgun; two ribs were fractured, the lungs lacerated and the stomach perforated. Although Beaumont assumed that St. Martin would not live the night, he miraculously survived. The wound in St. Martin's stomach, however, never healed completely, resulting in a permanent opening to the outside (a *gastric fistula*) through which Beaumont was able to observe the progression of gastric digestion under varying conditions over a period of years. It was these pioneering observations by Beaumont that stimulated much of the later interest in gastric physiology.

The accidental production of a gastric fistula in St. Martin provided the inspiration for a variety of surgical techniques, including production of gastric fistulas and gastric or intestinal pouches (isolated pouches no longer connected with the lumen of the gut). Transplantation of denervated pouches or pieces of digestive tract or pancreas to sites under the skin where revascularization can occur has enabled investigators to separate endocrine and nervous regulatory mechanisms. Finally the development of crossed circulatory systems between experimental animals was used to confirm the transfer of chemical factors (hormones) through the blood to target tissues. *In-vitro* studies

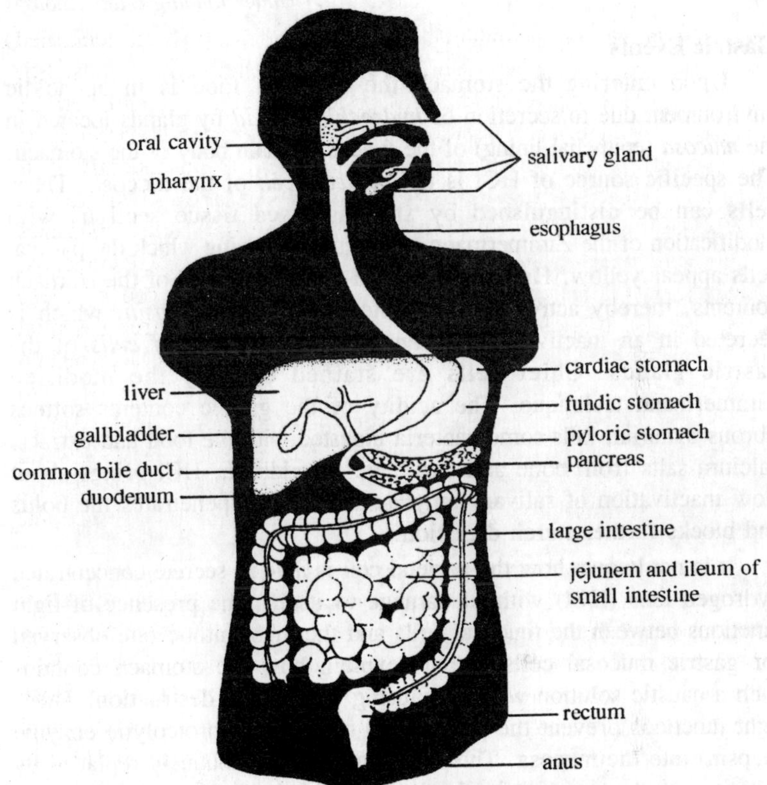

Fig. 2.1. The human digestive system.

of pancreatic slices or mucosal tissues from various regions of the gut have been employed profitably to ascertain details of the actions of the various GI peptides and factors regulating their release.

Oral Events

When food is ingested it is first torn and ground by the teeth and is mixed with *saliva* secreted from the exocrine salivary glands. Salivation is under direct neural control and can be stimulated by sight, smell or simply the thought of food or by the presence of food in the mouth. Saliva is a basic fluid containing the hydrolytic enzyme *salivary amylase*, which hydrolyses starches to disaccharides. The saliva and partially digested chewed food are mixed thoroughly to form a bolus that is pushed back into the pharynx by the tongue. After entry into the pharynx the bolus is swallowed by a reflexive series of events that propel it down the esophagus and into the stomach.

Gastric Events

Upon entering the stomach the bolus of food is in an acidic environment due to secretion of *hydrochloric acid* by glands located in the *mucosa* (epithelial lining) of the *fundus* or main body of the stomach. The specific source of HCl is the *parietal cell* of the mucosa. These cells can be distinguished by staining fixed tissue sections with modification of the Zimmermann technique, following which the parietal cells appear yellow. Hydrochloric acid reduces the pH of the stomach contents, thereby activating the proteolytic enzyme *pepsin*, which is secreted in an inactive form, *pepsinogen*, by the *chief cells* of the gastric glands. Chief cells are stained blue by the modified Zimmermann technique. The acidity of the gastric contents softens fibrous material, kills some bacteria ingested with the food and extracts calcium salts from bone and cartilage. In addition, HCl brings about slow inactivation of salivary amylase as the acid penetrates the bolus and blocks further starch digestion.

It is not known how the parietal cell is able to secrete concentrated hydrogen ions (HCl) without damage to itself. The presence of tight junctions between the mucosal cells and the high mitotic rate observed for gastric mucosal cells in part explain how the stomach contains such a caustic solution without causing irreparable destruction. These tight junctions prevent the penetration of acid and proteolytic enzyme (pepsin) into the mucosa. This outer layer is continuously replaced by mitotic activity to maintain the integrity of the mucosal barrier.

Gastrin theory and acid secretion

Edkins in 1905 showed that extracts prepared from the most posterior portion of the stomach, the *antrum*, stimulated acid secretion by the fundic glands, and he suggested the name of *gastrin* for the active substance in these extracts. He found no gastrin activity in extracts prepared from the fundic portion of the stomach. Edkin's gastrin hypothesis temporarily lost credibility with the discovery of *histamine*, a potent stimulator of gastric acid secretion, and the demonstration of histamine in extracts of the gastric mucosa. It was almost 30 years before it was shown that histamine-free extracts from the mucosa of the antral portion of the stomach possessed the ability to stimulate acid secretion by the parietal cells. Nevertheless it was not until gastrin was finally isolated and characterized chemically that the term gastrin theory was discarded.

At least two similar peptides with gastrin activity have been isolated (gastrin I and gastrin II) from the antral stomach. Both of the molecules are composed of 17 amino acids; their only difference is that the C-terminal tyrosine of gastrin II is sulfated. A molecule of 34 amino acids has also been found in the circulation; it consists of gastrin I or II together with a third 17 amino acid peptide component. This big gastrin constitutes only about 5% of the circulating gastrin. The preprohormone for gastrin is composed of 104 amino acids. This

	Secretin	Glucagon	GIP	PZCCK	Caerulein	Human Gastrin II
1	His	His	Tyr			
2	Ser	Ser	Ala			
3	Asp	Gln	Glu			
4	Gly	Gly	Gly			
5	Thr	Thr	Thr			
6	Phe	Phe	Phe			
7	Thr	Thr	Ile			
8	Ser	Ser	Ser			
9	Glu	Asp	Asp			
10	Leu	Tyr	Trp			
11	Ser	Ser	Ser			
12	Arg	Lys	Ile			
13	Leu	Trp	Ala			
14	Arg	Leu	Met			
15	Asp	Asp	Asp			
16	Ser	Ser	Lys			
17	Ala	Arg	Ile			(1) Glu
18	Arg	Arg	Arg			(2) Glu
19	Leu	Ala	Gln			(3) Pro
20	Gln	Gln	Gln			(4) Try
21	Arg	Asp	Asp			(5) Leu
22	Leu	Phe	Phe			(6) Glu
23	Leu	Val	Val			(7) Glu
24	Gln	Gln	Asn		(1) Gln	(8) Glu
25	Gly	Trp	Trp		(2) Gln	(9) Glu
26	Leu	Leu	Leu	Asp	(3) Asp	(10) Glu
27	$ValNH_2$	Met	Leu	Tyr (SO_3H)	(4) Tyr $((SO_3H)$	(11) Ala
28		Asn	Ala	Met	(5) Thr	(12) Tyr (SO_3H)
29		Thr	Gln	Gly	(6) Gly	(13) Gly
30			Lys	Trp	(7) Trp	(14) Trp
31			Gly	Met	(8) Met	(15) Met
32			Lys	Asp	(9) Asp	(16) Asp
33			Lys	$Phe\text{-}NH_2$	(10) $Phe\text{-}NH_2$	(17) $Phe\text{-}NH_2$
34			Ser			
35			Asp			
36			Trp			
37			Lys			
38			His			
39			Asn			
40			Ile			
41			Thr			
42			Gln			

Fig. 2.2. Comparison of amino acid sequences of polypeptide hormones from the digestive system.

molecule is cleaved enzymatically several times to release big and little gastrins. Most of the biological activity of these gastrins resides in the four carboxy-terminal amino acids consisting of *Trp-Met-Asp-Phe-*NH_2. Several peptides that possess this terminal sequence have been shown to stimulate acid secretion. A synthetic pentapeptide (*pentagastrin*) that incorporates the terminal tetrapeptide sequence is frequently used for experimental studies.

The gastrin theory for control of acid secretion combines the observations that parasympathetic stimulation, acetylcholine (ACh), histamine and gastrin all cause acid secretion, whereas atropine (an anticholinergic drug), procaine (an anesthetic), sympathetic stimulation or certain antihistamines tend to reduce acid secretion under some experimental conditions. Presumably parasympathetic stimulation through release of ACh causes release of gastrin from the *G cell* in the antral mucosa. Gastrin travels via the blood to the fundus of the stomach where it stimulates the release of histamine from cells situated in the mucosa. Gastrin activates the synthesis of histidine decarboxylase, the enzyme responsible for histamine synthesis. Histamine in turn stimulates release of HCl from the parietal cell, probably via an adenyl cyclase-cyclic adenosine 3´,5´-monophosphate (cAMP) mechanism, resulting in decreased pH of the stomach contents. Gastrin may stimulate parietal cells directly without employing histamine as an intermediate, and some investigators conclude that histamine is not involved in endogenous acid secretion. Vagal stimulation or application of ACh may stimulate the parietal cells directly to secrete HCl.

During active digestion the pH of the stomach may be between 1 and 2. Low pH in the antral portion of the stomach (especially near the pyloric sphincter) reduces gastrin release.

Secretion of pepsinogen

The major gastric enzyme is the protease pepsin that is secreted by the chief cells in an inactive form, pepsinogen. Conversion of inactive pepsinogen to pepsin is accomplished by the presence of an excess of hydrogen ions supplied by HCl secreted from parietal cells. The optimum pH for vertebrate pepsins lies between 1 and 2, the normal pH range observed in the stomach following stimulation of acid secretion. The presence of acid on the surface of the gastic mucosa may activate a cholinergic reflex that evokes pepsinogen release. Parasympathetic stimulation via the vagus nerve causes release of pepsinogen, but hormonal control of pepsinogen secretion has not been established. A duodenal peptide motilin, has been implicated in regulating pepsinogen secretion. Gastrin causes release of pepsinogen only when applied in doses great enough to *inhibit* acid secretion by the parietal cells, implying that gastrin is not the normal factor causing pepsinogen release from the chief cells. Several other GI peptides can invoke pepsinogen release but do so only when applied in pharmacological doses.

Phases of gastric regulation

Gastric secretion is controlled via two levels. The *cephalic phase* involves stimulation of secretion via parasympathetic discharges elicited by the same stimuli that cause salivation; that is, sight, smell, taste, thought or presence of food. In the *gastric phase* of secretory control the presence of food in the stomach elicits secretion through vagovagal reflexes and/or through the gastrin mechanism. It has not been possible to determine which of these mechanisms is more important in controlling gastric secretion; probably all of these mechanisms operate in the normal digestive process.

Intestinal Events

While in the stomach, the bolus of food become saturated with the acidic gastric juices. These substances are thoroughly mixed by peristaltic contractions of the stomach to form an acidic, viscous fluid called *chyme*. Motility of the stomach, which is stimulated by the parasympathetic system, is responsible for churning the food mass and mixing it with gastric juices to form chyme. If the pH of the chyme in the antrum is sufficiently low (that is, less than 4.5) and of proper viscosity, the pyloric sphincter opens, and acidic chyme is squirted into the first segment of the small intestine, the *duodenum*. The exact mechanism controlling ejection of chyme from the stomach is not understood. Secretions also enter the duodenum from the exocrine pancreas and liver, through the *common bile duct*, in response to events taking place in the duodenum.

The duodenal mucosa contains many secretory cells, including cells responsible for digestive enzyme secretion as well as a variety of hormone-secreting cells and mucus-secreting cells. The intestinal mucosa is organized into thousands of tiny finger-like projections called *villi*. The presence of villi greatly increases the total surface area of the small intestine for both secretion and absorption of digestive products. The duodenum possesses more villi and secretory cells than do the more posterior sections of the small intestine (*jejunum* and *ileum*). The degree of specialization in the mucosa decreases progressively from anterior to posterior.

Secretion and pancreozymin

The presence of acidic chyme (pH less than 4.5) in the duodenum directly stimulates the *S-type cell* in the duodenal mucosa to release the peptide *secretin* into the blood. Bayliss and Starling discovered the existence of this factor that stimulates the pancreas to secrete basic juice and helps to neutralize the acidity of the chyme that has entered the small intestine. Although secretin levels in the blood do not increase following ingestion of a meal, the action of secretin on the exocrine pancreas is potentiated by another intestinal peptide hormone, *cholecystokinin*, that increases the blood following ingestion.

Originally it was believed that secretin was also responsible for stimulating secretion of the digestive enzymes normally present in the pancreatic juice, including the proteases chymotrypsin and trypsin, pancreatic lipase, pancreatic amylase and nucleases (DNase and RNase). After 40 years of controversy following the demonstration of secretin it was confirmed finally by Harper and Raper that purified secretin stimulates secretion of pancreatic fluid that is rich in sodium bicarbonate and poor in digestive enzymes. A second duodenal peptide, named *pancreozymin*, was found to contaminate some secretin preparations but not others, depending on the methods of preparation. Since *zymogen granules* represent vesicles of stored enzyme within the acinar (exocrine) pancreatic cell, the peptide that causes extrusion of zymogen granules from pancreatic acinar cells is logically called pancreozymin (pancreas-zymogen). It was postulated that the release of pancreozymin into the blood in response to the presence of peptides and amino acids in the chyme is due to direct actions of these molecules on pancreozymin-producing cells, similar to the action of hydrogen ions on the S-type cell. Sometimes the term *secretagogue* is applied to substances present in food, products secreted from the mucosa into the gut lumen or products of digestion that induce gastric or intestinal secretions.

Cholecystokinin

It was proposed in 1928 by Ivy and Goldberg that the presence of fat in the chyme or some of the digestive products stimulated release of yet another intestinal peptide that they named *cholecystokinin* (*chole* bile + *kystis* bladder + *kinein* move). Cholecystokinin travels via the blood to the gallbladder where it stimulates contraction of the smooth muscles comprising the walls of the gallbladder. At the same time it causes relaxation of the sphincter muscle that controls exit of bile from the gallbladder (the sphincter of Oddi). As a result, bile is expelled from the gallbladder, enters the bile duct (bile-bladder-move) and is transported to the duodenum. Bile is a viscous, complex mixture consisting largely of bile salts and bile pigments. Bile salts are powerful emulsifiers of fats (detergents). Bile pigments are breakdown products of hemoglobins, and they provide to bile and feces their characteristic colourations. Many other substances produced by the liver are present in bile, including metabolites of steroid hormones and inorganic iodide from deiodination of thyroid hormones. When bile enters the small intestine the bile salts emulsify globules of fat, causing them to be dispersed as small fat droplets within the aqueous digestive fluids. The emulsification of fats allows for marked reduction in the volume-to-surface ratio of the fat droplets, facilitating hydrolytic attack by pancreatic lipase to release glycerol and fatty acids for absorption.

Pancreozymin-cholecystokinin

Isolation and purification of GI peptides was finally accomplished by Mutt and Jorpes in Sweden more than half a century after the discovery of secretin by Bayliss and Starling. Investigators have since succeeded in identifying many distinct GI peptides. Secretin was confirmed as a separate peptide consisting of 27 amino acids. Purified secretin stimulated secretion of pancreatic juice that was low in enzyme content. However, the biological functions previously ascribed to PZ and CCK were found to reside in the same peptide consisting of 33 amino acids. Consequently the rather cumbersome name *pancreozymin-cholecystokinin* (PZCCK or CCKPZ) has been proposed to designate the single peptide that has been described in the literature under separate names for these separate functions. The *I-type* cell in the intestinal mucosa has been identified as the synthetic source for PZCCK.

The separate functional roles of secretin and PZCCK have been demonstrated elegantly in vitro with slices of exocrine pancreas. Physiological levels of PZCCK do not evoke release of basic pancreatic juice except in the presence of secretin,"' implying a permissive role. The action of secretin is markedly enhanced by PZCCK. Addition of purified PZCCK causes extrusion of zymogen granules from the acinar cells. Acetylcholine also causes extrusion of zymagen granules, suggesting a role for parasympathetic control over enzyme release from the exocrine pancreas. Parasympathetic influence and PZCCK probably operate via separate mechanisms. Atropine blocks the action of ACh but not of PZCCK. Purified secretin does not induce zymogen granule extrusion in these preparations. These data support the hypothesis for direct vagal (parasympathetic) influence over pancreatic enzyme release as being part of the normal regulatory mechanism; however, PZCCK does not require neural factors for its action.

Glucose-dependent insulinotropic peptide

Another intestinal GI peptide *enterogastrone*, was proposed in 1930 by Kosaka and Lim to be released in response to arrival of fat from the stomach. Enterogastrone was believed to inhibit gastric motility (peristalsis) as well as reduce acid secretion by the parietal cells. These actions would slow the entrance of fat into the small intestine and allow more time for proper processing of fat.

A peptide has been isolated from the intestine that inhibits gastric function. It was named *gastric-inhibitory peptide* (GIP) and is produced by the K cell in the duodenal mucosa. Chemically, GIP is similar to glucagon and secretin but is larger (42 amino acids). This peptide blocks gastric peristalsis and secretion of acid and enzyme. However, these effects occur only in the experimentally denervated stomach. It also stimulates release of insulin from the endocrine pancreas. This insulin-releasing action is now believed to be the physiological role for GIP. Following glucose uptake, GIP is secreted into the blood and causes release of insulin. Because of this action, GIP's name has been changed to *glucose-dependent insulinotropic peptide* while retaining the same acronym.

Neither bile secretion by the liver nor the secretion of *Brunner's glands* in the intestinal mucosa are influenced by GIP. Brunner's glands secrete viscid alkaline mucus that is believed to neutralize gastric juice entering through the pylorus. These glands are concentrated in the region of the pyloric sphincter and gradually decrease in frequency posteriorly, only occasionally being present in the most anterior portion of the jejunum. These cells also may secrete mucus in response to the presence of acidic, chyme in the duodenum.

Motilin

A unique small peptide (22 amino acids) that stimulates both motility and secretion of pepsinogen by the stomach has been isolated from the duodenum. This peptide, *motilin*, is structurally unlike the other GI peptides. Motilin is released from the *EC-type* cells when alkaline conditions are present in the duodenum, but the actual regulatory mechanism(s) governing its release is (are) uncertain. The physiological significance of this effect has been questioned since very alkaline conditions (pH 8.5-10) are required for releasing motilin. Acidification may cause motilin release under certain conditions in humans.

Phe—Val—Pro—Ile—Phe—Thr—Tyr—Gly—Glu—Leu—Gln—
Arg—Met—Gln—Glu—Lys—Glu—Arg—Asn—Lys—Gly—Gln

Fig. 2.3. Amino acid sequence for motilin.

Vasoactive intestinal polypeptide

Another peptide isolated from the intestinal mucosa has been shown to relax smooth muscle, thereby increasing the flow of blood to the viscera. This *vasoactive intestinal polypeptide* (VIP) consists of 28 amino acids and is structurally similar to both glucagon and secretin as well as to GIP. Vasoactive intestinal peptide is produced in neurons and by a pyramidal-shaped *H cell* found in the small intestine and in the colon of some species. Sufficiently high doses of VIP have been shown to produce secretin-like effects on the pancreas as well as a hyperglycemic response. These actions of VIP on the exocrine pancreas and blood sugar levels may be pharmacologic. A major role for VIP is that of an inhibitory neurotransmitter produced by intestinal nerves. It inhibits vascular smooth muscle but stimulates glandular epithelia. It is the inhibitory action on vascular smooth muscle that increases blood flow into intestinal tissues. The immunoactive VIP found in the brain is structurally identical to intestinal VIP.

Enterocrinin

In 1938 intestinal extracts were shown to stimulate intestinal secretion itself, and a hormone immediately was postulated in the usual manner to account for this activity. The purported hormone was named *enterocrinin*. However, both VIP and GIP have been shown to be capable of evoking secretion by the small intestine. Secretin and PZCCK are ineffective. Enterocrinin is probably not a descrete peptide but it remains to be demonstrated that intestinal secretion during normal digestion is controlled by either VIP or GIP or both.

Enteroglucagon

In addition to those hormones isolated from the small intestine, the *EG_I-type cells* produce *enteroglucagon*, which is structurally and functionally like pancreatic glucagon. Enteroglucagon is extractable in a small and a large form. The latter has been named *glicentin* and consists of 69 amino acids. This may be a prehormone form since it contains the glucagon sequence enclosed within peptide fragments of 8 and 30 amino acids. The physiological role for enteroglucagon and its relationship to pancreatic glucagon and glucose metabolism are open for speculation. A relationship to release of calcitonin (CT) and calcium regulation has been proposed.

Other gastrointestinal peptides

A peptide isolated from the duodenum of the pig specifically stimulates release of the inactive protease chymotrypsinogen. When activated in the duodenum, chymotrypsin enzymatically converts trypsinogen into the active protease trypsin. This peptide has been named *chymodenin*. This finding may represent a specialized refinement in control of specific enzymes from the pancreas according to the particular composition of the food mass.

A *gastrin-releasing peptide* (GRP) has been isolated from the antral gastric mucosa of dogs. It is composed of 27 amino acids and is structurally similar to *bombesin*, a peptide that may function as an enteric neurotransmitter.

Somatostatin is produced by neurons and paraneurons (D cells) located throughout the small intestine. It seems to function in paracrine fashion by inhibiting release of all other GI peptides. Somatostatin has been demonstrated in the arterial circulation of dogs following a meal and may act as a hormone to inhibit release of gastrin, insulin and pancreatic polypeptide. Two forms of somatostatin have been isolated. The first is identical to the hypothalamic tetradecapeptide neurohormone. The second consists of somatostatin plus an additional 14 amino acids. This may be a presomoatostatin form. *Dynorphin*, a heptadecapeptide from the central nervous system and pituitary gland is also present in the intestine. *Neurotensin* (tridecapeptide) and *substance P* (onadecapeptide) are neurotransmitters of the central nervous system that are also found in the intestine. Substance P was the first peptide to be identified in both the intestine and brain. It is not clear whether substance P and neurotensin are neurotransmitters only or if they have paracrine functions in the intestine.

The new biochemical methodologies for isolating and sequencing of peptides are allowing the isolation of new GI peptides faster than physiological studies can be performed to find a function for them! Porcine histidine isoleucine heptacosapeptide (PHI) is such a peptide. It consists of 27 amino acids, is structurally similar to the secretin family of peptides and is in need of a good, stable function.

Some proposed GI peptides

Numerous different peptides have been postulated to influence intestinal and gastric physiology, including *coherin* from neurohypophysial extracts, *villikinin*, *duocrinin*, *bulbogastrone*, *urogastrone*, *oxyntomodulin*, *sorbin* and several others from the gut. However, none of these proposed regulators has been shown to be an important, integral part of the digestive control mechanisms.

Phases of intestinal regulation

Three stages or phases of intestinal regulation can be identified involving neural and endocrine mechanisms. There appears to be a distinct *cephalic phase* mediated via vagal stimulation that influences pancreatic secretion. The *gastric phase* involves vagal and vagovagal stimulation of gastrin release that appears to influence pancreatic secretion. Finally, and certainly the most important regulatory mechanism, the *intestinal phase* relies primarily on release of peptides stimulated by the composition of the intestinal contents.

Chemistry of Mammalian Gastrointestinal Peptides

A number of specific peptides have been isolated from the GI tract of man and several domestic mammals. As described previously, these peptides exhibit the biological activities classically associated with secretin, gastrin and PZCCK and more recently with GIP, VIP, motilin and others. The sequences of amino acids that have been worked out for some of these peptides suggest that three chemical classes of GI peptides have evolved. The first group includes the gastrins, PZCCK and some related peptides. Secretin, enteroglucagon, VIP, GIP and PHI form the second group. The third group is composed of bombesin, GRP, substance P and a number of related peptides that have been isolated from amphibian skin (physalaemin, phyllomedusin, etc.) and molluscs (eledoisin). This last group are known also as tachykinins. Motilin and somatostatin are each unique peptides and are not similar to the peptides in any of these groups.

Embryonic Origin of Gastrointestinal Endocrine Cells

Although the use of immunological and fluorescent techniques has enabled investigators to identify the actual cellular sources for many of the GI peptides, there is still considerable disagreement with respect to the embryonic origin or origins of these cells in mammals. Pearse proposed that all of these GI cellular types as well as calcitonin-secreting C cells of the thyroid gland, parathyroid chief cells, α- and β-cells of the pancreatic islets, melanin-containing cells, adeno-hypophysial cells and the chromaffin cells of the adrenal medulla belong to the so-called APUD cellular series (amine content and amine precursor uptake and decarboxylation). These APUD cells are derivatives of neural crest or other neural ectoderm cells, suggesting that GI endocrine cells are of ectodermal rather than endodermal origin and that they have migrated into the intestinal mucosa early during development. An alternative view might be that at least some of these different cellular types have independently acquired APUD characteristics subsequent to or coincident with their differentiation from endodermal cells. The occurrence of certain GI peptides such as VIP in neural tissue argues strongly for a neural origin for these peptide hormone-secreting cells. A single origin for GI endocrine cells is supported further by observations that immunoreactive gastrin, CCK and glucagon appear to be localized in a single cellular type in the invertebrate chordate amphioxus and in the cyclostomes.

Complex Interactions of Gastrointestinal Peptides

Many studies have been published in the past few years that involve observation of the effects of administering combinations of GI peptides as well as the influences of one peptide on the release of another. Studies of this type indicate considerable overlap in the functional roles of the various peptides (for example, glucagon-like activity in secretin) although pharmacological doses usually were employed. At the present time it is difficult to sort out interactions due to structural similarities, pharmacological doses or both from those interactions that might represent true synergisms, functional overlaps or inhibitions. Consequently there is considerable literature concerning such interactions that will not be discussed here. The reader is encouraged to study literature and to find order in the chaos.

Influence of Gastrointestinal Peptides on Other Endocrine Systems

The influence of GIP on insulin release has already been mentioned, and it is possible that enteroglucagon release would also stimulate insulin secretion. Consequently GI peptides might influence the metabolism of carbohydrates, amino acids and fats after absorption.

Pentagastrin administered in small doses stimulates release of calcitonin from the C cells of the mammalian thyroid. Since experimental hypercalcemia produced by systemic infusion of calcium results in elevated levels of circulating gastrin, it has been proposed that uptake of calcium from the gut acting through the release of gastrin as a mediator effects release of CT. This increase in CT levels would enhance deposition of calcium in bones and offset any effects of parathyroid hormone (PTH) on this target tissue. The early release of gastrin during the processing of a meal may represent an "anticipation" of the consequent uptake of calcium that will occur so that CT released by gastrin lowers plasma Ca^{++}, thereby stimulating PTH release, which in turn would influence Ca^{++}/HPO_4^{-2} management by the kidney and indirectly enhance uptake of calcium from the gut. The action of PTH on bone would be blocked by CT. Enteroglucagon has also been suggested to evoke CT release in a similar manner.

Comparative Aspects of Gastrointestinal Peptides

There have been few studies concerning endocrine regulation of GI physiology in submammalian vertebrates, and most of these studies have been performed since the advent of purified mammalian peptides. Obviously the investigation into comparative regulation was hampered by the lack of understanding and interest in the general physiology of digestion in submammalian species. For example, although extensive research in teleosean fishes of commercial importance has been accomplished with respect to diets, growth and feeding ecology, few experiments have been concerned with physiological control mechanisms. The following brief account should serve to further emphasize the neophyte status of this area of comparative endocrinology and, it is hoped, encourage some developments in the study of comparative aspects of GI peptides.

Invertebrates

Immunoreactive gastrin has been extracted from the GI tract of two molluscan species. The levels of extractable gastrin are comparable to those of mammals. Gastrin has also been demonstrated in the neuroendocrine cells of an insect, *Manduca septa*, supporting possible neural origins for this peptide. These data suggest a broad phylogenetic distribution of these peptides associated with digestive functions. Peptides characteristic of invertebrate systems, are turning up in vertebrate guts and nervous systems. For example, the head inducing substance of *Hydra* has been found in human brain and intestine. The implications of these isolated observations may become more meaningful through studies of other invertebrate groups.

Class Agnatha: Cyclostomata

Cytological studies by Ostberg and co-workers on the intestine of the Atlantic hagfish *Myxine glutinosa* have revealed the presence of primitive open-type endocrine cells. These cells extend from the basal portion of the intestinal epithelium to border on the lumen of the gut. Hagfish intestinal endocrine cells do not possess APUD characteristics, although APUD-type cells have been reported in the pancreatic islets that differentiate into insulin-producing B-cells. Antibody to human gastrin, pentagastrin and porcine glucagon binds to endocrine cells in the *Myxine* gut. These intestinal endocrine cells in the Atlantic hagfish do not resemble zymogen cells either, suggesting a separate origin for the endocrine and enzyme-secreting cells. In contrast, the intestinal epithelium of larval and adult lampreys (*Lampetra* spp.) contains APUD-type cells that react to antibodies prepared against mammalian glucagon and gastrin.

Secretin-like and PZCCK-like activities have been demonstrated in intestinal extracts prepared from river lampreys, *Lampetra fluviatilis*, and sea lampreys, *Petromyzon marinus*. Both secretin and PZCCK activities were assayed by monitoring pancreatic secretions in the anesthetized cat. Similar observations have been reported for *M. glutinosa*. Gallbladder strips prepared from a Pacific hagfish, however, did not respond in vitro with contractions to porcine PZCCK although ACh caused contractions. Secretion of intestinal lipase in this same species is stimulated by porcine PZCCK. These observations suggest that the evolution of gallbladder receptors for PZCCK occurred after the appearance of molecules in the intestine that possess PZCCK-like activities.

Somatostatin is not present in the intestines of hagfishes. It does occur in the pancreas.

Class Chondrichthyes

In their classical studies Bayliss and Starling reported the presence of secretin-like activity in extracts prepared from dogfish shark and skate intestines when these preparations were assayed in mammals. The activity they measured may have been due to secretin-like or PZCCK-like factors (or to both) present in these extracts. Intestinal extracts prepared from the holocephalan *Chimaera monstrosa* also possess PZCCK activity. Porcine PZCCK stimulates contractions in strips of gallbladder prepared from dogfish sharks, and the intensity of the response is proportional to the dose of PZCCK.

Class Osteichthyes: Teleostei

Only a few species of teleosts have been investigated within the entire class of bony Fishes. The first published observations are those of Bayliss and Starling who reported that intestinal extracts prepared from salmon (*Salmo salar*?) possessed secretin-like activity (possibly also PZCCK activity as well). Similar activities were reported for pike, *Esox lucius*, and cod, *Gadus morhua*, when intestinal extracts were assayed in either birds or mammals. The magnitude of these responses was similar to those induced by purified mammalian VIP. PZCCK activity has been reported in the intestine of the Atlantic eel, *Anguilla anguilla* and the pike. Isolated strips of gallbladder from Pacific salmon (*Oncorhynchus*) contract in the presence of porcine PZCCK, indicating sensitivity of the salmon gallbladder to the mammalian peptide.

A gastrin-histamine type of mechanism is present in the teleost stomach. Extracts from the gastric mucosa of sunfish (*Lepomis macrochirus*) stimulate acid secretion in bullfrogs, and large doses of histamine (10-15 mg/kg weight) induced acid secretion in the European catfish *Silurus glanis*. Histamine-induced acid secretion in cod (15 mg/kg) is blocked by certain antihistamines, supporting the existence of a mammalian-like regulatory system.

Somatostatin and motilin were not demonstrable in the intestine of *Gillichthyes mirabilis*.

Class Amphibia

Regulation of gastric mechanisms has been studied more extensively in frogs than in any other nonmammalian species. It appears that amphibians possess mechanisms very much like those of mammals, involving both neural and endocrine mechanisms. Stomachs of intact frogs or isolated gastric mucosa prepared from frogs (including *Rana pipiens*, *R. catesbeiana*, *R. temporaria* and *R. esculenta*) respond with acid secretion when subjected to ACh, histamine, pentagastrin or crude gastrin preparations from nonmammals or mammals. Similarly gastric mucosa isolated from a urodele, *Necturus*, secretes acid in response to pentagastrin. Treatment with atropine or surgical vagotomy reduces acid secretion in frogs as it does in mammals. Supposedly the release of pepsinogen in *R. esculenta* can be effected by increasing parasympathetic activity. Caerulein, a peptide previously thought to be present only in anuran skin and which is known to stimulate acid secretion from stomach mucosa in a variety of vertebrates, appears to be the endogenous "gastrin" in *R. temporaria*.

Bayliss and Starling reported that extracts from frog intestines would evoke pancreatic secretion in dogs, providing evidence for the presence of secretin-like or PZCCK-like factors or both. Frog gallbladders will contract in the presence of porcine PZCCK in vitro, which supports the possible existence of a PZCCK-like factor in amphibians as well as a role for PZCCK in regulation of gastric processes.

Class Reptilia

The only observation with respect to GI regulation in reptiles dates back to the observation by Bayliss and Starling that a factor or factors capable of causing pancreatic secretion in mammals is present in the intestine of a tortoise. Immunoreactive somatostatin is present in the intestine of the lizard *Anolis carolinensis*, but motilin is not. It is remarkable that additional studies have not bee reported or if reported have escaped recognition. Certainly the field is open for some careful comparative studies.

Class Aves

Mammalian gastrin can stimulate acid secretion in birds, and large amounts of PZCCK cause release of enzymes from the avian exocrine pancreas. However, no good evidence has been presented for a physiological role for these hormones in birds. Glucagon and GIP have no effects on pancreatic secretion. Extracts prepared from chicken intestines are strong stimulants of pancreatic secretion when assayed in turkeys, but these extracts are only weak stimulants in mammals (cat, rat). Purified porcine secretin only weakly stimulates exocrine pancreatic secretion in turkeys, but purified mammalian VIP is a potent stimulator. These data suggest that secretin-like activity in birds may reside in a molecule that is more like mammalian VIP than it is like secretin. However, chicken VIP has been isolated and differs structurally from porcine VIP at only four positions. Somatostatin and motilin have been demonstrated in the intestines of Japanese quail, but nothing is known about their functions.

3

HORMONES IN BODY COLOUR

Ten years ago G. H. Parker surveyed the studies (in the control of animal colour change, made by himself and others during the period from 1910 to 1943, and concluded that "the distinction between nervous and humoral activation for chromatophores seems to be disappearing". During the past decade the discovery that the greater part of endocrine control in crustaceans is neurosecretory in nature has done much to support the views of Parker by lessening still further the distinction between control by nerves and control by hormones, in the field of animal colour change. It is now known that the blood-borne chromactivators in crustaceans are, in fact, manufactured and released by cells in the central nervous system, which in other respects resemble typical neurones. The control of colour change in crustaceans, once thought to be effected by nervous control of a gland (the sinus gland) is now revealed as a more diffuse neuroendocrine mechanism in which certain nervous elements function as glands by manufacturing blood-borne chromactivators.

Advances in our knowledge of the control of pigmentary effectors in vertebrates have by comparison been less striking than the evolution of our understanding of crustacean endocrine mechanisms, but here also research, by pointing to a more intimate relationship between nervous and hormonal control than had hitherto been visualized, has clarified some of the inconsistencies of work on vertebrate metachrosis.

At one time the demonstrated innervation of the chromatophores of certain vertebrate animals, notably teleost fishes and reptiles, was accepted as evidence that these effectors were under nervous, not endocrine, control, and that it was local electrical disturbance rather than chemical stimulation that was responsible for the observed pigmentary movements. During the past few decades, however, biologists have begun to realize that the ability to conduct electrical disturbance is not an exclusive property of nerve cells and that the ultimate effects of nervous stimulation may in fact be predominantly chemical rather than electrical. The need for distinction between nervous and hormonal control of animal colour change, therefore, becomes less important, and the general thesis advanced by Parker, namely that animal colour changes are essentially under neurohumoral control, is seen to be acceptable.

We are, however, still confronted by problems: Why should the chromatophores of certain animals (e.g., teleost fishes) be directly innervated, while those of others (e.g., crustaceans) are not? And why should the control of colour change in some groups (crustaceans) be predominantly neurosecretory, in others predominantly nervous (some reptiles), and in amphibians mainly hormonal? The concepts of neurosecretion and of neurohormones do much to explain the otherwise puzzling inconsistencies in the control of pigmentary movements, but they do not explain why chemical activation at a distance by blood-borne hormones should be predominant in certain animals whereas local chemical stimulation should be preferred in others. Possibly the answer to this problem may be found by studying the relation between the colour of an animal and its environment. Local stimulation made possible by direct innervation permits greater speed of activation and subsequent check of the activator.

This is in contrast to the prolonged and widespread stimulation made possible by steady release of comparatively stable neurosecretory substances into the blood stream. It is certainly no accident that nervous control of pigmentary effectors is predominant in the swiftly moving cephalopods, teleosts, and reptiles, that a hormonal control is characteristic of the sluggish amphibians, and that a more elaborate neurosecretory control effects the sudden adaptive movements, superimposed on the slow rhythmic and prolonged changes, which characterize the colour changes of crustaceans.

If we accept the hypothesis that a fundamental activity of nerve cells is secretion, and that, therefore, the ultimate control of pigmentary effectors is humoral, we become less concerned by some of the contradictions which disturbed the earlier workers in the field of animal colour change. It then becomes possible to construct a logically satisfying general theory of neuroendocrine control of pigmentary effectors. In such a scheme many immediate pigmentary movements are due to activation by relatively small molecules (e.g., adrenalin-like compounds) that are released locally from nerve endings close to the pigmentary effectors and are rapidly destroyed after they have exerted their effect; more prolonged adjustments of pigmentary position are effected by larger and more persistent molecules (probably peptide in nature), produced and released by specialized neurosecretory elements of the central nervous system, or by endocrine glands, and which circulate in the blood stream for considerable periods of time. In this brief survey of recent advances in our understanding of the control of pigmentary effectors I wish to stress the importance of the study of animal colour change as a means of elucidating certain fundamental aspects of neuroendocrine control. There is undoubtedly a more ecologic aspect of colour-change study which deals primarily with the relationship between the chromatophore system and the changing environment of the animal. In the hands of the experimental physiologist, however, the chromatophore has another significance as the most sensitive indicator of neuroendocrine disturbance known.

Unit of Colour Change

Colour changes in animals are brought about by the dispersal or concentration of pigment granules along finely branched processes of colour cells called *chromatophores*, which may each constitute either a single cell containing one type of pigment, or an aggregation of cells to form chromatophores which are *dichromatic*, *trichromatic*, *tetrachromatic*, or *polychromatic*.

The general structure of animal chromatophores has been surveyed by Parker (1948), Knowles (1955), Knowles and Carlisle (1957), and others. Recently, use of the electron microscope to study the chromatophores of the teleost *Lebistes* has revealed a previously unsuspected organization within the chromatophore and has indicated that some of the previously held views on chromatophore organization and structure may need revision.

It was generally accepted that pigmentary movements in chromatophores were accompanied by alterations in the sol-gel balance of the cytoplasm, and that this alteration was, in fact, the mechanism for pigmentary movement. High-power resolution, however, has shown that the pigment granules in a *Lebistes* chromatophore are contained in an inner sac, separated from the nucleus by a nuclear membrane of about 300 Å, and from the outer cytoplasm by another membrane about 80 Å in thickness. The outermost layer of cytoplasm, beyond this latter membrane, appeared to contain fibrils each about 80 Å thick, and Falk and Rhodin have suggested that the contraction of these fibrils constricts the inner pigment-containing sac and so "concentrates" the chromatophore. The author has carried out some preliminary studies, on crustacean chromatophores, and has demonstrated that fine constrictions may be seen with the optical microscope under appropriate illumination. This would be in accordance with the view that in these chromatophores also, pigment granule concentration may be due to an active constriction of part of the cell possibly accompanied by a gelation of the cytoplasm. Further studies using electron microscopy on the chromatophores of various animals are needed before we can formulate a general theory of chromatophore structure and action. Recent discoveries, however, have pointed out the way to a better understanding of this problem.

Separation and Identification of Chromactivating Substances

Within recent years the principal vertebrate chromactivating hormone has been separated, identified, and termed *MSH* (the *melanocyte-stimulating hormone*). A detailed account of this discovery is given elsewhere in this volume, and it will suffice here to mention that MSH seems to be a peptide of low molecular weight, in the form of an octadecapeptide containing 18 amino acids in linear arrangement.

Separation by paper electrophoresis has revealed a number of chromactivating substances in neurosecretory organs of crustaceans. Subsequent tests by enzymatic destruction have shown that one of these chromactivating substances, the A-substance, has peptide links which are essential for its activity. Comparable results have been obtained for another crustacean pigment-concentrating substance, by Ostlund and Fange (1956), who have reported destruction by chymotrypsin; and Perez-Gonzalez (1957) has enzymatically destroyed a *Uca*-darkening substance, also by using chymotrypsin. It seems evident that many crustacean hormones are, like MSH, polypeptides.

The problem of how many hormones may interact to bring about the elaborate colour changes of animals and whether these hormones may be closely related chemically still awaits solution. Close chemical resemblance between ACTH and MSH, and the fact that the former may, when injected, bring about effects similar to those of the latter, indicate that crude extracts of substances with basically different functions *in vivo* may, because of a general molecular resemblance, have chromactivating effects when injected into experimental animals. The possibility of a range of precursor substances all with chromactivating effects must also be considered. These, as well as some other difficulties in the interpretation of experiments on chromatophores, have been reviewed by Knowles (1955).

The possibility of mutual masking; by antagonistic hormones, present together in crude extracts, has received experimental support. Recently Enami (1955), using a method of differential solubility in alcohol, has shown that a melanophore concentrating principle (MCH) could be separated from MSH, and that although it was present in crude pituitary extracts it seemed to have a hypothalamic origin. This work, and other papers on the role of the pituitary in the colour changes of fishes, has been reviewed recently by Pickford and Atz (1957).

Antagonistic chromactivating hormones in the neurosecretory organs of crustaceans have been separated by paper electrophoresis, and it may be that both pigmentary dispersal and concentration are active processes, under hormonal control.

That direct innervation may play a part in chromatophore alterations has received experimental support. This question was fully reviewed in 1948 by Parker and little has occurred since to modify his general conclusions. Chromatophores in the main are sensitive to adrenalin and respond to it by pigmentary concentration. In a number of vertebrate animals, notably *Teleost* fishes and reptiles, the chromatophores are richly innervated by autonomic fibres and stimulation or interruption of these fibres brings about local pigmentary movements. Acetylcholine also has been shown to have an effect, antagonistic to adrenalin, in some forms.

To sum up, we find that two physiologically distinct systems operate in animal colour change. One of these is essentially hormonal in the classic sense and seems to exert its effect through peptide substances circulating in the blood stream. There is evidence, however, that this may be supplemented or supplanted, in some forms, by a system of direct innervation in which active compounds of relatively low molecular weight are released by nerve fibre terminations close to the chromatophores.

Neurohaemal Organs in Crustaceans

The discovery that histologically distinguishable secretory droplets in nerve fibres could be correlated with chromactivating substances that effected colour-changes when injected into the blood stream, was independently made by Enami (1951) working on the sinus gland and Knowles (1951) who studied the post-commissure organs. Since then the general basic structure of these organs and the role they play in the neurosecretory systems of crustaceans have been further elucidated, mainly by Welsh and his pupils who worked on the sinus gland and by the observations of Knowles on the post-commissure organs.

These two important neurohaemal organs in crustaceans consist essentially of terminations of neurosecretory fibres, filled with granules or droplets of secretion and separated from a blood-sinus by a thickened layer of epineurium connective tissues in which some cellular elements may be observed.

The author recently had an opportunity to study the sinus glands of *Squilla mantis* and the post-commissure organs of *Leander serratus* with the electron microscope, and has thereby been able to supplement histological observations made by him and other workers, with the optical microscope.

Electron microscope studies show that each neurosecretory fibre termination is bounded by a layered membrane approximately 75-100 Å thick. These fibre terminations are separated from an adjacent blood-sinus by an epineurium wall into which the neurosecretory fibres of the post-commissure organs penetrate.

Whether any of the neurosecretory fibres in the crustacean neurohaemal organs examined eventually terminate at the surface of the epineurium, or whether a wall of ground-substance separates all the fibre terminations from the blood sinuses into which they liberate their secretory products, has not yet been ascertained by the author. In the sections so far observed, an amorphous wall generally about 5000 to 8000 Å in thickness has been seen, and so far fibre terminations penetrating this wall have been discerned in only a few instances. Beneath the surface of the epineurium there are many cells, lying between the neurosecretory fibres, which may be interpreted as supporting tissue probably engaged in synthesis or repair of the epineurium. An abundant and clearly apparent endoplasmic reticulum is characteristic of these cells. In nerve fibres leading to the sinus glands and to the post-commissure organs a great quantity of secretory material may be seen. This material is present as round or ovate units which differ greatly in size. In the sinus gland some fibre terminations characteristically contain inclusions about 1500 Å in diameter which resemble in dimensions inclusions in the fibres of the post-commissure organs; other fibres seem to contain larger droplets, about 2500 Å in diameter. The larger the droplets are, the more electron-dense they seem to become after permanganate fixation. A careful examination has revealed that each "droplet" is bounded by a many-layered membrane which appears to consist essentially of two electron-dense layers with a less dense layer between them; each layer measures approximately 30 Å, a figure which conforms approximately to the measurements of cell membranes described by Robertson (1957) and others. The concept that each droplet is surrounded by a membrane is also in accord with the results given by Perez-Gonzalez (1957), who remarked that various treatments which disrupt cell membranes (e.g., heating, freezing, and thawing and detergents) accentuate the strength of sinus gland extracts and with the recently published observations by Hodge and Chatman (1958).

The concept, however, of discrete spherical or ovate droplets of secretory material, each entirely surrounded by a many-layered membrane, is not one which provides a completely satisfying explanation of the observed facts, for two reasons:

1. Within certain droplets still more membranes may be seen.
2. In certain sections the droplets are not spherical but extremely elongated, and have the appearance of beads of secretory material lying in tubules within the cytoplasm.

A hypothesis that the secretory material in neurosecretory fibres is contained in fine tubules, and not in spherical units, helps to explain the otherwise puzzling variation in size among the droplets observed in electron micrographs of crustacean neurohaemal organs; in sections of some of these, droplets of many varying sizes have been observed in a single cell. If, however, we interpret the smallest droplets as transverse or oblique sections of tubules (about 300 Å in diameter) almost devoid of secretory material, and the larger units (750 to 2000 Å) as sections of "beads" of engorged portions of a tubule, the variation in size becomes easier to understand. One cannot fail to be impressed also by the genera] resemblance of this system of tubules in the cytoplasm of neurosecretory cells to the appearance of the endoplasmic reticulum system described in a review by Palade (1956), and said to be especially characteristic of actively secretory cells.

It is perhaps noteworthy that the largest droplets have been observed in the sinus gland only, and that this organ alone among those examined is suspected of a control of metabolism. In this organ it seems likely that the different sizes of the inclusions may represent different hormonal substances; this view would be in accordance with the studies of Potter (1954), who was able to distinguish different fibres containing different secretory material on the basis of distinct tinctorial affinities.

Electron microscope studies of crustacean neurosecretory cells are still in progress by this author, and at this stage it would be imprudent to do more than indicate some of the present trends of this research. There are indications, however, that electron microscope examination of crustacean neurosecretory systems and of the chromatophores they activate, correlated to investigations of the chemical nature of the chromactivating substances, may contribute to our general knowledge of the mechanism of neurosecretory control.

4

Corpuscles of Stannius

The corpuscles of Stannius have been, from time to time, described as the adrenocortical homologue of the Teleost fishes. Despite the considerable volume of work on these enigmatic bodies, mostly in the latter part of the nineteenth century, their real function is still in doubt. In addition to the adreno-cortical hypothesis they have also been held homologous with the Mullerian duct and latterly, in a persuasive paper, as osmoregulatory organs.

This paper, whilst recalling previous work, adds some comments on the nature of these bodies, their development, histological picture, and mode of function in some of the salmonid fishes of the West coast of North America.

Embryology

The earliest work on the development of these bodies deserving of attention is that of Giacomini (1912, 1920, 1922), in conjunction with information on the origin of the adrenocortical tissue in salmonids and lophobranchs. In a later paper Garret (1942) describes their development in *Amia calva* and in *Salvelinus fontinalis* as well as several other Teleosts and he confirms the work of Giacomini and others.

In the Pink salmon of the Pacific coast (*Oncorhynchus gorbuscha*) the first appearance of the corpuscles takes the form of groups (two pairs) of neutrophilic cells in the dorsal aspect of the pronephric duct, readily distinguishable from the basophilic cells of the duct itself. The duct stage is followed by a period of rapid cell division producing a swelling in the wall of the duct. It is attached to the duct by a stalk for a short period and then migrates into the kidney substance. Following the stalk period, and as the body moves off into the kidney, it becomes invested by venous sinusoids but it is itself still avascular. It is situated, however, close to the dorsal aorta which lies immediately dorsal to the developing mesonephros.

As the body increases in size the venous sinusoids penetrate it, coursing between undifferentiated and rapidly dividing cell groups. As it reaches the dorsal surface of the kidney it receives an arterial supply either from the dorsal aorta direct or via a renal artery. In the Steel-head trout the bodies lie lateral to the kidney, and the ultimate site varies from species to species.

The body of the corpuscle now assumes its definitive form but no cell types are distinguishable although the cells contain many basophilic granules. In the Pink salmon two pairs are the normal complement but occasionally one pair or three corpuscles may be found. The attempt to homologize these bodies with the Mullerian duct is not a valid thesis since, as de Beer says, the proof of homology lies in the adult structure not in ontogenetic stages.

Histology

In the Pink salmon the corpuscles of Stannius are large bodies 5 × 8 × 3 mm, yellow pink in colour, deeply, although loosely, buried in the dorsal surface of the kidney. They have a dorsal hilus in which arteries and veins of vascular supply and drainage can be clearly seen. The arteries are of medium size. The gland is invested in a thin collagenous capsule which penetrates and divides it into lobes and lobules. The vascular supply follows the collagenous septae.

In the Pink salmon there appear to be two types of morphological arrangement for the cells. Peripherally the cells are arranged in cords into which the sinusoids pass, often with a marked central vessel. The cells apposed to the sinusoids are regularly arranged with their proximal borders filled with secretory granules. The nuclei, large, and ovoid to round, basophilic, and somewhat vesicular, lie below the cytoplasm distal to the sinusoid. In some of the cords the centre is filled with irregular masses of cells whose nuclei are often intensely basophilic and may be shrunken, indicative of degeneration.

Toward the centre of the gland the cells are arranged in anastomosing plates, commonly two cells thick, with the sinusoids flowing between the plates. The cells have the same disposition as those in the peripheral cords and do not exhibit any difference which may be related to function. The plates do not show the pycnotic cells of the cords.

Extensive attempts, using a variety of polychrome and other stains failed to reveal any cell types in the gland and it is suggested that only the one type of cell is present. In addition to the secretory granules the cells contain many mitochondria.

The sinusoids are lined with reticuloendothelial cells, some of which are modified into large phagocytes of the same type as the cells of von Kupfer in the liver.

Experimental Results

Samples of blood were taken by ventricular puncture from a mature population of Sockeye salmon (*Oncorhynchus nerka*). The samples were collected directly into sterile dextrose-citrate. From the same fish from which the blood was taken, corpuscles were collected into suitably diluted methanol.

The citrated plasma and the methanol preserved Stannius corpuscles were extracted by the methods of Axelrod and Zaffaroni, and Zaffaroni. Chromatographic separation was carried out on paper beside a 10-gamma standard of cortisone and hydrocortisone. An appreciable concentration of hydrocortisone can be seen in the plasma but is entirely lacking in the corpuscles. The results of the chromatograms would seem to indicate conclusively that the corpuscles are not responsible for the production of corticosteroids of the C^{21} type.

5

Ultimobranchial Gland

Among derivatives of the embryonic pharyngeal entoderm are certain small, usually paired glandular structures best known today as the ultimobranchial bodies. These "bodies" were originally regarded as "lateral" or accessory thyroids because in typical mammals they were found to join with the lateral lobes of the thyroid gland. They were also known as suprapericardial bodies, because of their position in elasmobranchs and in certain amphibians, and as postbranchial bodies or as telobranchial bodies.

The term "ultimobranchial body" is to be preferred, since it may be used in any class of vertebrates and implies simply an origin similar to that of the pharyngeal pouches but does not imply that this outpocketing represents any particular pouch. The varied development, differentiation, and fate of this structure make it of peculiar significance and render it subject to different interpretations. No functions have been proven, although several have been hypothesized.

Occurrence

The comparative embryology of the ultimobranchial bodies has been extensively investigated. They are present among all the orders of each class of vertebrates except the cyclostomes. With some exceptions in teleosts, ultimobranchial bodies have been noted in fishes. According to Giacomini (1908) and Watzka (1933), the ultimobranchial bodies can be considered homologous in all teleosts and with similar structures in selachians also, but they are difficult to render homologous throughout the various, entire series of vertebrates.

In amphibia, these structures are almost universally present among anura, and Wilder (1929) found them without exception in urodeles.

They are present in virtually all reptiles, birds, and mammals, being reported absent only in some snakes and in the bat and horse. They are always paired in mammals but may be unpaired in other vertebrates. When unpaired, it is usually the left that is present (urodeles; some reptiles and birds). In most nonmammalian forms, the fourth and fifth pouches and the ultimobranchial anlagen develop as united blind outpocketings on each side of the posterior part of the pharynx. In typical mammals, however, it appears uniformly on both sides behind the fourth pouch, except in the rat and mouse where the third pouch is the last.

If these structures are truly ultimobranchial, they cannot be considered directly homologous throughout the different classes of vertebrates, where the number of arches and pouches in the series may vary. However, there seems to be no reason to force a homology too far. Whatever the factors involved may be, they affect the caudal branchial region and in most vertebrates a structure develops that is comparable if not homologous.

Character in Nonmammalian Forms

Ultimobranchial tissue has a vesicular structure in fishes. In elasmobranchs (*Squalus acanthias*), Camp (1917) found that this so-called gland consists of large, distended vesicles which frequently intercommunicate and contain mucous secretion. In teleosts, Giacomini (1908) found granular substance mixed with degenerated nuclei, and in both groups Watzka (1933) reported a thyroid-like colloid. According to Maurer (1888), a single large follicle or a complex of smaller ones represents this body in anura. During certain developmental stages, this sometimes contains a serous product but never colloid. In urodeles, although variable in size, form and position, it occasionally exhibits considerable secretory activity of variable quality and may bear a superficial resemblance to thyroid tissue. It remains throughout life as an epithelioid or epithelial structure which is frequently vesicular. The differentiation of its cells, however, is frequently so slight that there is the suggestion of persistence of an essentially embryonic structure. In urodeles, there is little or no evidence to support the assumption that the structure in question possesses physiological significance, either as an internally secreting gland or as an exocrine gland that has lost its duct. Its inactivity or inhibition during development suggests no correlation with metamorphosis in amphibia. However, contrary to Maurer's statements (1899, 1902) that colloid is absent in all nonmammalian vertebrates, appearing first in *Echidna*, Watzka (1933) noted colloidal substance stores in vesicles of the turtle, lizard and snake.

The ultimobranchial body in reptiles may have an endocrine function, at least in the lizard (*Lacerta*) where it is well developed. During the early postembryonal period, this structure, at first a compact mass of gland-like epithelial cords, becomes well vascularized. With growth and differentiation, there is evidence of secretory activity in numerous small thyroid-like vesicles. These enlarge and produce a fine homogeneous colloid in adult animals. According to Eggert, the activity of ultimobranchial tissue in the lizard seems to parallel seasonal changes. This is especially noticeable in June and July, when secretion begins and there is evidence of mitotic activity. This activity decreases in the fall, and during hibernation the colloid becomes denser and the structure is quiescent. Spontaneous involution occurs with age in the second and third summers and is associated with large secretion-filled cysts. Connective tissue and lipoid material usually replace these follicles and there is infiltration of lymphocytes.

Eggert has subjected this tissue to experimental analysis. No change occurs in animals in which ultimobranchial tissue has been removed. Neither extirpation of the thyroid nor involution of the testes cause demonstrable change in ultimobranchial tissue. Presumably, thyrotropic hormone acts directly on this tissue because prolonged injection of TSH causes increased activity with tremendous secretion and cyst formations in the absence of thyroid tissue. In this respect, ultimobranchial tissue may resemble thymus tissue. Furthermore, observations dealing with metamorphosis following extirpation of the thyroid gland suggests no thyroid-like activity in this ultimobranchial tissue, but simultaneous injection of TSH with thyroxin decreases cyst formation (normally, thyroxin produces no essential change). This author's concept of the nature of ultimobranchial tissue in reptiles may be open to question and has been ignored for years.

In birds the disposition of the last three pharyngeal pouches (IV, V and VI) closely resembles that in reptiles. In these animals the rudimentary fifth pouch is a transitory structure and the sixth pouch (found in sequence) gives rise to the ultimobranchial body. It is also quite likely that these pouches become incorporated or combined as a "caudal pharyngeal complex."

Although Lillie (1919) called attention to the development of thymus and parathyroid tissue from the ultimobranchial bodies in the chick, Dudley (1942) has shown that parathyroid tissue and thymus-like areas occur in addition to typical ultimobranchial parenchyma. The formation of one parathyroid in each ultimobranchial body would agree with a branchiomeric interpretation of the pharynx (which attributes to a posterior rudimentary pouch potentialities of one or two well developed pouches ahead of it—typically pouch III and IV), but the number of glands found is variable, and they may even be multiple. If not a matter of embryonic induction—the presence of parathyroid tissues being linked with the proximity of a parathyroid III and IV—this may mean that entoderm of the last diverticulum has ultimobranchial and parathyroid potentialities. In the presence of numerous lymphocytes, a "lympho-epithelial reaction" similar to thymus tissue may also occur. Verdun noted the "distinctly glandular character of this parenchyma" as long ago as 1898.

Based on the premise that all organs are of use or have been functional in ancestral forms, the evidence presented by Eggert (1938) for an endocrine function in reptiles is the most impressive. In birds, Terni (1927) and Watzka, working independently, both agreed that the ultimobranchial body had all the "morphological requisites of an actively functioning endocrine gland and Terni even believed it to be cyclic or seasonal. However, after an extensive review of the problem, Dudley (1942) has more recently found no reason to regard this tissue as having an endocrine function in birds. She stated that only rarely, in the fowl, are these structures reminiscent of secretory follicles, thyroid or otherwise. This tissue may possess certain secretory attributes, but as Kingsbury (1915) pointed out, "colloid may signify no more than a retained secretion of high protein content . . . and vesicles or cysts may be expected to arise in epithelial masses originating from a surface epithelium." If it is not a gland, with specific functions, there remains the interpretation offered by Kingsbury: "continued growth activity in the branchial entoderm."

In most mammals, the ultimobranchial body is so intimately associated with the fourth pouch that the two may be appropriately called the fourth or caudal pharyngeal complex. Although in birds and lower vertebrates generally no real transformation into thyroid tissue occurs, the developmental relations of this "complex" and the thyroid become very intimate in mammals, and there is even evidence that this tissue differentiates into colloid-filled vesicles. Echidna is the only mammal thus far reported whose ultimobranchial body never becomes embedded into the thyroid, but differentiates into glandular tissue, which, while differing from the thyroid in structure, nevertheless possesses colloid-containing follicles. Its similarity with that of birds has been noted.

There are many arguments as to the fate of the ultimobranchial body, especially in mammals (including man). After it becomes intimately incorporated within each lateral lobe of the thyroid gland, its further history varies in different species, as has been reviewed. The question naturally arises as to whether or not ultimobranchial bodies take part in the formation and functions of the thyroid gland.

MAMMALS

Concepts

Since the fate of the ultimobranchial body may be variable not only in different species but also among individuals, interpretations of its significance have been both numerous and diverse.

The *alleged fates* of this structure in mammals are as follows: (1) That it differentiates as an inherent "*lateral thyroid*" component and contributes to the general thyroid parenchyma. This was the early concept.

(2) That it is an "*organ of unknown function.*" Verdun (1898) described cysts and cell strands which persisted as "special rests" (rabbit, rat, dog, and mole) and after further study thought the ultimobranchial body might represent a gland of unknown function. In the bird he regarded it as a vestigial gland. Grosser (1910) working with human material stated that "the ultimobranchial body clearly represents a ductless gland which has become rudimentary" but suggested (1912) that it might be a *thyroid growth center* which ultimately disappears. After examining a variety of animals, Watzka (1933) considered the ultimobranchial body as representing in phylogeny an organ of internal secretion of apparently increasing significance up to and including birds. De Winiwarter (1933) held the opinion that in the guinea pig it could form a so-called parathyroid V and thymus V as well as an "assemblage of canals and cavities"; in the cat (1935) he considered it "an enigmatic organ of unknown function." In the developing human thyroid gland, Politzer and Hann (1935) could not decide whether gland-like remnants of the ultimobranchial body indicated developing thyroid parenchyma or "functionally different gland tissue" within the thyroid. They did not believe the ultimobranchial body degenerated. Godwin (1937) thought that in the dog, at least, ultimobranchial tissue might become scattered to form the so-called "parafollicular cells" ("interfollicular" cells or "macrothyrocytes"),

and in 1940 he suggested that "the idea that it may be an 'organ' in the process of appearing is a possibility to be kept in mind."

(3) That it represents "*more or less indifferent tissue*." In the pig and guinea pig, Simon (1896) thought the ultimobranchial body degenerated and disappeared. Rabl (1907) was of the opinion that it degenerated completely in the mole, but in the guinea pig (1922) it persisted without transformation and with little change in postnatal thyroids. Recent work by Klapper (1946) indicates that it finally undergoes degeneration and completely disappears. In a few instances it persists as irregular cystic masses which on occasion may be converted to a thymus IV. Kingsbury reported that this structure usually undergoes degeneration in mammals, disappearing entirely or resulting in the formation of epithelial cysts; in man this occurs "typically without trace." In the calf (1935) he could find no evidence for the origin of true thyroid parenchyma from ultimobranchial tissue, but he stated that in some instances vesicles indistinguishable from thyroid follicles might develop.

Kingsbury has interpreted the ultimobranchial body only as "the expression of a continued growth activity in the pharyngeal entoderm apparently associated with the gill-forming potentialities of the region." He has created a premise for considering the ultimobranchial body as relatively indifferent tissue which may be influenced in its development by variable factors of environment. Upon the basis of this assumption Rogers (1927), studying the rat, considered the partially differentiated ultimobranchial tissue as a natural transplant which, composed of indifferent material, becomes influenced in its further development by the thyroid parenchyma, "differentiating not according to its origin, but in harmony with its surroundings." This induction theory appears to explain the transformation of ultimobranchial tissue into thyroid-like follicles observed in many mammals. According to Badertscher (pig), Rogers (rat), and Godwin (dog), the amount of thyroid tissue derived from the ultimobranchial body is relatively slight.

Relation to the Thyroid Gland

Since the degree of incorporation and intimacy of fusion of the ultimobranchial body with thyroid parenchyma sometimes varies depending upon the growth mechanics involved during development, it may be intimately and extensively associated with thyroid parenchyma, remain in a "vasculostromal" hilus or lie wholly outside the thyroid gland. According to Watzka (1933), considerable structural variation accompanies the development of ultimobranchial tissue, depending upon its position relative to thyroid parenchyma. In the rat, pig, and dog, the ultimobranchial body becomes so intimately associated with the developing thyroid parenchyma that it undergoes reticulation and transforms into vesicles that are indistinguishable from thyroid follicles. In many mammals, however, a variable portion of the ultimobranchial body remains outside the thyroid; there it "shows a different and not so uncharacteristic and apparently meaningless cystic structure as one occasionally finds within the thyroid gland of many mammals". Its architecture sometimes resembles that of mucous acini. Watzka further states that these structures exhibit well the innate growth of potentialities of the ultimobranchial bodies when they are not restricted by incorporation within the thyroid gland. The writer, from his observations on postnatal thyroids, can readily subscribe to such a belief.

Observations and Experimental Studies

My own work with young sheep indicates that the ultimobranchial bodies are usually represented by large, multiple cysts lined by stratified squamous epithelium in the thyroid gland. They have been regarded as arising congenitally due to the failure of the ultimobranchial body (in sheep and cattle) to fuse intimately with the thyroid parenchyma during embryonic development. This cystic manifestation of metaplasia may reflect a deficiency of vitamin A, for sheep and cattle are grazing or pasture animals. The epithelium of these cysts frequently produces cords or clumps of glandlike cells which invade the neighbouring thyroid parenchyma in a manner that often suggests the growth of glandular neoplasms. Depending presumably upon the state of activity of the thyroid gland, these cysts can be the source of two peculiar outgrowths. These may resemble either typical thyroid parenchyma or adenoma-like nodules which may be solid or cystic, and they originate from the basal (germinative) and/or the suprabasal (*degenerating*) cells in this stratified squamous epithelium. The evidence here was interpreted as indicating that this cyst epithelium, indicative of ultimobranchial tissue which lacked inherent thyroid-forming potentialities, is nevertheless capable of producing undifferentiated cells. These can become induced by adjacent "active" thyroid parenchyma (and possibly other factors) to transform into thyroid-like vesicles; some of which ultimately may develop true neoplasms.

Despite the uncertainty as to whether or not the ultimobranchial body differentiates into typical thyroid parenchyma, the problem is more than an academic one. Dunhill (1931), Ward (1940), Cohn and

Stewart (1940), and others have suggested that aberrant thyroid tissue, possibly a derivative of the ultimobranchial body, is the site of origin for so-called *lateral aberrant thyroid tumors*.

According to Dunhill, as a result of imperfect growth and differentiation, misplaced bits of this tissue whether in the lateral neck or within the thyroid gland itself "may foredoom the host to carcinoma from before the day of birth." In this connection it is of interest to note that in the baboon, ectopic ultimobranchial tissue (associated with a thymus IV) apparently can give rise to accessory thyroid tissue. This spontaneous aberrant thyroid tissue originates from the wall of cysts lined by stratified squamous epithelium. It is a manifestation of ultimobranchial tissue which is not incorporated within the thyroid gland and, therefore, not induced by the proximity of thyroid parenchyma to transform into thyroid tissue. As far as the author is aware, this is the only example of accessory thyroid tissue from a portion of the caudal pharyngeal complex left outside the thyroid gland in mammals, and this strengthens the concept that ultimobranchial tissue is a "lateral thyroid." However, since the main thyroid mass showed marked hyperplasia suggesting overactivity, the stimulus for the appearance and growth of this tissue may be due to an increased demand for thyroid function during a previous stress period (adolescence?). As Ward (1940) has suggested regarding the growth of lateral aberrant thyroid tumors, such differentiation of ultimobranchial tissue may well lie dormant until the proper stimulus is furnished (e.g., perhaps increased endocrine demand during pregnancy).

On the other hand, that ultimobranchial tissue represents "more or less" indifferent tissue, which may, on occasion, be stimulated to produce thyroid tissue even outside the thyroid gland has also received support. Thymic adenomata, resembling accessory thyroid tissue, have been reported in a large number of rats following prolonged administration of thiourea. Although these "new-growths" frequently occurred coincidentally with tumors in the thyroid gland, little similarity existed between neoplasms in the two sites. They may be derived from portions of ultimobranchial tissue carried into the thymus during embryonic development, or even from hyperplastic foci of thymic epithelium. Such nodules probably develop as compensatory aberrant goiters following excessive thyrotropic hormone stimulation caused by the goitrogen. In the rat, at least, they suggest that the potentiality for ectopic thyroid tissue in the mediastinum is considerable. Similar manifestations, less marked in character but receptive to radioactive iodine, develop in thymus tissue following extirpation of the thyroid gland; and these indicate that the thymus may be a source of thyroid hormone following thyroidectomy, owing to excessive stimulation by thyrotropic hormone. There is evidence too that the ultimobranchial bodies can differentiate into thymus tissue, and possibly even parathyroid tissue, as suggested by experimentation.

In the rat, a useful animal for experimentation, ultimobranchial cysts occasionally develop spontaneously. They are most often seen in atrophic glands, in females, and in thyroids of older animals. As in sheep, the basal cell layers of these squamous cysts also produce gland-like cells. In the presence of hyperplastic parenchyma, these cells appear to transform into essentially typical thyroid tissue. In atrophic glands, these new cells sometimes form vesicles which resemble mucous acini, and staining by special method indicates a mucus reaction. Following either this cell outgrowth or separation from the basal cell component of these stratified epithelial cysts, thin-walled remnants of this epithelium persist. These represent largely only the more superficial or "suprabasal cell" layers of the original cyst. Such densely staining, thin-walled cysts characterize thyroid glands in old animals. Although these keratinizing cells are undergoing degeneration, some presumably remain temporarily viable. They are often contiguous to spontaneously developing cystadenoma; and the origin of some such neoplasms in the thyroid was assigned to these degenerating cystic masses in the past.

A number of factors (intrinsic and extrinsic) can modify ultimobranchial tissue in the rat. Ultimobranchial tissue here normally becomes indistinguishably transformed into thyroid-like tissue during embryonic development. At birth, it usually cannot be recognized as an entity separate from surrounding areas of the thyroid. According to some investigators, a similar situation holds for the development of the human thyroid gland.

In *young* rats, ultimobranchial tissue apparently functions as thyroid tissue, for storage of iodine following radioautograph techniques shows it to be little different from other thyroid areas. But this is not true when it has become cystic as in older animals. Ultimobranchial tissue is "labile," and, depending upon the age of the animal and the species, it may be represented either as a squamous cell or secretory cell type of metaplasia. In mice, for example, it is common to find mucus-secreting cysts, which may also possess cilia. Gorbman (1947) has noted that in mice these atypical follicles or cysts do not store iodine. It seems clear that ultimobranchial tissue may react differently in various species.

In view of the similarity of rat thyroid embryology to that of man, many studies have been undertaken to clarify the factors that may influence this thyroid tissue component.

Apparently, prolonged administration of a goitrogen, such as propylthiouracil, will induce ultimobranchial tissue to undergo neoplasia, but adjacent areas of thyroid tissue may also be involved, and there may even be multiple nodules. Ultimobranchial tissue is typically found at this site. Squamous cell metaplasia has also been associated with repeated injection of estrogenic hormones; but it is not certain whether these hormones induce this type of lesion in the thyroid or augment a pre-existing metaplastic site. Nevertheless, the position of these lesions is also consistent for ultimobranchial tissue. Prolonged administration of the carcinogen, methylcholanthrene also appears to be responsible for ultimobranchial tissue metaplasia, especially in mice; this may be due to the reputed estrogenic action of certain carcinogens.

Thyroids from vitamin A deficient rats *invariably* show sites of keratinizing, cystic metaplasia. Such lesions are usually in the center of both lateral lobes. Thus, they develop in the precise site occupied by the ultimobranchial body and are *predictable*. Although these changes are reversible following vitamin A therapy, certain remaining cells or proliferations from these ultimobranchial cysts may persist indefinitely after periods of repair and subsequently become tumorigenic. Chronic vitamin A deficiency will induce these lesions in animals of all ages, and this has been the basis for many experimental studies designed to determine the role of epithelial metaplasia in the evolution of certain neoplasms. It is of particular interest to note that squamous cysts arising through metaplasia can be markedly augmented by estrogenic hormones, especially during periods of reparation.

Feeding the goitrogen, allylthiourea, and the carcinogen, 2-acetylaminofluorene, simultaneously, during chronic vitamin A deficiency, in rats has resulted in the development of two rather common types of thyroid neoplasms. These can be attributed to the ultimobranchial body. One kind of experimental tumor (a solid, anaplastic or undifferentiated type) can be traced back to its source, the basal or germinative cell layers in stratified squamous epithelium, of these induced metaplastic cysts. It develops precociously and can become highly malignant. Another type (cystadenoma or cystadenocarcinoma) can be related to the so-called "*suprabasal cell*" component of these persisting epithelial lesions. These tumors have been interpreted as arising from certain densely staining cells which have become *re-vitalized* while undergoing degeneration in the wall of these squamous cell cysts. Since they represent cells that were also "differentiating toward keratinization," some viability is presumed to be retained. Under the conditions of this experiment, these tumors required a prolonged period of time for development. Both types of neoplasms may occur in the same thyroid gland.

Few investigators have described morphological precursors for thyroid neoplasms. Thyroid-incorporated ultimobranchial tissue, however, can be considered one source of tumorigenesis within the mammalian thyroid gland. Experimentally, this tissue may be "manipulated" in various ways during chronic vitamin A deficiency to demonstrate the origin and sequence of development of certain thyroid tumors.

Finally, in order to better illustrate foci of incipient tumor formation (especially the cystadenoma) which may be related to ultimobranchial tissue metaplasia, young rats (after an initial period of vitamin A deficiency) were alternately fed a normal (reparative) diet for seven days, followed by a vitamin A deficient diet for 21 days. This alternation was continued for six to eight months, during which time the animals received thiourea continuously.

Thyroid tumors stimulated by goitrogens *alone* were rarely associated with ultimobranchial cysts. Such tumors are usually of the cystadenoma type and, in these "morphologically activated" thyroid glands, they are often *incompatible* with "vestiges" suggestive of origin. Nevertheless, animals subjected to recurring periods of vitamin A deficiency *during* goitrogenesis frequently showed neoplasms contiguous to cystic remnants of ultimobranchial tissue. During this periodic epithelial repair, basal cells become separated away from cyst walls, and apparently contribute to the thyroid parenchyma (basal cells may require the action of a carcinogen for tumorigenesis). Consequently, in some instances, these cystic tumors could be *traced* directly in sections to "partially degenerated," yet differentiating, suprabasal cells which persisted in remnants of this stratified squamous epithelium.

The experimental evidence acquired here substantiates the concept that certain cells may become *re-vitalized* during degeneration and subsequently become stimulated to develop in an aberrant manner. Here, derived from ultimobranchial tissue within the thyroid gland, these cells formed *cystadenomata* under the influence of thyrotropin.

6

Regulatory Hormones

This chapter contains discussions of a number of vertebrate regulatory substances that do not fit easily into any of the previous chapters. They are either found only in mammals or are too general to classify with respect to their origins, actions or both. A kidney hormone, *erythropoietin*, several growth factors, the ubiquitous *prostaglandins* and related compounds, and the *thymus hormone* or hormones are discussed as they occur in mammals. *Kallikreins* and *kinins* are discussed, although they are often not included as hormones per se. Nevertheless their formation is similar to those events involved with the renin-angiotensin system, and they are included here. Chemical substances that function as chemical messages between organisms and influence physiology and behaviour are known for most vertebrate groups. These chemical substances are known collectively as *semiochemicals*. Finally the *epiphysial complex*, including the pineal gland, is discussed.

Erythropoietin

Erythropoietin is a glycoprotein hormone (mol wt 60,000-70,000) produced in mammals in response to hypoxia (reduced oxygen availability). The hormone was so named for its ability to stimulate erythropoiesis (erythrocyte or red blood cell formation) in bone marrow, and it is an essential factor in both fetal erythropoiesis and in adaptation to respiratory distress in adult mammals. These adaptations include increases in hematocrit, in life span of red blood cells and in hemoglobin content of these red blood cells. All of these changes contribute to greater oxygen-carrying capacity of the blood.

Erythropoietin is produced from a plasma substrate under control of the kidney via a mechanism similar to the renin-angiotensin system.

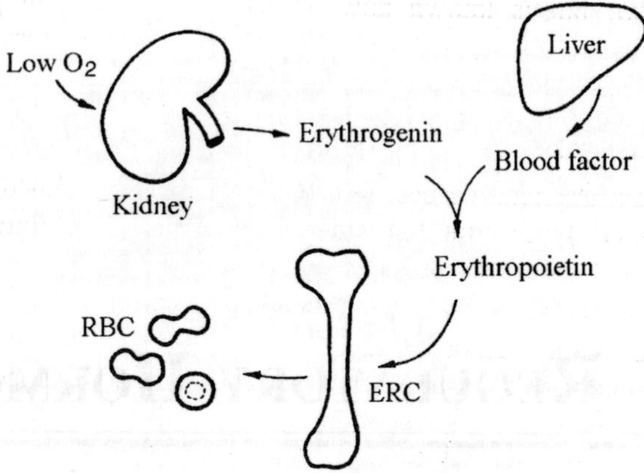

Fig. 6.1. Erythropoietin. Under conditions of hypoxia (low O_2) the kidney produces a polypeptide, erythrogenin, that interacts with a blood factor from the liver to release erythropoietin. Erythropoietin stimulates bone marrow cells (ERC) to produce more erythrocytes (RBC) which enhance O_2 transport.

Erythropoietin acts on erythropoietin-responsive cells (ER cells) in the bone marrow, causing them to differentiate into erythroblasts (erythrocyte-forming cells), which differentiate into reticulocytes and erythrocytes. At the molecular level erythropoietin induces synthesis of hemoglobin in the ER-cell, which marks the conversion of the ER cell to an erythroblast.

The kidney produces an erythropoietic factor termed *erythrogenin* that contains erythropoietin in an inactive form. An unidentified factor in the serum causes erythropoietin to be released from erythrogenin. The fixed macrophages (Kupffer cells) in the liver may also produce erythrogenin especially during liver regeneration.

Regulation of erythropoietin production is determined by availability of oxygen to kidney or liver cells. Hypoxia increases production and hyperoxia reduces production of erythropoietin. Similarly agents that normally increase oxygen consumption, such as thyroxine and dinitrophenol, increase erythropoietin production. Factors that bring about a reduction in oxygen consumption (hypophysectomy, goitrogens, starvation) decrease its production.

Growth Factors

Somatomedins

Early studies on the actions of growth hormone (GH) indicated that GH stimulated the liver to produce a blood-borne factor which mediated the action of GH on cartilage. This product was named *somatomedin*, but is known now to be several different peptides: somatomedin A, somatomedin C, insulinlike growth factors I and II (IGF I, IGF II), and multiplication stimulating activity (MSA). All of these peptides (1) are dependent upon GH for their synthesis, (2) enhance incorporation of sulfate into cartilage and (3) exert insulinlike effects on extraskeletal tissues. (Another peptide, somatomedin B, has been dropped from this list since it does not stimulate sulfate incorporation.) The four peptides comprising the MSA complex are the most potent of these factors for stimulating cell division (mitogenic).

The insulinlike activity of these peptides is a consequence of their structural similarity to insulin. They are all small proteins (5000-9000 daltons) consisting of two peptide chains with as much as a 50% overlap in amino acid sequences with insulin. Somatomedin C and IGF I are similar and in fact may be identical peptides. In plasma the levels of IGF II far exceed those of IGF I. The activity of IGF II is more like insulin than IGF I and does not support cartilage growth as well as IGF I.

Insulin is not responsible for all of the insulinlike activity observed in the blood. After removing insulin by treating blood with antibodies prepared against insulin, most of the insulin's biological activity

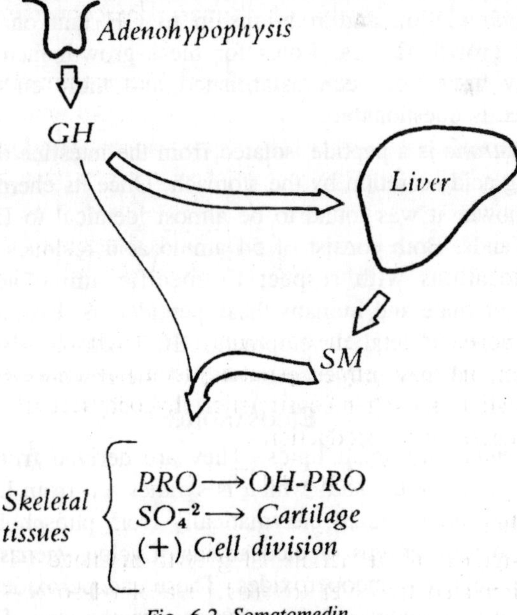

Fig. 6.2. Somatomedin.

remains. The name for the proteinaceous preparation responsible for this residual activity is *nonsuppressible insulinlike activity soluble peptide* or NSILAS. It is a heterogenous collection of peptides that has been difficult to characterize since it probably includes some or all of the other insulinlike peptides.

At least some of these "somatomedins" are essential intermediates in the actions of pituitary GH. They cooperate with insulin to produce mitogenic effects sometimes attributed to insulin action.

Other Growth Factors

Growth factors comprise a family of proteins that stimulate various aspects of embryonic development. They are also known to cause proliferation of cells or tumors in culture. These growth factors include *nerve growth factor* (NGF) and *epidermal growth factor* (EGF) isolated from the mouse submaxillary gland. *Ovarian growth factor* (OGF), *fibroblast growth factor* (FGF) and *myoblast growth factor* (MGF) have been isolated from the bovine pituitary gland. A *platelet-derived growth factor* (PGF) is produced by platelets. All of these compounds are named for their source or for a major cellular type they stimulate. They vary in size from about 6,000 to 35,000 daltons. Nerve growth factor actually exhibits structural similarity to insulin but lacks the cartilage stimulation and relationship to GH that characterize the insulinlike growth factors. Roles for these growth factors in normal physiology have not been established and their endocrinological significance is questionable.

Urogastrone is a peptide isolated from the intestine that is capable of blocking acid secretion by the stomach. Once its chemical structure became known, it was found to be almost identical to EGF from the salivary glands. Both consist of 54 amino acid residues and differ at only 12 locations with respect to specific amino acids. During pregnancy of mice and humans these peptides increase and may play important roles in fetal development. EGF activity also appears in human milk and may influence growth of the newborn as well.

EICOSANOIDS

Eicosanoids are small lipids. They are derived from a common precursor, *arachidonic acid*, which is synthesized from linolenic acid. Arachidonic acid liberated enzymatically from phospholipids can be converted to four groups of eicosanoids. Cyclooxygenase transforms arachidonic acid into endoperoxides. These endoperoxides are used to synthesize *prostaglandins*, *prostacyclin* or *thromboxanes*. Drugs such as aspirin and indomethacin inhibit cyclooxygenase and block their syntheses. A separate enzyme, 5-lipoxygenase, forms the *leucotrienes* from arachidonic acid. This enzyme is not inhibited by aspirin and indomethacin. Elucidation of these compounds and their synthetic pathways resulted in the awarding of the 1982 Nobel Prize in Physiology or Medicine to three principal researchers; Sune Bergstrom, Bengt Samuellson and John Vane.

Prostaglandins

In 1933 Maurice Goldblatt in England discovered some lipids in human seminal plasma with some peculiar properties. At about the same time, U.S. von Euler in Sweden found similar substances in extracts prepared from sheep vesicular glands, and he later named these compounds *prostaglandins* (PGs) based upon what he believed to be their source in man, the prostate gland. In 1956 Sune Bergstrom succeeded in elucidating the structures for 16 PGs. They are all related to the basic structure of prostanoic acid, and they can be separated into four classes on the basis of structural differences. The most commonly occurring PGs are PGE_1, PGE_2 and $PGF_{2\alpha}$. Prostaglandins are not restricted to the male genital tract, and they have been found in most tissues of both males and females.

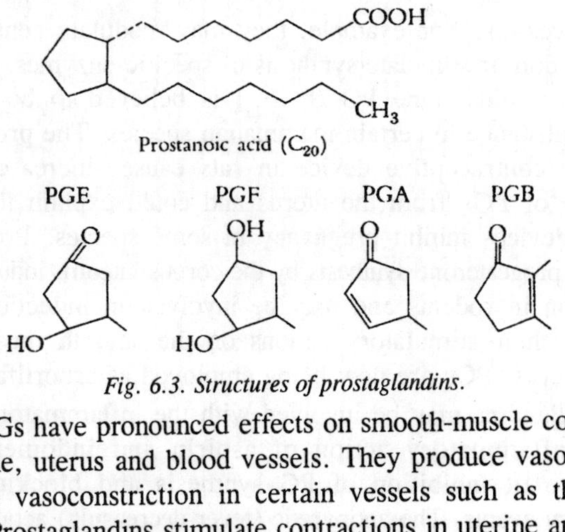

COOH

CH_3

Prostanoic acid (C_{20})

PGE PGF PGA PGB

O OH O O

HO HO

Fig. 6.3. Structures of prostaglandins.

The PGs have pronounced effects on smooth-muscle contraction in the intestine, uterus and blood vessels. They produce vasodilation but may cause vasoconstriction in certain vessels such as those of the placenta. Prostaglandins stimulate contractions in uterine and intestinal smooth muscle. Bioassays for PGs capitalize on these smooth-muscle effects or on vasodepressor effects.

Prostaglandins at first seem to be involved in a wide variety of unrelated physiological processes in addition to effects on smooth muscles

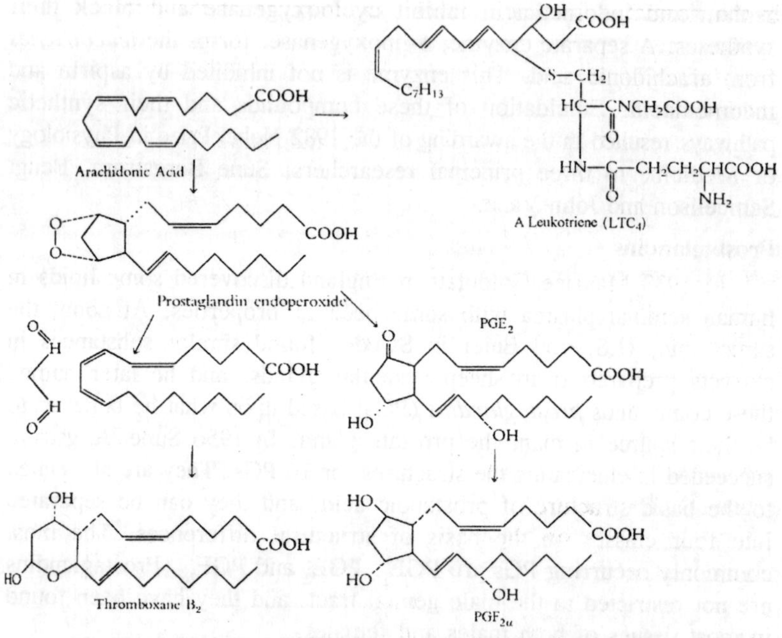

Fig. 6.4. Relationship of leukotrienes, prostaglandins and thromboxanes.

and blood vessels. For example, they may modulate central nervous system function or stimulate synthesis of specific enzymes, testosterone and corticosteroids. One PG ($PGF_{2\alpha}$) is believed to be the uterine luteolytic substance in certain mammalian species. The presence of an intrauterine contraceptive device in rats causes increased synthesis and release of PGs from the uterus and could explain the basis for how such devices inhibit pregnancy in some species. Prostaglandins also reduce progesterone synthesis by the corpus luteum, induce ovulation and lactation in rodents and may be involved in induction of labor. Because of their stimulatory actions on the smooth muscles of the pregnant uterus, PGs are now being employed as arbortifacients.

Prostaglandins may be involved with the inflammatory response. The anti-inflammatory action of aspirin and indomethacin is a consequence of inhibition of PG synthesis and blocking of early inflammatory events. The antipyretic (fever-decreasing) action of aspirin is also related to inhibition of PG synthesis. Consequently, these drugs have proved to be useful tools in studying the physiology of PGs. Glucocorticoids are also anti-inflammatory, but they produce their effects by interfering with the participation of leukotrienes and kinins in the normal inflammatory response. Because of the differences in mechanisms involved, aspirin and indomethacin are termed *nonsteroid anti-inflammatory drugs* (NSAID) to distinguish them from glucocorticoids.

A different role for PGs may be as mediators in the mechanism of action of peptide hormones that stimulate cAMP production in target cells such as for luteinizing hormone (LH), thyrotropin (TSH) and parathyroid hormone. Prostaglandins may also mediate the effects of estrogens on release of pituitary LH. One must not generalize too broadly on the role of PGs with respect to cAMP formation since the effects may be highly tissue specific. For example, PGE_1 mimics corticotropin (ACTH) and TSH in the adrenal and thyroid gland respectively by stimulating cAMP formation. In adipose tissue where cAMP formation and lipolysis are stimulated by epinephrine and glucagon, PGE_1 inhibits cAMP formation and blocks lipolysis.

Prostacyclin

Prostacyclin (PGI_2), a compound closely related to the prostaglandins of the E, F, A and B series, is generated by the walls of many blood vessels. This eicosanoid is a potent inhibitor of blood platelet aggregation and inhibits blood clotting. Prostacyclin is synthesized from the same precursors as the prostaglandins.

Thromboxanes

Researchers discovered the thromboxanes during studies on prostaglandin metabolism and action. Thromboxane A_2 causes translocation of free calcium ions to bring about changes associated with the shape of the platelet. It is this change in platelet shape that allows platelets to aggregate and facilitate clotting. Thromboxanes also may be released from the platelet and cause local constriction of vascular smooth muscle. This might enhance clotting by reducing the diameter of the arterioles and slowing blood flow through the capillary beds.

Leukotrienes

Leukotrienes are a novel group of at least 15 related compounds occurring in five structural classes. Their formation from arachidonic acid was discovered in 1979, and there is still much to be learned concerning their physiological roles. They are synthesized and released by white blood cells in response to injury or invasion of foreign antigens. Leukotrienes contribute to inflammatory or allergic response by causing contraction of vascular smooth muscle and increasing vascular permeability. Glucocorticoids may produce their anti-inflammatory effects by limiting the availability of free arachidonic acid and thereby blocking synthesis of the leukotrienes. Synthesis of leukotrienes is not influenced by NSAIDs.

Thymus Gland

The function of the mammalian thymus was unknown until after the middle of the twentieth century when its role in the immune response system was recognized. Prior to 1961 it was presumed to play some ill-defined role in juveniles but not in adults, since its rapid deterioration began at puberty. Jacques F.A.P. Miller discovered that thymectomy of newborn mice was followed by a short period of normal growth after which the animals suddenly became ill and soon died of a wasting disease. Autopsy revealed these mice were severely deficient in lymphocytes, which normally account for 70% of the mouse white blood cells. It was soon established that the thymus serves as the source or "seed bed" of lymphocytes normally found in other lymphoid tissues, including the spleen, lymph nodes and the small concentrations of lymphoid tissue in the wall of the intestine (Peyer's patches). Not only are thymectomized mice deficient in lymphocytes, but their ability to produce circulating antibody is also impaired. Furthermore, thymectomized mice readily accept foreign tissue grafts. Inoculation of thymectomized mice with foreign lymphocytes may result in a "graft-versus-host" reaction in which the foreign cells immunologically "reject" the host. These observations led to the conclusion that lymphocytes participate in immunological responses. Such interactions are referred to as cell mediated.

Thymectomized mice that receive grafts of thymus from mice of the same inbred strain (to reduce occurrence of graft-versus-host reactions) do not develop the wasting disease, and their lymphoid tissues produce lymphocytes at normal rates. When the grafted thymus came from a mouse with a cytologically distinct abnormal chromosome (chromosome marker), it was discovered that the lymphocytes being produced were genetically like cells of the host and did not contain the marker chromosome of the graft donor. These results suggested that the thymus does more than just produce lymphocytes mitotically and laid the basis for suspecting existence of a thymus hormone that controlled differentiation and maturation of lymphocytes.

We know now that the thymus stimulates differentiation and maturation of *T-lymphocytes* in the thymus as well as in other lymphoid tissues. These T-lymphocytes are responsible for cell-mediated immunity including transplant rejection, graft-versus-host reactions, delayed hypersensitivity to foreign antigens, resistance to viruses and immune surveillance of tumor cells. Humoral immunity (circulating antibody) is also influenced by T-lymphocytes. Circulating antibody is produced by *plasma cells*, derived from another type of lymphocyte, the *B-lymphocyte*, which resides in the bone marrow. There are different populations of T-lymphocytes that can aid or suppress antibody formation. Failure of T-lymphocytes to differentiate in the absence of thymus hormone explains observations originally made for thymectomized mice.

The thymus actually produces a variety of stimulatory and inhibitory humoral agents that seem to be important in differentiation, immune response, endocrine functions, calcium regulation, cell growth and metabolism. Ten agents have been isolated from thymus tissue that might be considered candidates for thymic hormones, and at least eight more factors are suspected. Only five of these agents have been reasonably characterized chemically. All of these agents are present in thymus, and all disappear from the circulation following thymectomy.

The most promising preparation for hormone status is one termed *thymosin*. There are actually at least two thymosin peptides. Thymosin-α_1 consists of 28 amino acids (M.W. = 3108) and stimulates mitosis in T-lymphocytes. The gene for thymosin-α_1, has been synthesized. Thymosin-β_4 is a slightly larger peptide (43 amino acids, 4982 daltons).

One of the other candidates isolated from thymus is *lymphocyte-stimulating hormone* (LSH), which occurs as a small heat-labile protein (molecular weight 8000) and as a conjugated form (mol wt about 15,000). Both of these LSH substances increase the proportion of lymphocytes among the white blood cells and confer the ability to synthesize antibody lo a foreign antigen. *Thymosterin* is a steroid derivative isolated from thymus that inhibits tumor growth and stimulates antibody synthesis and lymphocytopoiesis. The *homeostatic thymic hormone* is a relatively small molecule (mol wt about 2000) that contains amino acids, amino sugars and possibly a nucleotide. This substance, appears to play a permissive role in the negative feedback effects of several hormones on the hypothalamo-hypophysial axis as well as being immunologically active itself. *Thymopoietin* has been characterized chemically as a peptide (49 amino acids, 5562 daltons) which stimulates bone marrow to produce thymic lymphocytes.

KININS

Kallikreins are serine proteases that bring about release of small peptides called *kinins*. All kallikreins are glycoproteins with molecular

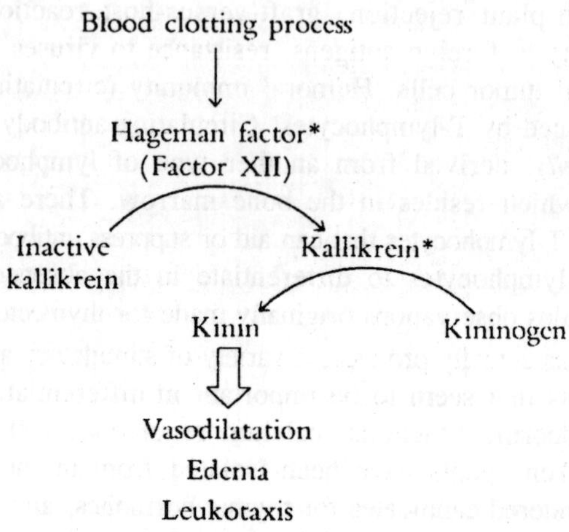

Fig. 6.5. Kinin production and action.

weights between 25,000 and 40.000. They were first found in pancreas and blood, and were named from *kallikreas*, the Greek word for pancreas. The name is a misnomer because kallikreins have been characterized from pancreas, urine (kidney), blood and salivary glands and also reported in lung, intestine, brain and nerve. Kallikreins are structurally and functionally similar to the proteolytic enzymes of snake venom and to trypsin. All of these enzymes can cause release of kinin from the substrate *kininogen*. There are different kininogens that correspond to specific kallikreins. These kininogens are large glycoproteins (about 120,000 daltons). Kinins are hypotensive agents. A specific kinin, the nonapeptide *bradykinin*, is produced by renal kallikreins and is thought to relax smooth muscle of arterioles in the renal vascular bed.

Pancreatic kallikrein is found in zymogen granules and is released into pancreatic juice rather than blood. Kallikreins may activate digestive enzymes (trypsin, chymotrypsin, etc.). They are also believed to convert many prohormones to active hormones (e.g., prorenin, proinsulin, proglucagon). In addition, kallikreins are implicated in fertilization (proteolytic penetration by spermatozoa), implantation, cell growth and possibly in embryonic development and differentiation. Kallikreins are found in secretory granules containing growth factors and may be responsible for release of peptides such as EGF and NGF.

Kinins have been reported from mammals, birds and most reptiles but have not been demonstrated in snakes, amphibians, teleosts, holocephalans or elasmobranchs. The absence of kinins in anamniote vertebrates has not been explained.

Chemical Communication

The phenomenon of chemical communication among animals includes a variety of compounds or semiochemicals that are secreted externally and affect the physiology or behaviour or both of other individuals. Such compounds may function as intraspecific signals, as interspecific signals or both. The first category includes conspecific semiochemicals termed *pheromones*, and the second category consists of the interspecific (or transspecific) *allelomones*.

Table 6.1. Classification of vertebrate semiochemicals

	Pheromones (Intraspecific)
Primer pheromones	Reproductive maturation, growth inhibitors/stimulants
Signal pheromones	Courtship, mating; territorial and trail marking; recognition: sex, species, parent/young, dominant/subordinate; alarm; location of food; use of garlic by man to keep his house free of vampires
	Allelomones (Interspecific)
Allomones (advantageous to emitter)	Repellents; venoms; growth inhibitors
Kairomones (advantageous to recipients)	Attractants from prey organisms; alarm

The intraspecific category of pheromones includes conspecific semiochemical signals that influence the physiology and/or behaviour of the recipient. These chemical substances are involved with sexual attraction, group aggregation, avoidance or dispersal (alarm) responses, synchrony of gamete maturation, induction of mating behaviour, trail or territory marking, individual recognition and food location. Pheromones may be subdivided into *primer* and *releaser* pheromones. Primer pheromones initiate a chain of physiological events in the recipient animal such as the hormonal events that lead to ovulation in female mammals. Releaser pheromones elicit a more or less immediate behavioural response in the recipient such as the induction of copulatory behaviour in male monkeys. A more appropriate term for these substances might be *signal pheromones* in that the recipient may be made aware of the sexual state of the emitter, for example, but may choose not to respond at that moment to the signal. In general, pheromones are highly species-specific molecules, and in some cases even two closely related species produce chemically distinct molecules for the same purpose that do not elicit responses by the other species.

There are two subcategories of allelochemic substances or allelomones. The first consists of *allomones*, which are substances that change the behaviour of the recipient so that the emitter benefits. The "perfume" emitted by skunks is a good example of an allomone. Interspecific substances that benefit only the recipient and not the emitter are termed *kairomones*. The attraction of mosquitos by L-lactic acid, a component of human sweat, is an example. These functional definitions cause some confusion, for one molecular species may be classified as more than one type of semiochemical.

All of these semiochemicals are compounds of small molecular weight that readily diffuse through water or air. These substances are extremely potent, and only a few hundred molecules arriving at an appropriate chemoreceptor are necessary to elicit a response. This sensitivity is not unexpected when one considers the problem of directing a chemical into the environment appropriately to reach the intended recipient in sufficient concentration to elicit a response.

Although the concept of semiochemicals has achieved scientific respectability only in recent years, their existence was suggested in early writings. Reproduction of the partridge *Perdix* was described in the 12th century Bestiary as follows: "Desire torments the females so much that even if a wind blows towards them from the males, they become pregnant from the smell." Even earlier, a Roman, Pliny the Elder, reported a variety of substances that could function as abortifacients such as menstrual secretions of other women or castoreum secreted by the beaver. These agents reportedly were effective when a pregnant woman came near. They were also burned in lamps during orgies. It is difficult to see how such substances would naturally function in chemical communications as do currently recognized pheromones, allomones and kairomones.

Much of the research with semiochemicals has been performed with insects and vertebrates, but this phenomenon is widespread in the animal kingdom as well as in plants, protistans and monerans. The discussion here will be limited to vertebrates.

Mammalian Pheromones

For centuries scientists have known that mammals use glandular secretions or urine to mark territories. Rabbits possess a chin gland that secretes a pheromone used for marking territory boundaries, and the behaviour of male dogs in marking upright objects with their urine is known to scientist and layman alike. Musk-secreting glands are found in many sexually mature mammals and may be involved in a variety of behaviours, including territory marking, defense and mating. Pheromones have also been implicated in alarm signaling, individual or group recognition, including maternal-young interactions, and in promotion of aggression.

Sex pheromones in mammals

Many reports have appeared with respect to sexual pheromones in laboratory mammals, including mice, gerbils, guinea pigs and hamsters. Crowded female mice enter anaestrus when no males are present (Lee-Boot effect). However, simply the odor from a male mouse can cause them to synchronously reenter estrus (Whitten effect). The endocrinological basis for these effects is suggested by observations that pheromones from female mice suppress pituitary release of follicle-stimulating hormone (FSH), whereas male pheromone stimulates FSH release that is followed in normal sequence by LH release and ovulation. A newly mated female will abort if placed with a "strange" male (not the previous mate), and the incidence of abortion increases with genetic dissimilarity of the strange male to the male with whom she was mated. If offspring result, they are always from the second mating (Bruce effect). This effect has also been observed in voles and may not be peculiar to laboratory mice.

Table 6.2. Some mammalian responses to primer pheromones

Bruce effect	A newly impregnated female will abort and return to estrus if exposed to odor from a strange male.
Whitten effect	An odor transmitted via male urine accelerates any synchronizes the estrous cycles in females.
Lee-Boot effect	Crowding of large numbers of females causes suppression of estrous cycles; smaller groups tend to exhibit pseudo-pregnancies.
Ropartz effect	Adrenal enlargement and increased production of corticosterone occurs when isolated mouse is exposed to the odor of other mice.
Vanderberg effect	Odor from a male accelerates sexual maturation in a female.

The sexual pheromones involved in the Lee-Boot and Bruce effects are probably modified steroids (steroid metabolites) and are transmitted via the urine of the male to the olfactory apparatus of the female.

Male mouse urine induces and accelerates estrous cycles of females (Whitten effect), and the effect is most pronounced on Lee-Boot groups of females. The time of vaginal closing in females is also influenced by male urine. Anosmic females (animals whose nostrils have been blocked or whose olfactory bulbs have been removed surgically) do not respond to male urine.

Pregnant and lactating rats produce pheromones that influence other females. Odors from pregnant females shorten the estrous cycle of nonpregnant females so that more females will be pregnant at the same time. Presumedly it is advantageous to have many females give birth at the same time so that the females can cooperate in pup care. In contrast, odors from a lactating female with pups lengthens estrous cycles of nonpregnant females. Thus a socially dominant lactating female can suppress fertility of other females until she is again in estrus herself. A similar lactating pheromone may be produced by gerbils.

Males may also be influenced by female pheromones. Pairing of a previously paired male mouse with a strange female results in elevation of plasma testosterone, indicating that endocrine responses of both males and females may be influenced through bisexual encounters. Proximity of ewes in estrus increases plasma testosterone levels in mature rams, implying that a female estrogen-dependent pheromone is responsible. The display of estrous odor preference in male beagles is an androgen-dependent behaviour induced by female pheromones. Removal of the olfactory bulbs has no effect on sexual behaviour of male hamsters, but removal of the vomeronasal organ as well as the olfactory bulbs abolishes copulatory behaviour. Removal of only the vomeronasal organ blocks copulatory behaviour in only about one third of the hamsters, indicating an important interaction between this organ and olfactorily influenced sexual behaviour.

The precise sources of many pheromones are not known. Those transmitted via the urine may originate in either the liver or kidney or both. Some steroids are converted by apocrine epidermal glands to pheromones in rabbits, and these pheromones readily diffuse into the air from the surface of the animal. Musk is a steroid derivative that is produced by specialized epidermal glands. However, not all semiochemicals produced by mammals are derived from steroids. For example, the distinctive allomone of skunks consists of several mercaptans. Phenylacetic acid has been identified as the scent-marking pheromone secreted by the ventral scent-marking gland of male

Musk odor of human urine

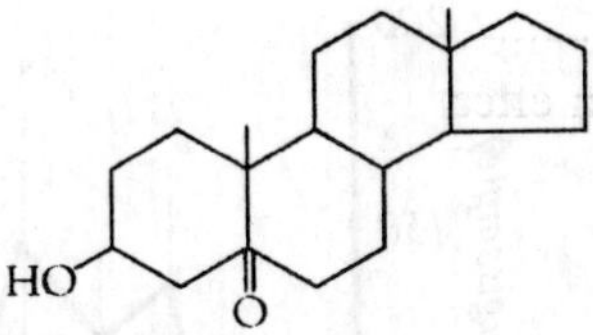

Territory-marking pheromone from black-tailed deer

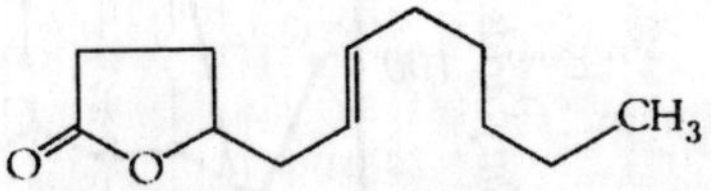

Antelope scent-marking pheromone — $(CH_3)_2CHCH_2COOH$ Isovaleric acid

Gerbil scent-marking pheromone — O—CO—CH_3 Phenylacetic acid

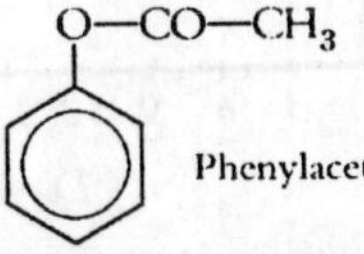

Sex attractant complex of volatile fatty acids from vaginal secretions of female rhesus monkey

CH_3COOH Acetic acid

CH_3CH_2COOH Propionic acid

$(CH_3)_2CHCOOH$ Isobutyric acid

$(CH_3(CH_2)_2COOH$ Butyric acid

$(CH_3)_2CHCH_2COOH$ Isovaleric acid

Fig. 6.6. Structure of some mammalian pheromones.

Mongolian gerbils, and the active substance secreted by the male pronghorn antelope to mark territory is isovaleric acid, a small fatty acid. The secretion of the tarsal gland of black-tailed deer is also a nonsteroidal lipid. Dimethyl disulfide has been identified as the attractant pheromone of vaginal smears obtained from hamsters, and its production may be estrogen dependent. Many of the pheromonal secretions of males described above appear to be influenced by androgens. Although all sexual pheromones may not be steroids or steroid derivatives, it is clear that secretions of most male or female sexual pheromones or the responses to them are dependent upon androgens or estrogens.

Primates also produce semiochemicals utilized in territorial marking and reproduction. Female rhesus monkeys produce a mixture of fatty acids of low molecular weight in vaginal secretions that stimulates sexual interest of males and may induce mounting behaviour and

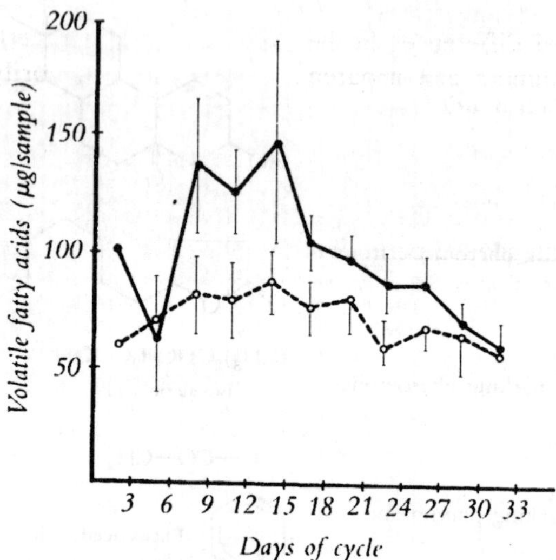

Fig. 6.7. Oral contraceptives and effects on composition of vaginal secretions collected over 3-day intervals.

ejaculation. The major volatile fatty acids produced are acetic acid, butanoic acid, propanoic acid, methylbutanoic acid and methylpropanoic acid. Synthetic mixtures of these fatty acids in appropriate ratios stimulate male interest in females. Males are not interested in females receiving steroidal contraceptives, suggesting that the releasers of overt male sexual behaviour are hormone-dependent agents associated with the normal ovulatory cycle. Estrogens stimulate fatty acid secretions, and progesterone is inhibitory; observations that correlate well with levels of fatty acids observed in vaginal secretions throughout the menstrual cycle. Human vaginal discharges exhibit a similar variation in fatty acid composition, although human females produce a much greater percentage of acetic acid than do rhesus monkeys. The use of oral contraceptives effectively obliterates the preovulatory increase in volatile fatty acids. The behavioural implications of these observations are not clear.

Although studies with humans are complicated by a number of psychological and social considerations, some evidence exists for production of pheromones and their roles in reproduction. A "dormitory effect" of menstrual synchrony has been described for all-female living groups. Even though it is generally accepted that the olfactory sense in humans is limited as compared to most mammals, several studies have shown definite sensitive olfactory discriminations, including sexually based differences in the abilities to perceive certain odors. Trained perfumers can apparently distinguish olfactorily between different skin and hair types, and some psychiatrists claim to be able to smell schizophrenics because of abnormal production and elimination of trans-3-methylhexanoic acid. The ability to detect some odors is sex dependent, such as the greater sensitivity of women to "boar taint" associated with spoiled pork. Whether these differences are related to possible pheromone production and behaviour in humans remains to be shown.

Chemical Communication in Teleostean Fishes

Intraspecific chemical communication is employed by teleosts for sexual and individual recognition, sexual attraction, induction of courtship behaviour, parent-young interactions, aggregation (schooling) and dispersal (alarm) behaviour. Purified semiochemicals have not been isolated and verified under natural conditions, and it must be recognized that the use of such terms as pheromone requires verification.

Individual chemical recognition has been shown in the minnow *Phoxinus*, the blind goby *Typhlogobius californiensis* and the yellow bullhead *Ictalurus natalis*. Studies of the last species demonstrated that a dominant bullhead could recognize subordinates and distinguish them from intruders. When threatened, subordinate fish sought out shelter controlled by the dominant bullhead, who would tolerate their presence until the danger was past. Intruders were not tolerated even under these conditions. Anosmic fish (olfaction blocked experimentally) could not make these distinctions.

Evidence for sexual attracting substances and substances that elicit courtship behaviour have been suggested for sticklebacks, shad, salmon, brown bullhead, several blennies and a goby. Experimental studies with *Bathygobius soporator* have confirmed production by the female of a pheromone that elicits courtship behaviour in males. The pheromone appears to come from the ovary. Segregated anosmic males show no courtship response to introduction of ovarian pheromone to their aquarium. A similar response occurs in the goldfish *Carassius auratus*. The early claim for production of a pheromone (copulin) by male guppies, *Poecilia reticulata*, which stimulated copulatory responses in female guppies has not been confirmed.

Several species including *I. nebulosus*, *T. californiensis*, *Heterochromis bimaculatus*, *Nannacara anomala* and *Cichlasoma nigrofasciatum* can distinguish their own broods of young fish from other young of the same species on the basis of olfactory cues. In the jewel fish *H. bimaculatus*, the young presumably produce a pheromone that stimulates certain aspects of parental behaviour that is recognized by only the parents. Fry of *Cichlasoma citrinellum* also exhibit chemical recognition of parental odors.

Although schooling behaviour includes a very strong visual component, experimental evidence supports a role for an aggregating pheromone in at least one species. This *aggregating pheromone* reduces swimming activity and causes individuals to remain near one another. Evidence for dispersing or *alarm pheromones* has been found. There is some lack of specificity among closely related species in responses to these alarm pheromones, and they might be more appropriately termed kairomones. However, interfamilial responses are slight, as are responses of species that are closely related phylogenetically but are geographically isolated. Certain species do exhibit a flight reaction when presented with the scent of a common predator. Brown bullheads have been observed to guard their nests in response to scent from a potential predator of their young.

Chemical Communication in Amphibians

Evidence for chemical communication in amphibians comes primarily from studies performed with urodeles. Salamanders and newts utilize semiochemical as well as tactile information, especially in courtship. Urodeles possess *hedonic glands* or *cloacal glands* or both that appear to be sources for pheromones. Anurans on the other hand lack hedonic and cloacal glands, and there are no convincing data to support the use of pheromones or allelomones in anuran reproductive behaviour.

Plethodontid and desmognathid salamanders rely largely on tubular *mental* (chin) hedonic glands for stimulation of courtship behaviour. Most of these salamanders are nonmigratory, largely terrestrial species. Migratory or nonmigratory semiaquatic salamanders (for example, Ambystomidae, Salamandridae) rely more on cloacal glands, particularly the abdominal gland of males that supposedly produces hedonic-type secretions. The other cloacal glands are responsible for production of spermatophores. Many behavioural observations on salamanders (*Ambystoma*) and several newts (*Taricha*, *Triturus*, *Notophthalmus*, *Cynops*) have implicated involvement of cloacal secretions in courtship.

Attractant pheromones have been reported for *Taricha granulosa*. Males are attracted upstream to a sponge soaked in female odor. Laboratory experiments employing a simple olfactometer device demonstrate that male *T. granulosa* exhibit directed movements toward air passing over either males or females, but females exhibit random movements with respect to "newt air" versus "non-newt air".

Evidence for territorial marking has been reported for *Plethodon cinereus* and *P. jordani* involving chin-touching and nosetouching behaviours. Scent production may be related to either the epidermal hedonic mental glands or to unsaturated lipids secreted by *nasolabial glands*. Only plethodontid salamanders possess nasolabial glands, and the use of nosetouching to transfer chemicals is probably unique to these salamanders.

Many amphibians have granular integumentary *poison glands* distributed over various areas of the body that probably discourage would-be predators. These glands secrete alkaloid toxins that readily meet the definition of allomones. Some of these toxins are extremely potent, and the simple act of handling the animal may produce irritation to the skin. Attempted consumption of a western newt (*Taricha*) may be a terminal event in a dog, although other organisms, such as garter snakes, do not appear to be affected by this toxin. Ingestion of these newts is fatal to humans, too. Certain South American frogs (for example, dendrobatids) secrete powerful neurotoxins used by Indians to tip their poison arrows.

A curious relationship has been documented for a possible growth-inhibiting pheromone in anurans. Crowding of tadpoles causes reduced growth rates, presumably due to secretions from the largest tadpoles. This growth retardation was first attributed to the presence of an algal cell in the feces. However, it is not clear whether the algal cell produces the growth-inhibitor pheromone or whether the tadpole does, since the phenomenon has been observed in the absence of any algal cells. Conditioned medium (that is, water that has contained crowded tadpoles) prepared from cultures of *Rana pipiens* tadpoles inhibits growth of tadpoles of ten other anuran species, indicating this pheromone could function as an allomone as well. Other studies have suggested that these effects are only consequences of crowding, and resultant behavioural interactions. Additional research may resolve these conflicting interpretations.

Anuran larvae may produce recognition pheromones. The ability of tadpoles to distinguish siblings from nonsiblings is thought to involve detection of chemical cues.

Chemical Communication in Reptiles

The use of semiochemicals (at least pheromones and allomones) for communication can be inferred from experimental data gathered for all major reptilian groups. However, visual information appears to be the primary mode for communication as it is in their feathered descendants, the birds. This is especially true in certain families of lizards (for example, Iguanidae, Agamidae) in which visual signals are primary determiners of both courtship and territorial interactions. Nevertheless, pheromones may be important chemical signals for eliciting male to male aggression and other behaviours. '

Lizards of several families (Scincidae, Lacertidae, Teidae, Gekkonidae) possess holocrine *femoral glands* on the midventral surface of each thigh and apocrine glands associated with the proctodeum (cloaca). These glands are prominent in males and show an androgen-dependent increase in secretory activity during the breeding season. Many behavioural observations suggest that olfaction plays an important role in courtship and territorial behaviour in lizards possessing femoral glands, and the femoral glands or the proctodeal apocrine glands or both may be sources of pheromones. Considerable research must be done before a pheromonal function of these glands can be confirmed, however.

Snakes, unlike their legged relatives, are well known for the role olfaction plays in their life histories and generally have well-developed glandular structures associated with the cloaca. Emission of a somewhat disagreeable material (allomone?) under the stresses of handling are known to even small children who have attempted to collect garter snakes (genus *Thamnophis*). Numerous studies have demonstrated involvement of odors in snake behaviour, including trail-following behaviour, sex recognition and preference for (aggregation) or avoidance of conspecifics. Distinct chemical differences for the lipid portion of cloacal emissions have been found for 25 species of snakes representing three families. These observations provide a chemical basis for species-specificities in cloacal emissions and a potential for species-recognition mechanisms.

When rattlesnakes are given a choice of a freshly envenomed mouse or a mouse killed by cervical dislocation, the snakes prefer the envenomed mouse. These observations suggest that either the venom itself or some substance released by the envenomed mouse attracts the snake and acts as a releaser of swallowing behaviour.

Holocrine mental glands of terrestrial and semiterrestrial chelonians produce saturated and unsaturated fatty acids that appear to have communicative significance. Models coated with mental gland secretions from male *Gopherus belandieri* attracted both males and females.

Moreover, the model elicited aggressive behaviour from the males. Apparently sex recognition can be accomplished through mental gland secretions. The size of these glands and the quantity of secretion may be an expressin of dominance in male tortoises. Inguinal and axillary glands (collectively termed Rathke's glands) in chelonians secrete musk and appear to be involved in defensive behaviour. Furthermore, Rathke's glands are involved with sexual behaviour in mud turtles and musk turtles. Chelonians also have cloacal glands that have been implicated in trail marking and other behaviours.

Crocodilians have been suspected of employing pheromones dispersed via secretion of musk glands. Both male and female alligators, *Alligator mississippiensis*, release musk scents during excitement and courtship. Semiochemicals may also be involved in territorial displays. Similar behavioural observations have been reported for the Nile crocodile *Crocodylus niloficus*, although no pheromones have been identified experimentally.

Chemical Communication in Birds

Birds are known to be highly oriented toward visual displays as well as vocalizations. Much of their physiology and resultant behaviours are influenced by photoperiod and temperature. Although a few studies have established the role of olfaction in location of prey, there is little evidence for either pheromones or allomones in birds. A possible sexual attractant has been reported for the duck, *Anas platyrhynchos*. Considering the importance of olfaction in nonavian vertebrates with respect to intraspecific and interspecific chemical communication, it is unexpected to find an entire vertebrate class without even suggestive evidence for these substances. But then, birds are certainly unique in many other respects, so may be it is not so surprising after all. Obviously the apparent absence of pheromones and allomones has not hindered their evolutionary success.

Epiphysial Complex

Almost all vertebrates exhibit one or two epithalamic structures that constitute the epiphysial complex. The components of this complex are the *pineal organ* and a more anterior projection, the *parapineal organ*. In fishes, amphibians and some reptiles (lizards), these organs are basically sac-like diverticula that are more or less open to the third ventricle of the brain. They consist of a basal portion composed of sensory and ependymal (supportive) cells and may have an attached stalk with a distal end vesicle that contacts the dorsal brain case. These structures probably arose in primitive fishes as a pair of

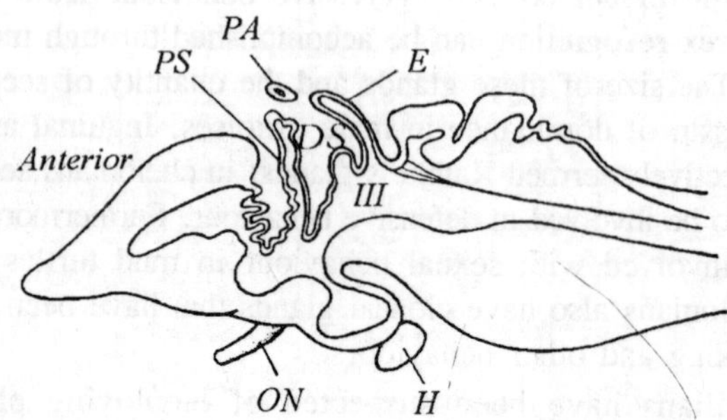

Fig. 6.8. The generalized epiphysial complex. E, Epiphysis cerebral or pineal organ; PA, parietal or parapineal organ; PS, paraphysis; DS, dorsal sac; III, third ventricle; ON, optic nerve; H, hypophysis.

diverticula that later changed positions relative to one another. A well-developed parapineal organ has been retained only in cyclostomes and lizards, whereas the pineal organ is found in all vertebrate groups with the exception of crocodilians. In anamniotes as well as in lizards the pineal organ has retained its sensory functions, but in other reptiles, birds and mammals the pineal appears to be only an endocrine structure and is often termed the *pineal gland*.

Two additional prominent dorsal evaginations of the brain occur in this same region: the paraphysis and the dorsal sac. The anteriormost evagination that actually develops from the telencephalon is known as the *paraphysis*. The paraphysis is best seen in amphibians as a highly vascularized, sac-like diverticulum; it may function similarly to the choroid plexus in producing cerebrospinal fluid. The *dorsal sac* arises in most vertebrates as a diencephalic (epithalamic) evagination just posterior to the paraphysis but anterior to the epiphysial complex. It is especially prominent in the ganoid fishes (Chondrostei, Holostei) and becomes less conspicuous in teleosts. In most vertebrates the dorsal sac contributes to formation of the choroid plexus.

The pineal complex is connected to an adjacent ependymal structure, the *subcommissural organ* of Dendy (SCO). The ependymal cells of the SCO produce an aldehyde-fuchsin positive secretion rich in disulfide bonds and cysteine similar to that observed in the pineal ependyma. The major secretory product of the SCO is a noncellular fiber that in some species extends into the central canal of the spinal cord for its entire length. This structure is known as *Reissner's fiber*. Its significance is not clear. The SCO and its Reissner's fiber have been described in vertebrates from cyclostomes to mammals. Originally it was supposed that Reissner's fiber was involved in regulating posture

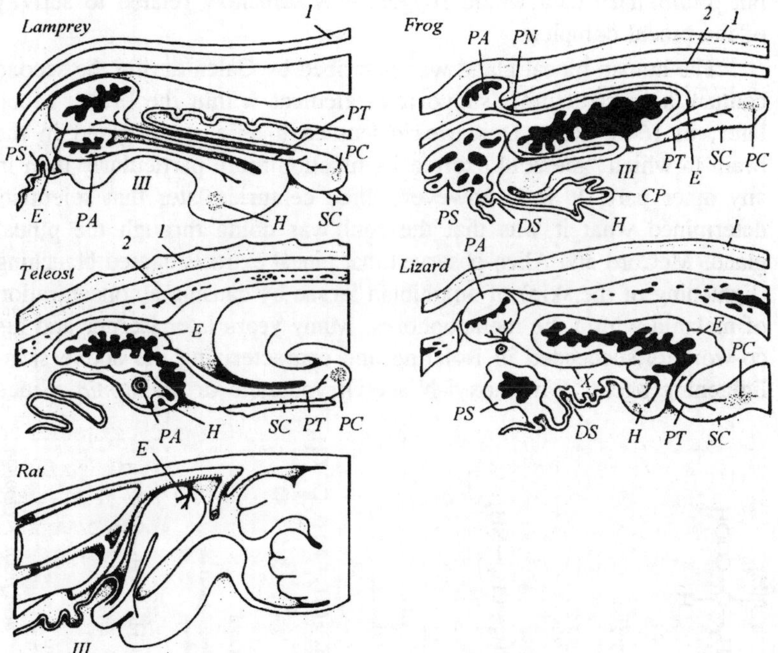

Fig. 6.9. Selected epiphysial complex. 1, skin, 2, skull; III, third ventricle; CP, choroid plexus; DS, dorsal sac; E, epiphysis cerebri or pineal organ; H, habenular commissure; PA, parietal or parapineal organ; PC, posterior commissure; PN, pineal nerve; PS, paraphysis; PT, pinal tract; SC, subcommissural organ; X, parietal nerve.

through tension produced in it by flexion of the body. This tension presumably operated through influences of Reissner's fiber on pressure-sensitive neurons. A more plausible suggestion is the possibility that Reissner's fiber contributes to formation of cerebrospinal fluid. Formation and dissolution of the fiber into the cerebrospinal fluid has been documented as a temperature-dependent process in the frog *Rana esculenta*. Reissner's fiber also binds biogenic amines (epinephrine, norepinephrine) present in the fluid in both *R. esculenta* and in mammals (cow, cat), suggesting still another role. Studies with mammals and reptiles imply a relationship among the pineal complex, the SCO and the adrenal cortex, but the nature of that relationship remains somewhat obscure. Cytological activation of the SCO in the lizard *Lacerta s. sicula* has been correlated positively with seasonal activities of adrenal cortical cells and of testicular Leydig cells. The actual role or roles for the SCO and its secretory products must await further research, but preliminary data would suggest it is somehow related to activity of the pineal complex.

The human pineal gland was described by Galen during the second century as a structural (supportive) element within the brain. Much later, in 1646, Rene Descartes discussed it as a small gland in the brain in which "the soul exerted its function more particularly than in any other part. It was, however, three centuries later that scientists determined what it was that the soul was doing through the pineal gland. McCord and Alien observed that pineal extracts caused blanching (lightening of the skin) of amphibian larvae by causing a concentration of melanin within the melanophores. Many years later Lerner and his co-workers succeeded in isolating and characterizing the active skin-lightening agent, 5-methoxyl-N-acetyltryptamine or *melatonin*. Since

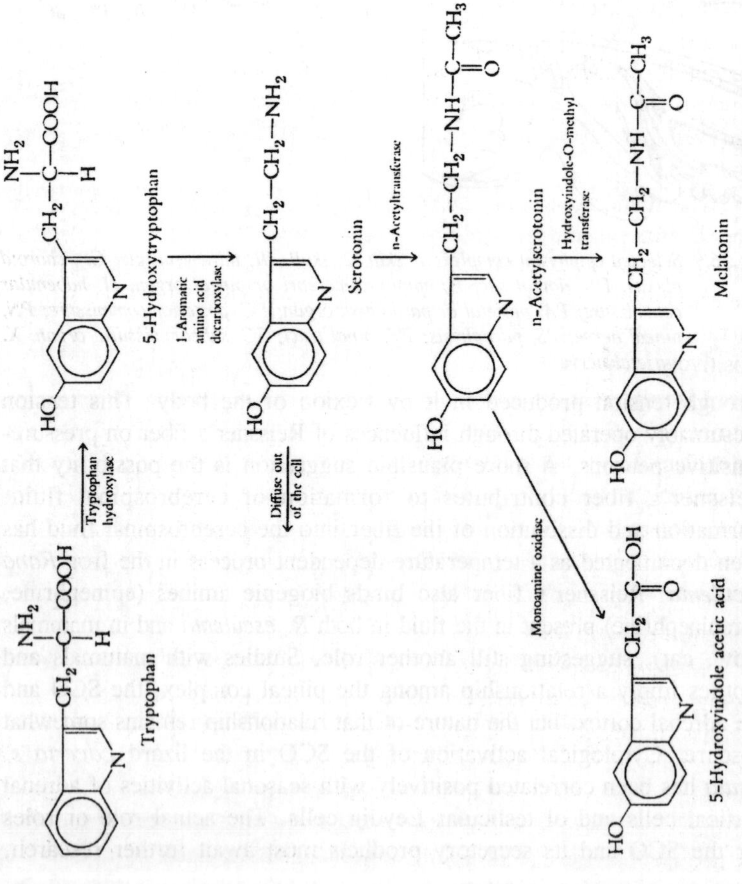

Tryptophan → (Tryptophan hydroxylase) → 5-Hydroxytryptophan → (1-Aromatic amino acid decarboxylase) → Serotonin → (n-Acetyltransferase) → n-Acetylserotonin → (Hydroxyindole-O-methyl transferase) → Melatonin

Serotonin → (Diffuse out of the cell)

Serotonin → (Monoamine oxidase) → 5-Hydroxyindole acetic acid

Fig. 6.10. Synthesis of serotonin, melatonin and 5-hydroxyindole acetic acid from tryptophan.

that time a number of biologically active agents and related compounds have been isolated from pineal tissue including serotonin (5-HT), *N-acetylserotonin*, *5-methoxytryptophol* (5-MTP) and *5-hydroxytryptophol* (5-HTP). The initial substrate for their synthesis is the amino acid tryptophan. Two enzymes have been studied extensively with respect to their role in regulating melatonin synthesis: *N-acetyltransferase* (NAT) and *hydroxyindole-O-methyltransferase* (HIOMT). N-acetyltransferase converts 5-HT to N-acetylserotonin, which in turn is converted to melatonin by HIOMT. The conversion of 5-HTP to 5-MTP is also catalyzed by HIOMT. Another enzyme complex, *monoamine oxidase* (MAO), acts upon 5-HT, converting it to *5-hydroxyindoleacetic acid*. Thus, two enzymes, NAT and MAO, compete for the same substrate, 5-HT, and their relative activities could represent mechanisms for regulating melatonin synthesis.

Early observations suggested that HIOMT was the rate-limiting enzyme for melatonin synthesis, and several studies describe correlations between HIOMT activity and melatonin synthesis. More recent studies, however, have established NAT as the rate-limiting enzyme in melatonin synthesis.

The mammalian and avian pineals are innervated by sympathetic fibers, and variations in norepinephrine synthesis are correlated with pineal functions. Circadian rhythms in tyrosine hydroxylase activity, the rate-limiting enzyme for norepinephrine synthesis, have been reported in rats. Moreover, the rhythm associated with this enzyme correlates positively with observed rhythms in HIOMT activity and NAT activity.

Like the PGs, melatonin has been shown to produce a rather limited number of cellular effects (Table 6.3). The major enzymes influenced by melatonin are those involved in steroid transformations, for example, 5α-reductase and MAO. The actions of melatonin on MAO activity and consequent effects on tissue 5-HT levels may explain effects observed on pancreas, pituitary, testes and the pineal itself following administration of melatonin.

Pineal and Rhythms

Melatonin levels in the blood exhibit a distinct diurnal rhythm being greater at night than during the day. This circadian rhythm persists under constant dark conditions. Plasma melatonin rhythm is a consequence of a circadian rhythm in NAT activity. This enzymatic rhythm is controlled by neural signals from the suprachiasmatic nucleus of the hypothalamus. Information on photoperiod detected by the retina is responsible for entraining the suprachiasmatic nucleus to light dark

Table 6.3. Summary of nonreproductive actions of melatonin and the pineal in mammals

Target	*Description*
Melanophores (melanocytes)	Melatonin implants in weasels, *Mustela erminea*, causes them to grow white coats (typical of winter) in the spring instead of brown coats.
Hair	Melatonin inhibits hair growth of intact or pinealectomized mice.
Connective tissue	Pinealectomy reduces permeability of subcutaneous connective tissue.
Adrenal cortex	Pineal substance, adrenoglomerulotropin, claimed to stimulate aldosterone release.
	Pineal may alter release of ACTH from adenohypophysis.
Parathyroid	Pinealectomy of rat caused hypertrophy of parathyroids, which was reduced by administration of pineal extract or melatonin.
Cardiovascular system	Vasopressor activity reported for pineal extracts, probably due to presence of AVT.
Immune response	Chronic administration of pineal extracts caused leukocytosis, lymph node hypertrophy and an increase in mitotic activity in the spleen. Probably it was a simple immunological response to antigens in the extract.
Thyroid	Melatonin or pineal extracts inhibit thyroid function, possibly through regulation of TSH release from the adenohypophysis.

cycles. The suprachiasmatic nucleus probably controls a number of circadian rhythms in mammals.

Pineal Secretion and Reproduction

So many potential physiological roles of the pineal gland have been reported that it is second only to prolactin (PRL) and the PGs in the number of possible roles. The major effects of the pineal gland in mammals probably relate to reproduction. In 1941 Fiske reported that keeping rats under conditions of constant light increased the frequency of estrus. Several years later Wurtman discovered that pinealectomy also increased the frequency of estrus in rats maintained under normal photoperiods, and a surge of investigation was launched into possible roles of photoperiod, the pineal gland and melatonin in controlling sexual maturity and reproductive cycles in mammals. A mass of data appeared that suggested melatonin released from the pineal gland acted through either the blood or cerebrospinal fluid or through both on the hypothalamus or directly on the pituitary to lower circulating LH levels. The early morning increase of circulating PRL in male rats has also been correlated with the release of some agent from the pineal. Presumably light inhibits sympathetic input to the pineal, resulting in decreased melatonin synthesis and increased levels of LH. Increased LH was considered the basis for increased estrus. The mechanism governing PRL has not been elucidated.

Recent studies have provided a different explanation for the effects of light on estrus. Pinealectomy or injection of massive doses of melatonin never produces marked effects on rat reproduction, and some workers have not found any effect of melatonin on rat reproduction. One study, in fact, reported stimulation of rat gonads by melatonin treatments. The golden hamster *Mesocricetus auratus* exhibits marked gonadal collapse when subjected to short photoperiods (less than 12 hours of light per day). Pinealectomized hamsters do not exhibit gonadal collapse when subjected to short photoperiods, and subdermal melatonin implants (in Silastic capsules) cause testicular atrophy in hamsters maintained on long photoperiods. If rats are made anosmic (olfaction blocked either mechanically or surgically) and are blinded, more marked gonadal atrophy occurs than was seen following pinealectomy. Furthermore, pinealectomy or melatonin treatment negates the effects of blinding and anosmia in rats as it does in hamsters. Additional support for a stimulatory gonadal role for melatonin has been reported for the ferret, in which melatonin may be responsible for bringing the animal out of photorefractoriness. Melatonin treatment restores the gonadal growth response to long photoperiod in postbreeding ferrets maintained on long photoperiods.

If melatonin is not an antigonadal agent as previously thought, how are the data which indicate an antigonadal role for the pineal explained? Early observations on the nature of pineal extracts showed that, in addition to melatonin and related compounds, peptide fractions could be prepared that exhibited a variety of biological activities. It is known that ependymal cells of fetal human and rat pineals synthesize arginine vasotocin (AVT) which has also been found in adult mammalian pineals. Arginine vasotocin may not be present in all mammalian pineals. Furthermore, the AVT sequence identified immunologically in the pineal could be a fragment of a larger peptide. If AVT is administered to neonatal mice during the period when the brain is undergoing sexual differentiation, increased growth of reproductive organs upon entering adulthood is observed. In contrast, if AVT is administered after the brain has undergone sexual differentiation the growth of accessory organs and in some cases the gonads themselves is inhibited.

The compensatory hypertrophy of the remaining ovary after unilateral ovariectomy is a response to increased gonadotropin levels caused by an effective reduction in circulating estrogens. Arginine vasotocin administered intraperitoneally or directly into the third ventricle of the brain prevents compensatory ovarian hypertrophy (COH). Much less AVT is required if it is administered through the third ventricle than if it is given intraperitoneally. Several related octapeptides including arginine vasopressin, lysine vasopressin and 4Leu-AVT also inhibit COH, although oxytocin does not. All of the active compounds have an identical ring structure and a basic amino acid at position 8. Treatment of these active molecules with mercaptoethanol disrupts the disulfide bridges necessary for maintenance of the ring structure. Such reduced octapeptides no longer prevent COH; in fact, they enhance it. Melatonin is not as effective as melatonin-free preparations in preventing COH.

Arginine vasotocin exhibits effects on 5α-reductase and MAO activities similar to those reported for melatonin. In fact, AVT or a similar peptide may prove to be the physiological regulator. An antigonadotropic peptide has been isolated from bovine pineals that prevents LH release from the pituitary. This peptide is not AVT since it possesses no oxytocin-like activity in biological assays and has no arginine residues. Release of LH or FSH from rat adenohypophysial cells in culture, however, is not influenced by AVT over a concentration range of 10^{-18} to 10^{-7} moles/liter culture medium. These data would support a role for AVT at the level of the hypothalamus. The presence of other hypothalamic peptides including oxytocin, arginine vasopressin, thyrotropin-releasing hormone and somatostatin has been demonstrated in human pineals. Any of these might be candidates for the illusive pineal inhibitory peptide.

Reiter et al. proposed a scheme that relates the actions of melatonin and pineal antigonadotropic peptides (PAGs) with respect to the observations discussed above as well as to the presence of calcium deposits (*corpora arenacea*) characteristic of some mammalian pineal glands. Since the pineal peptides appear to be structurally similar to

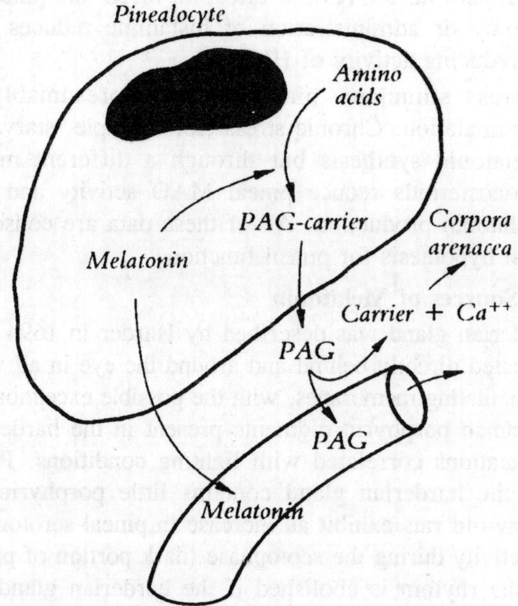

Fig. 6.11. Role proposed for pineal melatonin and pineal antigonadotropic peptides (PAGs) on reproduction.

the neurohypophysial octapeptides they suggest that PAGs are also stored bound to a carrier protein. Neurophysin-like molecules have been obtained from bovine and human pineals. These peptides may be related to the neurohypophysial octapeptides present in the pineal or to other pineal peptides synthesized in a similar manner. Consequently the mechanism for secretion of PAGs may also be similar. They suggest that PAGs are released into the interstitial fluids by exocytosis, which also deposits the walls of the secretory vesicles as "exocytotic debris." The PAG-carrier complex presumably interacts with calcium ions, producing a Ca^{++}-carrier protein complex that results in release of free PAGs that can diffuse into the blood. The Ca^{++}-carrier protein complex interacts with exocytotic debris to produce corpora arenacea. Melatonin would act by inhibiting synthesis or release or both of PAGs in the pineal cell.

Other Factors Affecting the Pineal

In addition to the well-known actions of light on the pineal gland, hypophysectomy, stress and gonadal steroids all influence pineal function. Androgens (testosterone, dihydrotestosterone) inhibit MAO activity in the pineal, which in turn causes increased melatonin synthesis. Estrogens have a reverse effect on MAO and pineal activity. Hypophysectomy or administration of histamine reduces melatonin synthesis by reducing activity of HIOMT.

Acute stress stimulates pineal activity, presumably through sympathetic stimulation. Chronic stress (for example, starvation) also increases melatonin synthesis but through a different mechanism. Elevated corticosteroids reduce pineal MAO activity and allow for increased melatonin production. All of these data are consonant with an antigonadal hypothesis for pineal function.

Extrapineal Sources of Melatonin

The harderian gland was described by Harder in 1694 in the red deer. It is located directly behind and around the eye in all vertebrates that possess nictitating membranes, with the possible exception of higher primates. Reddish porphyrin pigments present in the harderian gland undergo fluctuations correlated with lighting conditions. Prior to 12 days of age the harderian gland contains little porphyrin pigment. Blinded 12-day-old rats exhibit an increase in pineal serotonin as well as HIOMT activity during the scotophase (dark portion of photoperiod cycle), but this rhythm is abolished if the harderian glands are also removed.

Melatonin has been demonstrated in the rat harderian gland, and continuous illumination causes enlargement of the rat harderian gland and an increase in HIOMT. Harderian HIOMT differs from the HIOMT found in the pineal gland and from that found in the retina of the eye. In contrast, continuous illumination decreases pineal weight and pineal HIOMT activity. The importance of these observations to overall involvement of pineal indoles or harderian indoles to observations on reproduction or other pineal-influenced processes remains to be determined.

The retina of the eye may be another viable source for melatonin, but the nocturnal increase in circulating melatonin is of pineal origin. Melatonin has also been found in the colon of rats, but the significance of this observation is unknown.

Comparative Aspects of the Epiphysial Complex

Class Agnatha: Cyclostomata

The epiphysial complex of cyclostomes consists of a pineal organ and a parapineal organ, although the latter may be lacking in some species. Both organs possess variably shaped end vesicles that project dorsally against the roof of the brain case. The end vesicles contain sensory and ependymal cells structurally organized to suggest that they are photoreceptors. Efferent neural fibers from the pineal organ end at the posterior commissure, whereas those from the parapineal terminate at the habenular commissure. These organs probably relay photoperiodic information to other regions of the brain.

Nocturnal blanching has been observed in ammocetes larvae, and epiphysial levels of HIOMT are correlated with this nocturnal lightening in *Geotria australis*. Removal of the epiphysial complex causes persistent expansion of melanophores in *Lampetra planeri* and *G. australis* but not in *Mordax mordacia* which lacks a parapineal organ. This melanophore response may be mediated via the parapineal. Hypophysectomy of cyclostomes results in blanching due to melanophore contraction, and the action of the parapineal principle may be to inhibit release of melanotropin.

There is evidence for a possible antithyroid effect of the pineal organ. Pinealectomy may inhibit metamorphosis of the ammocetes in all of the above-mentioned lampreys, but confirmatory data are needed.

Class Chondrichthyes

The shark pineal organ contains photosensory and supportive cells. No experimental data, however, have been reported, and the functional importance of the pineal is not known for these fishes. A parapineal organ has not been described for any species in this group.

Class Osteichthyes: Teleostei

The epiphysial complex of teleosts consists of a pineal organ that is extremely variable in both size and degree of development. There is usually a prominent lumen, which in some cases is open to the third ventricle. Photosensory cells and ependymal cells are present in the pineals of several species. A reduced parapineal has been described in some teleosts.

Pigmentation, responses to light and thyroid changes are influenced by the pineal organ. Melanophore changes have been reported in several species of teleost, including rainbow trout, following injection of pharmacological amounts of melatonin. Some of these species also respond in a similar manner to epinephrine treatment. Other species respond only to epinephrine with no reaction to melatonin. Circulatory melatonin levels exhibit no correlations to adaptation by rainbow trout to different backgrounds, indicating that melatonin may not be a factor influencing normal pigmentary responses in this species. Furthermore, levels of HIOMT in rainbow trout pineals are not altered by either continuous light or darkness.

The white sucker *Catostomus commersoni*, is responsive to light intensity and/or thermal gradients. This sensitivity is mediated through the pineal. Shielding the sucker's pineal from light causes the fish to choose the warmer portion of horizontal temperature gradients or the better illuminated portions of chambers maintained at a constant temperature. When the shield is removed, the fish returns to its original preference. The pineal may provide information on light intensity for use in mediating thermal behaviour. When there is no temperature gradient, retinal receptors alone determine the location of the fish.

Structural correlations have been described between the pineal organ and phototactic responses by fishes. Species with a translucent covering over the pineal organ (a definitive pineal spot) exhibit predominantly positive phototaxis, whereas species with a pigmented, opaque skeletal covering do not show phototaxis. Species that have pigment cells located so that dispersal and concentration of pigment granules could regulate the intensity of light reaching the pineal organ exhibit responses varying from positive phototaxis to no response. The pineal organ might be involved in some other functions for the species showing no phototaxis. The presence of a pineal spot is more common in deep-sea fishes than in freshwater or shallow-water marine species and may relate to the influence of light on vertical migrations performed by deep-sea fishes.

Pinealectomy of *Poecilia reticulata* (guppy) causes pituitary enlargement and hyperplasia of the thyroid. This effect also occurs in *Fundulus heteroclitus* if pinealectomy is performed during the winter months (December through March). Pinealectomy between February and June produces no effect on the thyroid, however. Pituitary and thyroid of the characin, *Astyanax mexicanus*, are not affected by pinealectomy, but both stimulation and inhibition of the goldfish thyroid have been described. A possible influence of the pineal on gonadal development has been reported for *F. heteroclitus* and *F. similis*, but no relationship was found in goldfish or in *A. mexicanus*. Additional studies performed on a seasonal basis involving a large number of species are needed before any definitive statements can be made with respect to the pineal and reproductive or thyroid functions. Differences in photoperiod regimens could explain these differences.

Melatonin has been measured in the retina of rainbow trout, and HIOMT is present in the retinas of several teleostean species. Levels of melatonin in trout retina exceed levels reported in the pineal.

Class Amphibia

The epiphysial complex of amphibians, like that of most vertebrates, consists only of a pineal organ. The frontal organ has not been investigated for its capacity to synthesize melatonin. The proximal or basal portion of the amphibian epiphysial complex as well as the retina of the eye contain HIOMT activity and melatonin. Pinealectomy reduces circulating melatonin to daytime levels, suggesting that the retina may be responsible for basal levels of melatonin in the blood.

As mentioned earlier, the role of melatonin on melanophores in larval Amphibia was first suggested by the observations of McCord and Allen. Since that time it has been shown that melatonin is the pineal agent responsible for the blanching of tadpoles or larval salamanders when held in the dark. As little as 0.0001 μg/ml medium causes aggregation of melanin granules in melanophores of *Xenopus laevis* tadpoles.

Attempts to relate pineal function or melatonin with thyroid function have been equivocal. Pinealectomy of tadpoles of the midwife toad *Alytes obstetricans* accelerates metamorphosis, but a similar operation in larvae of the newt *Taricha torosa* was without effect. Earlier observations in *Bufo americanus*, however, indicated that feeding mammalian pineal to tadpoles accelerated metamorphosis. Pinealectomized larval tiger salamanders, *Ambystoma tigrinum*, exhibit decreased thyroidal uptake of injected radioiodide (^{131}I), but neither purified melatonin nor commercial bovine pineal powder influences iodide uptake of intact larvae. Certainly additional studies of the relationship of pineal factors to thyroid function would help to resolve some of these apparent contradictions.

Reproduction may be under inhibitory influence of the pineal organ, at least in anurans. Accelerated gonadal development follows pinealectomy of *A. obstetricans* and *Hyla cinerea*. Gonadotropin-induced ovulation from *Rana pipiens* ovaries in vitro is inhibited by addition of melatonin to the culture medium. Bovine pineal extract similarly inhibited human chorionic gonadotropin-induced spermiation in male *R. esculenta*, but purified melatonin had no effect. This dichotomy of melatonin's influence in male and female anurans warrants further investigation. The possible influence of pineal principles on reproduction in urodeles has not been studied.

Unlike the other gnathostomous vertebrates some anurans have retained a well-developed end vesicle known as the *frontal organ* or *stirnorgan*. Because of the presence of photosensory cells the frontal organ is often referred to as the parietal eye.

Class Reptilia

Reptiles can be separated into several groups on the basis of the anatomy of the epiphysial complex. Melatonin has been localized in the blood, pineal gland and retinas of snakes, lizards and turtles. Although a pineal is absent in alligators, melatonin is present in the blood. Presumably this melatonin is of retinal origin. Lizards possess an elaborate sac-like, pigmented pineal organ containing both sensory and ependymal cells. The lumen of the lizard pineal lies close to the third ventricle but does not join with it. A parapineal organ penetrates the skull, forming a parietal spot on the surface. The parapineal is often termed the *parietal eye*. Turtles and snakes have retained only the basal portion of the pineal organ and have lost the end vesicle and stalk of the pineal organ as well as the complete parapineal organ. Nevertheless the turtle pineal organ is the best-developed epiphysial structure in vertebrates. The crocodilians apparently "have seen fit to discard" the entire epiphysial complex.

The parietal eye of lizards has been examined with respect to a number of events, including thyroid function, thermoregulation and reproduction. The lizard parietal contains HIOMT activity suggesting that it synthesizes melatonin. Removal of the parietal eye stimulates thyroid hyperplasia and oxygen consumption. These data suggest that melatonin or some other pineal principle influences pituitary function, as reported for teleosts, although direct effects are not ruled out entirely. It has been proposed that the reptilian parietal eye is a photothermal radiation dosimeter that monitors solar radiation and, in turn, regulates activity patterns of lizards. Indeed, excision of the pineal or parietal eye alters thermal responses of lizards. This suggestion has been expanded and supported with extensive documentation, and it may well be the major functional role for the epiphysial complex in lizards.

A definite effect of the epiphysial complex on reproduction has been reported. Excision of the parietal eye of the lizard *Anolis carolinensis* stimulates ovarian development in reproductively quiescent animals. This effect is blocked by administration of melatonin. The onset of gonadal recrudescence in this lizard is induced by long photoperiod and warm temperatures. The parietal eye appears to be the transducer through which photoperiod influences reproduction.

Class Aves

The pineal organ of birds has been reduced to the glandular basal portion, the pineal gland. No parapineal or remnant thereof is present. Structurally the avian pineal exhibits considerable diversity, and it is composed of several cellular types. The avian pineal is innervated by sympathetic fibers as reported for mammals.

Avian pineals are biochemically like their mammalian counterpart. Variations have been reported in HIOMT and NAT activities with respect to lighting conditions, but the most dramatic effects involve NAT. Activity of this enzyme exhibits a marked increase with onset of the scotophase, a peak about the middle of the scotophase and a decrease rapidly following the onset of the photophase. Brief exposure to light at the peak of NAT activity causes a rapid reduction to photophase levels. Melatonin rhythms that correlate with rhythms in pineal NAT activity have been reported for brain, pineal, retina and serum of birds. The brain and especially the hypothalamus in birds may be the primary site of action for melatonin and may explain effects of melatonin on gonadal function, thermoregulation and locomotor activity.

The role of the pineal in the reproductive biology of birds may be progonadal. Pinealectomy inhibits androgen synthesis, whereas administration of melatonin stimulates androgen synthesis, presumably by altering gonadotropin release from the adenohypophysis. Pinealectomy of quail delays ovarian development, an observation that also supports a progonadal role. Melatonin injections cause a decrease in gonadal weight suggesting that the progonadal agent might be a peptide. Marked species differences may occur, as evidenced by depression of testicular function in the duck following pinealectomy. However, no effects of melatonin treatment on parameters of gonadal development and their relationship to photoperiod could be demonstrated in either white-throated sparrows or border canaries.

Pinealectomy abolishes endogenous body temperature rhythms as well as free-running locomotor activity rhythms in house sparrows, *Passer domesticus*. Effects on locomotor activity appear to involve two pathways, one of which can bypass the pineal organ. Pinealectomized birds that have lost free-running locomotor activity in the dark still exhibit entrainment to light-dark cycles, supporting the presence of a bypass system. Both systems can be entrained by light, but the pineal has control over the bypass system.

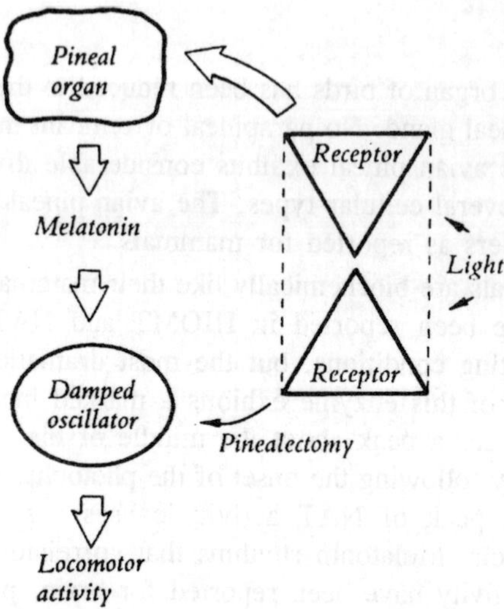

Fig. 6.12. Role of pineal in controlling circadian locomotor rhythm in birds.

Evolution of Melatonin's Functions

An intriguing hypothesis for the original function of melatonin and the evolution of other functions is based on the presence of melatonin synthesizing systems in retinas, parietal eyes and pineals and the observations that pineals of more primitive vertebrate groups are photoreceptors. Evidence suggests that both retinal and pineal melatonin exhibit night-time (scotophasic) peaks of synthesis. This hypothesis proposes that melatonin was initially a local hormone for regulating the distribution of melanosomes in the retina. During the day, melanosomes are dispersed in retinal pigment cells which protect the photoreceptors from intense light. At night, elevated melatonin causes concentration of melanosomes and allows dim light to maximally stimulate the photoreceptors. Similar mechanisms presumedly operate in the photoreceptive outer portion of the pineal, in the parietal eye and in the amphibian frontal organ. The increase in melatonin synthesis during the scotophase causes a greater proportion of melatonin to appear in the blood. Consequently, the scotophasic elevation in melatonin is a reliable internal cue for obtaining information about seasonal photoperiods. Information concerning length of the scotophase is reflected in circulating melatonin levels. Thus, according to this hypothesis, the diurnal rhythm in melatonin has been co-opted as the blood borne signal entraining a number of other internal events during the evolution of vertebrates.

7

AVIAN REPRODUCTIVE HORMONES

Reproduction is a cyclic phenomenon, and the majority of avian species are seasonal breeders whose physiological mechanisms regulating gametogenesis are synchronized by environmental stimuli, ensuring that young are produced at the time of year best suited for their survival. Only in some domesticated species, or in those species inhabiting an environment showing little seasonal variation in food availability, has this ancestral cyclic pattern sometimes been lost. Generally, the reproductive organs of most birds undergo a great annual variation in size and functional activity; the whole reproductive process from copulation through the fledging of the young is usually crowded into a few weeks, particularly in those birds that undergo long migrations to high-latitude breeding grounds.

The reproductive system of birds repeats the basic vertebrate pattern. Thus, the seasonal fluctuations in gonadal activity are produced by the environmental stimuli exerting their influence through a central nervous mechanism that leads to gonadotropin release from the adenohypophysis. This is mediated, as appears to be common throughout the vertebrates, by the liberation of neurohumoral substances from the median eminence of the neurohypophysis, these being transported by portal vessels to the pars distalis. The neuroendocrine system, therefore, constitutes a finely integrated and balanced coordinating link between the organism and its environment. The system has evidently had a long evolutionary history, since neurosecretory cells have been identified in primitive coelenterates, platyhelminths, annelids, mollusks, and arthropods, as well as in all the vertebrate classes. Therefore, we have restricted our dissertation to a consideration of the primary sexual organs, the accessory sexual structures, and the associated behavioural aspects that are concerned with the process of reproduction. We have also excluded any extensive consideration of the phenomenon of intersexuality, which is common in certain avian species.

In common with those of all other vertebrate groups, the avian gonads are responsible both for the proliferation of gametes and also for the secretion of the steroid sex hormones that control the development and functional activity of the accessory sexual structures and secondary sexual characteristics. In mammals, these two functions are known to be regulated by the secretion of two distinct gonadotropic hormones from the pars distalis of the anterior pituitary gland, namely, the follicle stimulating hormone (FSH), which is primarily responsible for the regulation of gametogenetic activity of the germinal epithelium, and the luteinizing hormone (LH), which in the male regulates the secretory activity of the interstitial Leydig cells and in the female induces ovulation and the formation of ovarian corpora lutea. In birds, it has sometimes been suggested that the gonads may be regulated by only one type of gonadotropic secretion, but pituitary cytology distinguishes two gonadotropic cell types as in the mammalian situation, and distinct FSH-like and LH-like fractions were earlier obtained by Fraps et al. (1947) from chicken pituitaries using a method previously applied to ovine glands. More recently, more purified samples of these two gonadotropic hormones have been separated from the chicken pituitary gland, and their specificity has been established by biological assay. There is also evidence of two anatomically distinct areas in the hypothalamus of the Japanese Quail responsible for the regulation of secretion of FSH and LH, respectively. Thus, electrolytic or radio frequency lesions in the anterior regions of the infundibular nuclear complex result in regression of the seminiferous tubules without any apparent depression of interstitial cell activity, whereas lesions placed in the medial, ventral, and posterior regions produce, in males, a regression of both gametogenesis and interstitial cell activity, and in females, a cessation of ovulation but no regression of the ovary or oviduct. Similarly, there is good evidence in cockerels, too, that the posterior infundibular region controls the release of an avian LH, whereas the release of FSH is associated with an anatomically separate region.

The male and female gonads are derived from a pair of sexually undifferentiated primordia associated with the intermediate mesoderm (nephrotome). The primordial germ cells within these structures are derived from the embryonic splanchnopleur and migrate in the blood to be housed in these locations and become the presumptive germinal epithelium. They sink below the surface into the connective tissue (stroma). The left presumptive gonad receives a greater compliment of primordial germ cells than the right, and thus establishes an asymmetrical gonadal development which generally persists throughout life. Initially, proliferation of the germinal epithelium in both presumptive gonads in either sex forms a potential testis (medullary tissue). In the female a second proliferation of cells gives rise to a cortex in the left gonad which then becomes the potential ovary. In some species, particularly the domestic hen, a few cortical cords may also sometimes be laid down in the embryonic right gonad as well.

In males, the embryonic gonadal primordia develop into paired testes, but in the female of many species only the left organ develops into a functional ovary, and the right generally remains in an ambisexual state. When the left ovary is removed, or when it becomes non-functional due to some pathological condition, a compensatory development of the rudimentary gonad may take place under the influence of the increased circulation of gonadotropin. In the large majority of cases in which this occurs the rudiment develops into a testis or ovotestis. The experimental induction of this condition in chickens has demonstrated that age can be a modifying factor on the result; full spermatogenesis can develop in the resultant ovotestis if the operation is done at an early age, but if performed in older birds full spermatogenesis is rarely achieved, but ovulations can take place. Under natural conditions, many old domestic hens suffering from senile changes become masculinized, assuming the cock plumage and the capacity to crow, because the rudiment of medullary (testicular) tissue becomes functional. Such sex reversal is also relatively common and often spectacular in pheasant species in which, for example, a somber coloured female Golden Pheasant (*Chrysolophus pictus*) can, as a result of such changes, assume the resplendent plumage of a male. A high incidence of intersexual individuals may also occur in strains of the domestic pigeon, in which apparently a delay in the degeneration of the embryonic cortical tissue causes genetic males to develop with an intact right testis but a left testis in which the cortical component persists and differentiates into ovarian tissue.

Occasionally, adult birds are found with two functional ovaries, particularly among members of the Accipitridae, Falconidae, and Cathartidae, but even in these specimens it is sometimes unaccompanied by a corresponding development of the associated right oviduct. Functional right ovaries are also frequently found in pigeons and in the Herring Gull (*Larus argentatus*).

Individuals without gonads are also sometimes found. Usually these lack both right and left Mullerian ducts and exhibit a general masculine appearance, thus resembling subjects experimentally castrated as embryos. Taber (1964) points out that in man gonadal agenesis or the Turner syndrome is accompanied by loss of one of the sex chromosomes, producing a neuter genotype resembling the female (genetic composition XO which resembles XX). Since males are the homogametic sex in birds, it may be significant that birds lacking gonads resemble the male.

Testis

The essential sex organs in the male bird are the paired testes which, unlike mammals in which they usually become housed in an extraabdominal sac, are located permanently in the body cavity just ventral to the anterior end of the kidneys. Each is attached by a short mesorchium to the dorsal body wall. They are supplied, together with their immediately contiguous ducts, by the spermatic arteries from the dorsal aorta; the testicular veins drain into the vena cava. Because of the asymmetry already established in the presumptive embryonic gonad, the left testis is commonly larger than the right, although in the pigeon and in the turkey, the reverse is frequent.

Each testis is an ovoid, encapsulated body surrounded by a substantial fibrous coat, the tunica albuginea, with a fragile serous outer sheath, the tunica vaginalis. In seasonally breeding species, the testis is subject to great annual variations in size, sometimes by as much as 400- to 500-fold. In the Japanese Quail, for example, testicular weights can be induced to increase from 8 mg to 3000 mg within 3 weeks when birds are artificially exposed to a stimulatory 20 hour photoperiod. Such gross extremes are common in seasonally breeding species and imposes a great strain on the ensheathing tunics, which are replaced annually by a proliferation of new fibroblasts rebuilding a new capsule from beneath the old weakened covering. This upsurge of fibroblasts usually occurs during the postnuptial (regeneration) phase of the testicular cycle, and Marshall (1961a) has suggested that this may be automatically initiated by the collapse of the old tunic. Occasionally, in birds living in captivity or under abnormal climatic conditions, the testes may collapse, with the concomitant formation of a new tunic, even though reproduction has not occurred and the tubules are still packed with spermatozoa. For some weeks after the postnuptial testicular regression, therefore, both the old and new tunics are seen in sectioned material, and the testis appears to be surrounded by a double coat. This transient stage can sometimes be a useful parameter for distinguishing between a juvenile bird, in which the sexual organs have not yet undergone any expansion into a breeding condition, and an adult showing a postnuptial testicular collapse.

Internally, each testis consists of a mass of convoluted seminiferous tubules, which in birds, unlike mammals, are anastomotic and not restricted by septa. They are lined by the germinal epithelium consisting of developing germ cells and nongerminal sustentacular or Sertoli cells. The latter have sometimes been described as forming a syncytium, but electron microscopy reveals them to be separate units whose cytoplasm becomes intimately wrapped around the germ cells, thus giving the appearance of a syncytium when viewed by the light microscope. Lofts (1968) has suggested that this surrounding of germ cells by folds of Sertoli cell might serve to maintain the integrity of a particular population of germ cells in a way analogous to the germinal cysts found in the testes of the anamniotes. During the period of sexual quiescence, the terminal epithelium generally consists of a single layer of stem (type A) spermatogonia and Sertoli cells. Because of the collapsed condition of the tubules, a lumen is sometimes not apparent during this stage because of the consequent compression of the Sertoli cells into the more confined area. However, with the advent of the breeding season, a recrudescence of mitotic activity in the stem spermatogonia causes the propagation of numerous germ cells which mature successively into secondary spermatogonia (type B), primary and secondary spermatocytes, spermatids, and spermatozoa, so that the germinal epithelium becomes several cells thick. Each germinal stage arises in a coordinated fashion, so that the germinal epithelium is not a mass of independently developing germ cells, but is characterized by the different germinal stages being organized into well defined cellular associations. Thus, the sectioned gonad presents an orderly appearance, with the spermatogonia close to the basement membrane and the successive cell types appearing progressively toward the central lumen. There are three clearly defined phases in the rehabilitation of the germinal epithelium to a full breeding condition: (1) the period of spermatogonial multiplication, during which new spermatogonia are formed continuously, and some start maturing into primary spermatocytes; (2) the period of spermatocyte division, when the cells undergo their meiotic division producing secondary spermatocytes which mature into spermatids; (3) the period during which spermatids start elongating and start their transformation into the mature spermatozoa.

In most single-brooded species, particularly those inhabiting temperate and high latitudes, once the gametogenetic resurgence begins, there is a rapid progression into the full breeding condition, and all seminiferous tubules appear uniform in cross section, with innumerable radiating bundles of spermatozoa. This is particularly evident in subarctic and arctic species in which the breeding season is particularly abbreviated and the birds migrate south before the onset of the harsh winter conditions. In continuously breeding forms, however, such as the domestic fowl and feral pigeon, and in multi-brooded species in environments which remain equable for longer periods (e.g., *Columba palumbus*, *Zonotrichia leucophrys nuttalli*, and *Passer domesticus*), the breeding season may be protracted for 3-4 months, and the testes remain spermatogenetically active for a much longer period. In these forms, spermatozoa become released in waves throughout the more protracted breeding season. In mammals, it has been established that the spermatogenetic condition of the germinal epithelium is not synchronous throughout the length of any given tubule but shows a longitudinal progression in the form of a wave of development passing along the length of the tubule. Since the particular stage of development reached in one segment of the tubule may differ from another segment further along, a transverse section of the testis will show an asynchronous pattern, with the germinal epithelium of adjacent sectioned tubules possibly in different stages of the epithelial cycle. Such an asynchronous pattern has sometimes been observed in birds, and Clermont (1958) distinguishes eight distinct stages in the epithelial cycle of the Pekin duck.

There is a tremendous variation in the morphological appearance of avian spermatozoa, but, like those of other vertebrates, they all consist essentially of the sperm head (with an apical cap above the acrosome containing the chromatic material), a midpiece, and a long propulsive tail. Once the spermatozoa are freed from the enfolding Sertoli cells, they migrate from the tubule lumina through the rete tubules and short vasa efferentia into the efferent ducts (vas deferens). These discharge into the expanded distal ends of the ducts, the seminal sacs, from which the spermatozoa are finally ejaculated into the urodeum.

The interstices between the convoluting seminiferous elements are packed with areolar connective tissue containing blood capillaries, lymph spaces, and the interstitial Leydig cells, which produce steroid sex hormone. Large numbers of melanoblasts may also be located in this tissue in some species, imparting a black or gray appearance to the organ. With the seasonal proliferation of the germinal epithelium and consequent expansion of the tubules, the interstitial tissue becomes dispersed and compressed into tight concentrations; fewer interstitial cells appear in testicular sections. Conversely, the interstitial tissue of the regressed testis of a wintering bird appears more conspicuous. These cells undergo well defined cyclic changes in their histophysiology. They show the general characteristics of other steroid-producing tissues. Thus, histochemical techniques have established the presence of the Δ^5-3β-hydroxysteroid dehydrogenase (3β-HSDH) enzyme system in these cells. This enzyme catalyzes the conversion of Δ^5-3β-hydroxysteroids to Δ^5-3-ketosteroids, which is known to be an important stage in the production of such steroids as progesterone and the androgenic steroid androstenedione. It has been found in all types of steroid hormone-producing tissues such as the adrenal, testis, ovary, and placenta, and its presence in the Leydig cells provides strong evidence of their steroid biosynthetic capacity. Ultrastructurally, too, they have the general characteristics attributed to steroid-producing cells, in that, in the breeding season at least, they possess a well-developed agranular endoplasmic reticulum and numerous mitochondria with tubular cristae. Generally, too, lipids are often found in their cytoplasm and react positively to tests for cholesterol.

The interstitial cells appear to be derived from fibroblast-like cells in the intertubular areas. In the young Japanese Quail, Nicholls and Graham (1972) have traced the evolution of the typical mature steroid-secreting Leydig cell from its fibroblast-like progenitor, by means of electron microscopy. It has been clearly established that quail subjected to short photoperiods from hatching exhibit no testicular growth or androgen secretion, but when exposed to a long photoperiod, rapid testicular development and androgen secretion take place. When the interstitial tissue of young quail maintained under a 6 hour photoperiod are examined, three to four layers of spindle-shaped cells with elongate nuclei and prominent nucleoli are observed in the areas between adjacent tubules, with the more centrally located cells being somewhat less elongate but still basically fibroblastoid in their ultrastructural appearance. Generally, granular endoplasmic reticulum is found in these cells, but when the birds are transferred to a highly stimulatory 20 hour photoperiod, the interstitial cells rapidly metamorphose into the typical secretory form; the cells and nuclei expand and develop the characteristic fine structure of steroid-secreting systems with the whole cytoplasmic area becoming filled with smooth endoplasmic reticulum in which are scattered rounded mitochondria with tubular cristae. Nicholls and Graham have recorded the first discernible changes in this metamorphosis at about 3 days after the transfer to the increased photoperiod. Interestingly, Follett and colleagues (1972) have found that plasma levels of LH, measured by means of a radioimmunoassay technique, begin to increase about 4 days after such treatment, which is therefore in good agreement with the ultrastructural evidence.

Histophysiology

General histophysiology

Testicular activity is cyclic, and in some species there is evidence that these cyclic changes may, in part at least, be endogenous. Thus, the testes of young Budgerigars (*Melopsittacus undulatus*), and Zebra Finches (*Poephila guttata castanotis*), become spermatogenetically active and produce spermatozoa even when the birds are kept in almost total darkness. A similar endogenous rhythm has also been recorded in an equatorial African weaver finch (*Quelea quelea*) kept on an unvarying 12 hour photoperiod and thermostatically controlled temperature for 2.5 years. Under these conditions, males show a seasonal testicular development and postnuptial regression similar to that observed in wild populations. In the Pekin duck, too, the testes of birds kept for a

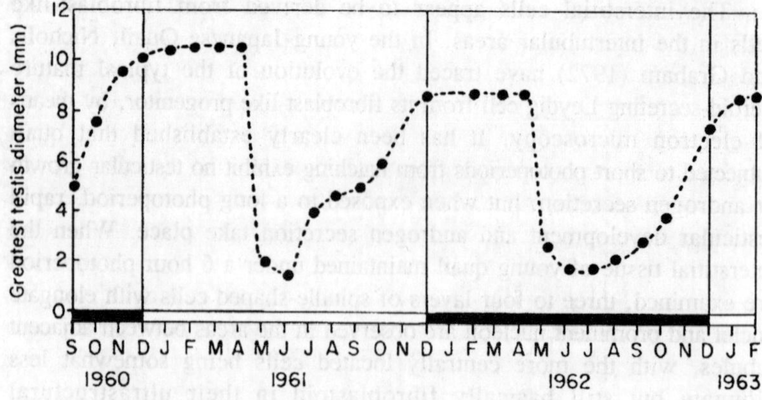

Fig. 7.1. Seasonal changes in the testicular size of the Red-billed Quelea (Quelea quelea) kept under an unchanging daily 12 hour photoperiod.

number of years under conditions of either continuous darkness, or continuous light, similarly undergo a seasonal enlargement and regression.

Marshall regards such an endogenous rhythm as the primary initiator of the seasonal reproductive periodicity. However, in many species such a rhythm is not apparent, and birds like the White-crowned Sparrow (*Zonotrichia leucophrys gambelii*), Brambling (*Fringilla montifringilla*), or Japanese Quail show no spermatogenetic recovery when maintained under winter light conditions and require the seasonal advent of a long photoperiod before a recrudescence of their gametogenetic activity occurs. Even those species possessing an apparently autonomous reproductive rhythm may still be influenced by environmental stimuli, and Marshall (1955) has likened the endogenous cycle to a cogwheel that is seasonally engaged by various environmental "teeth" that produce the final synchronization that ensures reproductive success.

Among various tropical species or species inhabiting a particularly equable environment with a continuous abundance of food, a precise annual breeding cycle may not occur. An equatorial population of *Zonotrichia capensis*, for example, has two complete cycles each year, and Miller (1959) considers this to be an expression of an endogenous 6 month rhythm which is only slightly affected by seasonal rainfall. Even in some temperate zone species, such as the populations of feral pigeons inhabiting our towns, the continuous supply of suitable food provided by a benevolent public and spillage in the docks has resulted in the evolution of a capacity for breeding throughout the year in a high proportion of the flocks. Generally, however, although birds display widely diverse breeding patterns, natural selection has favoured the establishment of annual breeding cycles, particularly in the avifauna of the subarctic and temperate zones. Thus, testicular recovery usually begins in the spring, and the reproductive period is usually terminated by early summer whereupon the testes rapidly regress and remain small throughout the winter months.

The annual cycle can broadly be classified into three basic phases:

Regeneration or preparatory phase

The regeneration or preparatory phase comes immediately after reproduction in single-brooded species, and after the final ovulation of the season in multi-brooded forms. Morphologically, it is marked by the rapid regression of the testes, which can be attributed to a changed and possible diminished gonadotropin release from the anterior pituitary. In many species, this takes place at a time of year when the environmental photoperiod is still highly stimulatory, but the neuroendocrine apparatus no longer responds. The bird is said to be in the refractory period and remains sexually quiescent even when artificially subjected to long photoperiods. Such postnuptial refractoriness is a feature of nearly all species so far submitted to controlled and critical photostimulation experiments, and it has been tacitly assumed or inferred that it is a phenomenon of universal occurrence among avian species. However, recent photoexperimentation on the Wood Pigeon (*Columba palumbus*) has shown this species, and probably others as well, to be without a photorefractory phase. The Wood Pigeon, like the Stock Dove (*Columba oenas*), is multibrooded with a breeding season extending into early autumn where agricultural grain harvests provide abundant food for rearing the later broods. Reproductive activity ceases when autumn day lengths fall below a stimulatory level, and the bird then enters its regeneration phase. Spermatogenesis can, however, be immediately reactivated by artifically prolonging the photoperiod. A refractory period is not therefore a necessity for gonad rehabilitation as previously supposed.

The regeneration phase is characterized by a rehabilitation of both the interstitial tissue and seminiferous tubules, and is a period of sexual quiescence during which the bird displays little or no sexual behaviour. In many species, infiltration of the interstitial tissue by large numbers of leukocytes takes place during the early stages, and the replacement of the weakened testis tunic, and sometimes also the postnuptial molt, occurs during this phase. In the White-crowned Sparrow, for example, a rapid and intensive postnuptial molt is inserted between the breeding period and the onset of autumnal migration. However, the Wood Pigeon molts throughout the breeding season when the gonads are fully active, while the European Turtle-Dove (*Streptopelia turtur*) exhibits a partial molt immediately as it enters the refractory phase but defers the molt of its primaries until it reaches its African wintering grounds. The duration of the regeneration phase varies from species to species, and its termination is probably marked by the completion of the interstitial cell rehabilitation, which may sometimes manifest itself in the form of an autumnal resurgence of sexual behaviour seen in many species. In birds that have a refractory period, the end of the regeneration phase is also heralded by the restoration of photosensitivity.

Acceleration or progressive phase

The acceleration or progressive phase succeeds the regeneration phase and is a period marked by the interstitial cells and seminiferous tubules responding to gonadotropic secretions from the adenohypophysis. The postnuptial molt is usually completed. A recrudescence of gametogenetic activity occurs under favourable environmental conditions but becomes retarded or stimulated by a variety of environmental inhibitors or accelerators that are mutually antagonistic. Thus, the cycle hastens, slows, or sometimes stops altogether, depending upon the factors currently presented to it by the changing environment. Temperature is perhaps the most important modifier of the testicular cycle during this stage, and its effect has been demonstrated experimentally by a number of investigators. In addition, there are many field data that indicate a correlation between temperature and speed of testicular development. Low temperatures will inhibit the cycle of most species, and an unusually cold spring will often nullify the accelerating effects of long sunny days. Among temperate zone birds, some species begin their acceleration phase in late summer or early autumn, but generally this becomes depressed or even halted by the onset of winter conditions until after the winter solstice. In some tropical or xerophilous species, the absence of rainfall may similarly retard the gametogenetic progress during this phase, whereas among waterbirds and many others, a frequent inhibitor is the lack of a safe and traditional nesting site.

The acceleration phase is a period marked by increasingly intensive sexual activity and song, during which gametogenesis leading to the production of spermatozoa occurs. In some species, the winter feeding flocks begin to disintegrate as individuals start territorial selection and defensive displays. The phase varies enormously in duration, both interspecifically and intraspecifically. An interesting example of this is shown by the differences displayed by the resident British population of European Starlings (*Sturnus vulgaris*) and the continental starlings that migrate to Britain for the winter. Spermatogenetic activity begins in the British population in late September, but does not usually progress beyond the proliferation of spermatogonia and occasional primary spermatocytes until February, when a burst of activity populates the seminiferous tubules with secondary spermatocytes and later stages. Continental birds, on the other hand, show no spermatogonial division in autumn, and the first mitoses are not seen until late December or early January, and primary spermatocytes not until early March. Thus, both populations respond to January and February photoperiods, though this marks a beginning of spermatogenesis for continental birds but a resumption of the acceleration phase already started in the British population. Unlike continental birds, the British population shows earlier and more intensive interstitial cell activity in the autumn and winter, which is evident by the earlier changes in bill colouration and development of the vas deferens. This early elevation of androgen titer is also responsible for autumnal sexual displays and even winter breeding in exceptionally mild winters, events virtually absent in the life history of the continental population.

Culmination phase

The culmination phase, during which actual ovulation and insemination occur, may be regarded as a distinct component of the annual cycle, since a species-specific requirement is generally necessary before the cycle can culminate in oviposition. By the end of the preceding phase, the male bird has reached a fully reproductive condition with expanded testes containing seminiferous tubules charged with masses of spermatozoa. Nevertheless, although the internal physiology may be now wholly prepared for reproduction, many complicated behavioural factors may influence the final culmination. The male generally assumes this reproductive state before the female, and the final timing of oviposition then depends on the female receiving the appropriate psychological stimuli, both from her mate and the environment. The appropriate habitat and interpair displays are often essential to stimulate final oocyte development, ovulation, and insemination. The action of stereotyped behaviour in causing specific hormone secretion is now well appreciated, and conditions of captivity, for example, have been shown to inhibit gonadotropin secretion in the Pintail (*Anas acuta*), thus preventing breeding.

Interstitial cells

There is an extensive literature concerning the seasonal changes observable in the avian interstitial tissue, but many of the early reports are contradictory, often being based on unsatisfactory and now outmoded histological procedures. Because of their dispersal by seasonal tubule expansion, the Leydig cells have sometimes been stated to be absent from the gonad at certain times of the year, and an inverse relationship between sexuality and interstitial (Leydig) cell activity has sometimes been claimed. This, of course, is not true and the more recent histochemical and electron microscopic observations have clearly established the close relationship between these cells and the androgen-dependent sexual structures. Furthermore, the selective destruction of the germinal epithelium of cockerels by roentgen radiation, which leaves the interstitium and secondary sexual characters apparently unaffected, also indicate this tissue as the site of androgen production. Tumors of the interstitial cells result in an increased production of androgens and 17-ketosteroids.

In birds, as in the majority of seasonal vertebrates, the interstitial cells undergo well defined seasonal secretory cycles involving a rhythmic accumulation and depletion of cholesterol-positive lipoidal material. The interstitial cells of young birds are generally heavily impregnated with such material. Then, at the approach of the sexual season and consequent buildup of spermatogenetic activity in the seminiferous elements, they become rapidly depleted of their lipids and cholesterol and become strongly fuchsinophilic. In a species such as the Northern Fulmar (*Fulmarus glacialis*), in which the young do not begin breeding until 7 years old, the interstitial cells of newly hatched birds are less lipoidal but become more heavily impregnated when the birds are just over 2 years old. In adults, the interstitial cells of the sexually quiescent winter gonad are generally small and often sparsely lipoidal with numerous fuchsinophilic elements that are more easily seen after dissolution of the lipids in wax-embedded material. During the acceleration phase, these cells rapidly increase in size and there is a buildup of the lipoidal inclusions, so that the interstitial tissue is seen to consist of compressed aggregations of heavily lipoidal and cholesterol-rich cells. They also react positively to tests for 3β- HSDH. Then, as injuveniles, the lipoidal content rapidly diminishes at a time when androgen-dependent sexual displays reach their maximum intensity. During this period, the cholesterol reaction also becomes weaker, and may disappear altogether, although 3β-HSDH activity remains strong. The nuclei of Leydig cells also attain maximum diameter at this time, reflecting an increase in secretory activity. In migratory waders (Charadriiformes), this depletion of interstitial lipids is often evident just before the birds leave their African wintering grounds on their north-bound migration.

With the advent of the regeneration phase, the now exhausted interstitial cells have reached the end of their secretory cycle, and it has been reported that they disintegrate so that ultimately their total number is drastically reduced. It may be that the massive invasion by leukocytes that takes place at this time clears the spent Leydig cells by phagocytic action. However, there is a need for further investigation by electron microscopy to confirm this point, and the possibility that the spent Leydig cells return to an inconspicuous fibroblast-like form should not be excluded. Concomitant with the atrophy of the exhausted generation, a new generation of juvenile interstitial cells begins to arise, presumably by their differentiation from fibroblast-like progenitors, and gradually begin to mature.

This seasonal replacement by new interstitial cells at the end of each breeding period is not unique to avian testicular cycles. A similar phenomenon has also been recorded in some snakes and in the common frog (*Rana temporaria*). Marshall (1961b) suggests that the sequence is part of an endogenous rhythm that can occur even in the absence of any gonadotropic rhythm. Thus, even after complete removal of the adenohypophysis, Coombs and Marshall (1956) have reported that the interstitial cells of domestic cockerels still renew themselves and develop a new generation of Leydig cells with some lipoidal and cholesterol-positive material. It is doubtful, however, whether such cells would ever become secretory in the absence of gonadotropins.

The length of time necessary for interstitial cell rehabilitation probably varies from species to species, but there is evidence that in some birds at least it may be a fairly rapid process. Thus, Lofts and Marshall (1957) have recorded that the newly arisen interstitial cells of fifteen different migratory species were already beginning to manufacture small cholesterol-positive lipid droplets in their cytoplasm when they were autopsied at the time of their departure from Britain on their southward migration.

The cyclical waxing and waning of cellular lipids, which although in itself is insufficient to implicate unequivocally a steroid secreting role, is a useful index of the functional activity of the tissue. The lipids are both cholesterol-positive and strongly birefringent, reactions that are probably indicative of precursor material involved in androgen biosynthesis. The prenuptial buildup in birds often precedes the hypertrophy of the accessory sexual apparatus and behavioural activities thought to be dependent on androgen secretion. In young chicks, for example, the increase in concentration of birefringent material in the interstitial tissue is in close agreement with the gradual hypertrophy of the comb, and in the House Sparrow (*Passer domesticus*) the level of 17α-hydroxylase activity in the interstitial tissue has been shown to increase rapidly between February and March, a time when the interstitial cells are losing the lipoidal material accumulated earlier in January and February.

The sudden depletion of lipids and cholesterol at the height of the breeding activity has also been noted in a number of reptilian species, and it has been suggested that it is probably indicative of a rapid utilization of precursor material at a time of high androgen release.

The observations of Jones (1970) that the height of the epididymidal epithelium in the California Quail (*Lophortyx californicus*) attains its maximum thickness at the time when the interstitial tissue is showing this phenomenon are in agreement with such an interpretation. Certainly 3β-HSDH activity is high at this time. In the Department of Zoology, University of Hong Kong, we have been attempting to correlate these histochemical events with the seasonal fluctuations in androgen production, measured by *in vitro* methods. For this, portions of testicular material have been taken at monthly intervals from the migratory Green-winged Teal (*Anas crecca*) throughout the year and incubated with radioactive pregnenolone as an added precursor. The biosynthetic activity of the tissue has been determined by standard chromatographic analysis of the steroids produced, and the percentage conversion of the added [16-^{3}H] pregnenolone to testosterone per unit weight of tissue has been calculated. Because of the seasonal dispersal of interstitial

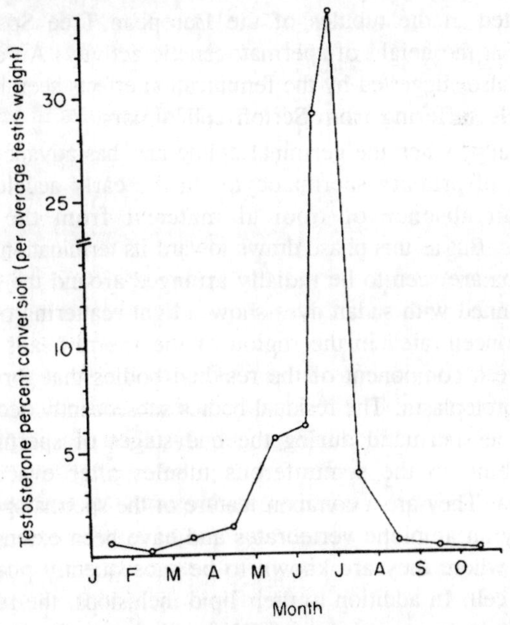

Fig. 7.2. The seasonal fluctuations in the in vitro production of testosterone from radioactive pregnenolone by the testis of the Green-winged Teal (Anas crecca).

tissue during testicular expansion, fewer Leydig cells will be contained in the tissue from the sexually mature testis compared with the same weight of tissue taken from the regressed testes of winter birds. The conversion figures have therefore been multiplied by the mean testicular weight to give a truer picture of the seasonal pattern of steroid production by the whole gonad. They support the hypothesis outlined above. Thus, the gradual elevation in androgen production from mid-February to early June coincided with the appearance of lipoidal inclusions in the interstitial tissue. Lipid depletion was observed in early July at a time when androgen biosynthesis reached a peak, and the subsequent rapid decline marked the bird's entry into the postnuptial refractory period.

Sertoli cells

As well as the interstitial tissue, the Sertoli (sustentacular) cells of the tubule also undergo cyclic seasonal changes involving an accumulation and depletion of cholesterol-positive lipoidal material. Furthermore, ultrastructurally these cells show the fine structure normally associated with a steroid-producing tissue and also react positively to tests for 3β-HSDH. The enzyme 17β-HSDH can also be demonstrated in the tubules of the European Tree Sparrow (*Passer montanus*) at the height of spermatogenetic activity. An endocrine role is perhaps also suggested by the feminization effects seen in the plumage of cockerels suffering from Sertoli cell tumors.

Generally, when the germinal epithelium has advanced beyond the production of primary spermatocytes in the early acceleration phase, there is an absence of lipoidal material from the intratubular components. But as this phase draws toward its termination, and bunched spermatozoa are seen to be radially arranged around the tubule lumen, sections stained with sudan dyes show a light scattering of fine lipoidal granules concentrated in the region of the sperm heads. These small droplets are a component of the residual bodies that form part of the spermatid protoplasm. The residual bodies subsequently become sloughed off from the spermatid during the end stages of spermateleosis and remain behind in the seminiferous tubules after evacuation of the spermatozoa. They are a common feature of the spermatogenetic process of probably all amniotic vertebrates and have been extensively studied in the rat, where they are known to be subsequently phagocytized by the Sertoli cell. In addition to their lipid inclusions, the residual bodies also contain a large amount of RNA, clusters of mitochondria and Golgi remnants, and, in the rat at least, it has been suggested that they may be instrumental in maintaining the radial coordination of the associated germ cell population.

In contrast to the slight scattering of intratubular lipids observed in the fully mature testis, a dramatic metamorphosis into a condition of heavily sudanophilic and strongly cholesterol-positive tubules takes place during the postnuptial regeneration phase. In sectioned material the seminiferous tubules appear to become filled with a dense amorphous mass of lipid and cholesterol, completely occluding the lumen, but with the better resolution provided by the electron microscope, the lipid is generally seen to be within the Sertoli cell cytoplasm. While some of this lipid might be contributed by the lipoidal inclusions of the phagocytized residual bodies, much of it is probably produced by the Sertoli cell cytoplasm. Steroid dehydrogenase activity ceases to be demonstrable in these cells at this time. The tubules remain in this heavily lipoidal state for some time; then, concomitant with the recrudescence of spermatogenetic activity in the adjacent stem spermatogonia, the sudanophilic material rapidly disappears. The duration between the sudden accumulation in the Sertoli cell cytoplasm and the start of lipid depletion varies from species to species and may be as long as several months. For example, in the migratory Whimbrel (*Numenius phaeopus*), the seminiferous tubules have already become heavily lipoidal by the time the animals leave Britain and head south to their African wintering grounds. The gonads remain in this condition for about 5-6 months, and the Sertoli lipid clearance and spermatogenetic recovery begin just before the birds leave the contranuptial area and return north to breed. By the time they arrive in Britain, tubule lipids are absent and testes are reaching sexual maturity. In columbid species, such a massive postnuptial tubule steatogenesis does not occur, and the quantity of Sertoli lipid is very much less.

In view of the fine structure and histochemical evidence, the question arises as to whether this seasonal waxing and waning of cholesterol-rich lipoidal material in the Sertoli cells is indicative of a seasonal endocrine function as is apparent in the adjacent interstitial tissue. Tentative evidence for this has been provided by Lofts and Marshall (1959), who chromatographically analyzed testicular lipids extracted from birds with regressed gonads containing heavily lipoidal tubules but a lipid-free interstitium. The results showed the presence of progesterone, which also correlated with a positive progestogenic reaction in the blood subjected to a parallel bioassay. A similar analysis on birds with gonads with expanded nonlipoidal tubules but heavily lipoidal interstitial cells resulted in an absence of demonstrable progestogenic activity in the blood, and only androgenic steroids were identified in the testis extracts. By present-day standards, the above data would require more rigorous criteria for specific identification of the steroids, but so far as the authors are aware, no further confirmation of these data has yet been attempted. However, there is strong evidence in favour of this hypothesis in mammals and also in reptiles, in which a separation of seminiferous tubules and interstitial tissue has been achieved by microdissection. By incubating seminiferous tubules with labeled precursors, their steroid biosynthetic capacity has been clearly established. Furthermore, in reptiles, Lofts (1972) has shown that testosterone is a major steroid being produced by this tissue, and that the production varies on a seasonal basis.

Control of Testis Function

Effects of exogenous hormones

(a) *Androgen*. There is an extensive literature dealing with the effects of exogenous androgens on the functional activity of the vertebrate testis, and injections of androgenic steroids have been shown to have stimulatory effect in hypophysectomized and intact fishes, reptiles, and mammals. In birds, administration of androgen has sometimes been reported to cause testicular regression, but has also been shown to induce testicular stimulation in adult Ring Doves, European Starlings, Red-billed Queleas, and House Sparrows. Furthermore, in the pigeon, testosterone administration maintains spermatogenetic activity and prevents testicular regression after hypophysectomy and even induces a reestablishment of spermatogenesis in birds in which the gonads have been allowed to regress completely after removal of the pituitary gland. This is in contrast with the situation in the quail, in which Bayle et al. (1970) report that injections of testosterone propionate into hypophysectomized birds fail to prevent the degeneration of the germinal epithelium, although there is evidence that this process is retarded in androgen-treated birds. In *Quelea quelea*, a recovery of spermatogenetic activity can be produced by exogenous testosterone proprionate in the fully regressed gonads of regeneration phase birds, and although Pfeiffer (1947) has reported that spermatogonial multiplication is not augmented in the regressed testes of House Sparrows subjected to similar treatment, it has been our experience that such gonads can be stimulated when sparrows collected in November and maintained under an 8 hour photoperiod are given daily androgen injections. Furthermore, an elevation of endogenous androgens induced by injections of mammalian LH stimulating the interstitial cells has an even greater effect than that produced by exogenous androgens.

Much of the above controversy can probably be attributed to the fact that androgens have been administered to birds without cognizance being taken of the precise reproductive condition of the animal at the time of injection or implantation of the steroid. Thus, androgens given to birds just starting their seasonal testicular enlargement apparently retard the rate of spermatogenetic recrudescence, but if injections are delayed until spermatogenetic development has proceeded beyond the point of primary spermatocyte production, the same androgen treatment can accelerate gonadal enlargement. In the sexually mature *Quelea quelea*, androgen therapy maintains the testes in breeding condition and prevents the postnuptial testicular collapse. The same is true of the European Starling and House Sparrow. In the pigeon, on the other hand, Chu (1940) has reported that daily injections of 2 mg of crystalline testosterone in sesame oil produces testicular atrophy in sexually mature birds. However, this observation may be somewhat speculative, since it was based on the results of only four specimens, and of these one killed after a week was normal and another killed after 30 days still contained spermatozoa. Furthermore, our own observations indicate that daily injection of testosterone propionate into feral pigeons does not cause a testicular regression but, on the contrary, appears to accelerate the maturation of spermatids into spermatozoa. This lends support to the suggestion of Kumaran and Turner (1949b) that androgens facilitate the transformation of secondary spermatocytes in the cockerel, and is also in agreement with its known effects in lower vertebrates.

The stimulatory effects of androgens on the testes of hypophysectomized birds and on the completely regressed testes of intact wild birds is, in both cases, probably a direct effect on the seminiferous tubules, since Lofts (1962b) has shown that the termination of such therapy is immediately followed by a rapid regression. However, although such evidence of a spermatokinetic effect of androgen has sometimes led to the suggestion that a seasonal resumption of interstitial cell activity may be responsible for the spring resurgence of gametogenesis, it does not explain the apparent contradiction that an inhibitory effect is produced during the early acceleration phase under similar experimental conditions. Lofts (1970) has suggested that there may be some antagonism between the exogenous androgen and gonadotropic hormones at a time when the germinal epithelium is

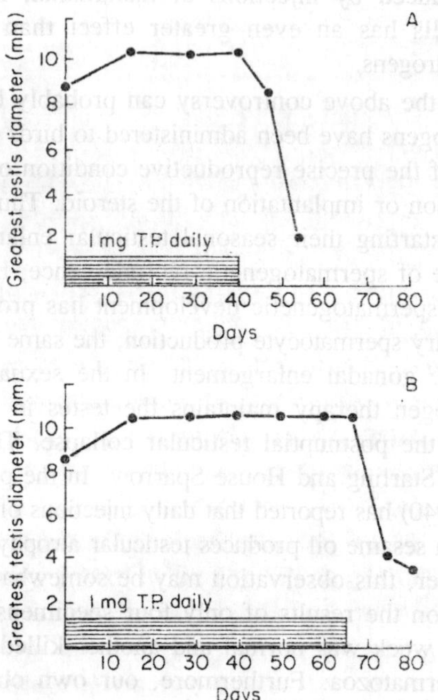

Fig. 7.3. These graphs show the effects on testis size of terminating daily androgen injections in the Red-billed Quelea (Quelea quelea). (A) This experimental bird was kept in full breeding condition by daily injection for 40 days. (B) This experimental bird was maintained in full breeding condition even though the uninjected controls entered and passed through a full refractory phase.

particularly sensitive to FSH, but it is perhaps unwise to extrapolate from experiments using large nonphysiological doses that endogenous androgens have a direct interrelationship with the germinal epithelium. The recent observations by Murton et al., (1969a) on Greenfinches (*Carduelis c. chloris*) throws doubt on such an hypothesis. In an experiment devised to find the photoinducible phase, male Greenfinches with regressed testes were kept in groups under a 6 hour daily photoperiod with an additional 1 hour light pulse given as an interruption in the night. In each group, the night interruption occurred at a different time. After 3 weeks of such treatment, the birds were autopsied for testicular development and the plasma was assayed for LH levels by a radioimmunoassay technique. The results demonstrate that birds of group B had a high plasma titer of LH, which correlated with a highly stimulated interstitium with numerous Leydig cells heavily charged with lipids and swollen prominent nuclei, yet the seminiferous tubules were spermatogenetically inactive. This indicates that, in Greenfinches

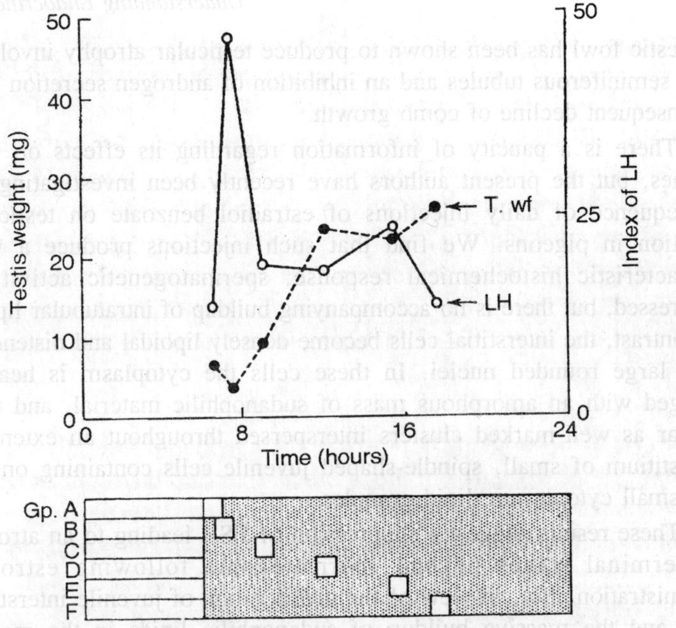

Fig. 7.4. The testicular weight (T. wt) and plasma LH levels in groups of Green-finches (Carduelis c. chloris) subjected to asymmetric skeleton photoperiods.

at least, high endogenous androgen levels produced by the interstitial cells do not stimulate a recrudescence of spermatogenetic activity in the sexually regressed testis, although the possibility of an influence on later germinal stages cannot be excluded. Furthermore, these data do not rule out the possibility that androgens from the Sertoli cells may also be involved.

(b) *Estrogen and progesterone*. There is scattered evidence that the male gonad may produce both estrogens and progesterone under both normal and pathological conditions. Chromatographic and bioassay techniques have indicated the presence of progesterone in testicular extracts of a number of species, and this is supported by the evidence of Fraps and his co-workers, who demonstrated the testicular source of plasma progesterone or a closely allied substance in cockerels. Production of estrogen and progesterone during the normal avian cycle is indicated, too, by the appearance of incubation patches in many species. These are produced by the synergistic action of these hormones with prolactin.

The suppressive effects of exogenous estrogen on the gonadotropic secretion in birds are well established, and estrogen administration to domestic fowl has been shown to produce testicular atrophy involving both seminiferous tubules and an inhibition of androgen secretion with a consequent decline of comb growth.

There is a paucity of information regarding its effects on wild species, but the present authors have recently been investigating the consequence of daily injections of estradiol benzoate on testicular function in pigeons. We find that such injections produce a very characteristic histochemical response; spermatogenetic activity is suppressed, but there is no accompanying buildup of intratubular lipids. In contrast, the interstitial cells become densely lipoidal and distended, with large rounded nuclei. In these cells the cytoplasm is heavily charged with an amorphous mass of sudanophilic material, and they appear as well-marked clusters interspersed throughout an extensive interstitium of small, spindle-shaped juvenile cells containing only a few small cytoplasmic lipid granules.

These results suggest a suppression of FSH leading to an atrophy of germinal stages beyond spermatogonia following estrogen administration. But, in view of the proliferation of juvenile interstitial cells and the massive buildup of sudanophilic lipids in the mature Leydig cells, an increased release of LH is indicated and would be in agreement with the known effects of estrogen in ovulating female mammals. Indeed, recent studies have confirmed that plasma levels of LH are markedly elevated following the treatment of male feral pigeons with estradiol benzoate. It seems likely, too, that there is a direct effect of estrogen on the interstitial cells causing an interference with the later stages of androgen biosynthesis resulting in an accumulation of precursor material, which would also lead to an atrophy of androgen-dependent secondary sexual characters such as comb growth. In the case of feral pigeons, new Leydig cells appear about the middle of the preincubation courtship cycle, and the displays typically occurring at this time are markedly enhanced by exogenous estrogens.

Little is known regarding the effects of progesterone in male birds, and the available data are somewhat contradictory. Farner, on the other hand, mentioned in discussion, that progesterone injections into *Zonotrichia leucophrys* produces a negative result. In pigeons, we have found a variable response to such injections, with some showing no effect on the testes, but others undergoing considerable testicular atrophy. However, it is doubtful whether endogenous progesterone production has any antispermatogenetic effect, since Lehrman (1965) has good evidence that this hormone is released just prior to egg laying, at a time when there is no obvious testicular regression.

(c) *Prolactin*. Prolactin is present in the pituitary gland of male birds and has been known for many years to retard spermatogenesis and lead to testicular collapse in pigeons, cockerels, and some passerine species. Apparently, exogenous prolactin does not induce a testicular regression in *Zonotrichia leucophrys gambelii*, *Carpodacus mexicanus*, *Zonotrichia albicollis*, and *Lophortyx californicus*, but it can inhibit photoperiodically induced gonad growth in *Z. l. pugetensis*, *Z. l. nuttalli*, the Song Sparrow *Melospiza melodia*, the House Sparrow, and Wood Pigeon.

The antigonadal effect of prolactin is suggested to be mediated by a suppression of FSH secretion and produces a buildup of intratubular lipids and cholesterol similar to the naturally occurring seasonal event. In *Ploceus philippinus*, feather regeneration, which is supposedly dependent on LH, is not affected by prolactin administration, but gonadal regression still takes place, again suggesting a suppression of FSH output. However, there is insufficient evidence to attribute a general antigonadal role to endogenous prolactin secretion under natural conditions. In multibrooded species, the seasonal testicular regression does not occur until after the completion of the last clutch, yet prolactin-induced brood patch formation and incubation behaviour take place from the time the first clutch is produced. Furthermore, as in *Z. l. gambelii*, a reduction in gonad size is not produced in prolactin-injected pigeons, although the same treatment is capable of terminating spermatogenetic activity in the closely related Wood Pigeon and has an antigonadal effect in other strains. In the California Quail, Jones (1969c) has recorded the highest pituitary prolactin content in males with testes in full breeding condition, and Gourdji (1970) has similarly recently reported that there is no antigonadal effect in domestic ducks, nor in Japanese Quail, and has further shown that an increase in hypophysial prolactin content under experimental photostimulation is accompanied by an increase in testicular growth. Prolactin does not inhibit the uptake of ^{32}P by the testes of cockerels *in vivo*. The assimilation of this material into the testis is stimulated by gonadotropins, and the measurement of its uptake into the testes of young cockerels is an established bioassay for gonadotropic hormones. Stetson and Erickson found that prolactin administration simultaneously with or up to 18 hours previous to the administration of either purified gonadotropins or pituitary extracts, not only had no inhibitory effect on ^{32}P uptake, but actually acted synergistically at high dose levels.

It seems likely that different levels of tolerance to prolactin secretion have arisen in different species, which would account for the variations in gonadal response. Certainly, there is no noticeable reduction in testis size in pigeons and doves when prolactin-dependent crop gland development and crop milk production is at its maximum, suggesting that columbid species must tolerate high levels of endogenous prolactin throughout the whole of their very protracted breeding seasons; there are, however, histological changes in the testes during the period of prolactin secretion. Hohn (1959, 1962) attributes a similar high testicular insensitivity to prolactin in the Brown-headed Cowbird (*Molothrus ater*) to an evolutionary adaptation to brood parasitism.

Pituitary-gonad axis

In birds, as in mammals, the hypothalamus exerts absolute control over gonadotropic secretion from the adenohypophysis, so that hypophysectomy, pituitary autotransplantation, or sectioning of the pituitary portal vessels produces a rapid testicular regression. The recent separation of avian FSH and LH fractions again shows a similarity to the mammalian situation in that testicular control is mediated by means of a dual hormonal regulation from the adenohypophysis. Lofts and Murton (1968) have proposed that the two gonadotropic hormones might not necessarily be secreted simultaneously, but rather, they may be secreted at different phases. Evidence in support of such a hypothesis appears to be provided by the Greenfinch experiments outlined earlier in this chapter. In these birds, testicular growth (and, therefore, presumably FSH secretion) was most strongly induced in birds (group F) that were given a light pulse 17-18 hours after dawn, whereas the LH peak occurred in response to a light pulse between 7 and 8 hours after dawn. These results imply that there may be two separate hypothalamic centers controlling LH and FSH release, each perhaps receiving information from discrete rhythms of photoperiodic sensitivity, and the results of Stetson (1969, 1971, 1972a,b) provide anatomical evidence of such a mechanism.

Differences in the photoperiodic response mechanism are known to occur in birds, and this has in turn led to a variety of annual breeding patterns. It has been suggested that this phenomenon, in terms of endogenous hormonal controlling mechanisms, must have its basis in differences in pattern in the release of the two gonadotropic hormones, and we have seen in the Greenfinch that a mechanism has evolved in which the photoinducible phase for LH is apparently entirely separate from that promoting FSH release. In *Coturnix*, Follett considers that LH and FSH may be secreted simultaneously at one circadian phase, and this may well also be true in the House Sparrow, where

Murton et al. (1970a) have been able to show that there is a greater overlap of LH and FSH release. In this species, a marked peak of LH release in relation to a photoinducible phase early in the 24 hour cycle does not occur.

Since much of our knowledge about pituitary regulation of avian testicular function has, up to the present time, been derived from studying the effects produced by the administration of mammalian gonadotropins, there is a need for caution and further experimentation with avian preparations before firm conclusions are drawn. However, much of the evidence to date suggests that LH stimulates interstitial cell activity in basically the same way as in mammals. Thus, Leydig cell hyperplasia has been noted in LH-injected young cockerels and was seen to be accompanied by parallel comb growth, and in pigeons, similar LH treatment produces a proliferation in the numbers of recognizable Leydig cells and a buildup in their lipoidal inclusions and steroid dehydrogenase activity. Such results have recently been endorsed by the first preliminary experiments using purified avian LH. Follett et al. (1972) report that daily administration of 50 μg of avian LH for 10 days to sexually regressed young Japanese Quail left under nonstimulatory photoperiods stimulates Leydig cell development into typical secretory form with abundant smooth endoplasmic reticulum.

The fact that exogenous androgens cause a regression when administered to early acceleration phase birds might indicate a negative feedback effect on the anterior pituitary, and the results of Kordon and Gogan (1970) showing that microimplants of testosterone within the ventromedial nucleus of the hypothalamus in ducks induces a testicular regression adds support to such a suggestion. The same investigators have also shown that the threshold of testosterone necessary to block light-stimulated gonad growth varies on a seasonal basis, being low at the time of testicular quiescence, increases when the gonads are stimulated, and decreases at the time of seasonal testicular regression. In *Spizella arborea* and Japanese Quail a similar inhibition of light-stimulated testicular growth is obtained by testosterone implants in the nucleus tuberis region. Evidence of a negative gonadal feedback mechanism is also suggested by the fact that the anterior pituitary of photostimulated castrated birds increase in weight more than the photostimulated intact controls, and that plasma levels of LH become up to five times higher in Japanese Quail subjected to such experimentation. Furthermore, there is a significantly higher acid phosphatase content in the supraoptic and median eminence regions of photostimulated castrated *Zonotrichia leucophrys* than in the intact controls. It would be informative to determine whether exogenous androgen administration would reduce the acid phosphatase content as it does plasma LH in quail.

Injections of purified preparations of mammalian FSH have a stimulatory effect on the germinal epithelium and can restore the spermatogenetic activity of a hypophysectomized bird or an intact bird during the postnuptial regressive stage. Such treatment also produces a depletion of the intratubular lipids, and an inverse relationship seems to exist between the Sertoli cell lipid cycle and the activity of the germinal epithelium. Thus, a termination of spermatogenesis (and presumably FSH output), either naturally during the postnuptial refractory period or artificially as a consequence of hypophysectomy is generally accompanied by an increase in lipids and sterols in the Sertoli cell cytoplasm. Conversely, a recovery of gametogenetic activity during the acceleration phase, or by the exogenous administration of FSH, is always paralleled by a depletion of these substances. Such an association has also been noted in other vertebrate groups, and in view of the steroidogenic capacity of this tissue, it has been suggested that the overall control of spermatogenesis by FSH might be mediated via a regulation of the secretory activity of the Sertoli cells. Such a mechanism would be more in line with the known actions of the other pituitary tropic hormones such as LH and ACTH, which are known to stimulate the secretory activity of other steroid-producing tissues. In the same way, FSH might thus stimulate the secretion of a steroid by the Sertoli cells, which, in turn, would have a spermatokinetic effect on the germinal epithelium. In view of the experimental evidence indicating that exogenous androgens can, under certain situations, stimulate spermatogenesis, an indirect stimulation of the germinal epithelium by FSH acting on the Sertoli cell would appear a more likely hypothesis than one that supposes that a protein hormone, on the one hand, and a steroid, on the other, can cause the same biological action, i.e., a stimulation of spermatogenesis. Evidence in support of such a mechanism is also beginning to appear in mammals and in lower vertebrates.

Refractory period

The development of photorefractivity in those species that have photoperiodically controlled breeding cycles may be regarded as an adaptive mechanism preventing propagation at times when stimulatory day lengths still occur, but when other environmental needs for successful rearing of young are becoming scarce. In migratory species, it causes a premature termination of the breeding period so that the birds are able to undergo their postnuptial molt and autumnal hyperphagia before their contranuptial flight and while an adequate food supply is still available. The length of time before such birds regain their photosensitivity varies, but generally in many temperate zone species the animals remain in this sexually quiescent period for some 3-4 months. By the time they once again become photosensitive it is usually late autumn, and stimulatory light conditions are no longer present in their environment and will not be experienced until the following spring.

The termination of photorefractivity has been shown to be advanced by experimentally subjecting birds to a period of short days, and it is only by exposure to a species-specific term of short photoperiods that these animals retain their capacity to respond to stimulatory day lengths. In some, such as *Junco hyemalis*, *Zonotrichia leucophrys*, *Z. albicollis*, *Sturnus vulgaris*, and *Anas platyrhynchos*, this requirement appears to be absolute, and refractory birds exposed to continuing long daily photoperiods remain sexually regressed. In others, on the other hand, the situation is less rigid, and even when experimentally exposed to continuous summer day lengths, they will spontaneously emerge from their photorefractory state. The domestic duck and *Quelea* are two examples of such a situation. A circadian-based mechanism may be involved in determining the duration of photorefractivity. Thus, when groups of photorefractory House Sparrows are maintained for 35 days under daily photoperiods of 6 hours and an additional 1 hour light pulse given as an interruption of the dark period at 8, 12, or 16 hours from "dawn," respectively, only sparrows receiving a light interruption at 8 hours from "dawn" recover their photosensitivity and show gonadal stimulation when transferred to a stimulatory 16L-8D light schedule. These results indicate that the total daily quantity of light is unimportant, since it was the same in each group. Hamner (1968) similarly claims a circadian component timing the refractory period in *Carpodacus mexicanus*.

The physiological basis for gonadal photorefractivity still remains enigmatic, and the hypothalamus, adenohypophysis, and the gonads themselves have all in turn been suggested as the possible site of the blockage. The early experiment by Benoit et al. (1950) demonstrating that a compensatory hypertrophy of the residual testis occurred in hemicastrated ducks when operated on during the nonrefractory period, but not when the operation was performed during the refractory period, provided evidence suggesting a cessation of gonadotropin output during the refractory phase and indicated that the hypothalamo-pituitary axis and not the pituitary-testis axis as the primary site involved. Furthermore, Stetson and Erickson (1971) have evidence that the gonadotropic content of the pituitary of castrated White-crowned Sparrows shows a spontaneous decline at a time when intact birds enter their refractory phase, again suggesting a hypothalamo-pituitary block independent of any gonadal influence. Nevertheless, it is perhaps still premature to exclude completely any gonadal involvement during this period of photorefractivity. The histological condition of the interstitial tissue of a refractory bird is quite distinct from that of a winter bird that has regained its photosensitive potential but is still sexually regressed, as has been observed in a number of species. This condition can also be induced experimentally. For example, the interstitial tissue of wild Mallards (*Anas platyrhynchos*) caught during their refractory phase in August and maintained in a protracted refractory condition by exposure to artificial summer photoperiods to the end of December has been shown to be much more extensive than that of birds caught at the same time but which have terminated their refractory state while being maintained under an 8 hour winter photoperiod. The testicular interstitial tissue of the latter appears to be much more lipoidal, although spermatogenetically both groups are identical. Hamner (1968) has recorded a similar accumulation of interstitial lipids at the termination of photorefractivity in *Carpodacus mexicanus*, and the same has been noted also in House Sparrows. On the basis of this and other cytological data, Lofts and Murton (1968) have speculated that a release of LH might continue during the refractory phase and be a relevant factor. It is during this time that the differentiation of new Leydig cells occurs, and as has already been noted elsewhere, this rehabilitation of the interstitium with a subsequent release of androgen toward the end of the period is probably responsible for autumnal sexual activity in a variety of species.

Evidence implicating a more positive involvement of the testis in the underlying endocrinological mechanisms of the refractory period is also provided by the recent work of Murton et al. (1970b) on the manipulation of photorefractivity in *Passer domesticus* by circadian light regimes. Although each of three experimental groups of birds were given the same daily ration of light (6 hours plus an additional 1 hour light pulse given as an interruption of the dark period), these investigators found that after 35 days quite distinct cytological differences developed in the testis of each group. Thus, in the two groups that had experienced a light pulse at 12 hours and 16 hours, respectively, from the experimental dawn, the testes had acquired an expanded interstitium with numerous spindle-shaped cells, whereas in the group that had its night interruption 8 hours from dawn the interstitium was less extensive and the Leydig cells remained small but were more heavily lipoidal and the nuclei more rounded. Subsequent exposure to a daily 16 hour stimulatory photoperiod demonstrated that only the latter group had regained their photosensitivity. Thus, the possibility of a feedback effect from the testis during the refractory period cannot be excluded at this time, although a positive conclusion will be reached only when the circulating testosterone levels of the photorefractory bird are eventually measured.

Ovary

Reproduction in birds, unlike the other vertebrate groups, is exclusively oviparous. The left ovary, which becomes the dominant and functional primary sex organ in most species, is attached to the body wall by a short mesovarium in close apposition to the kidneys. It is supplied with blood by the ovarian artery, which is usually a branch of the left renolumbar artery but occasionally is a branch of the dorsal aorta. The organ is drained by large veins that anastomose and converge into an anterior and posterior ovarian vein, both of which empty into the vena cava. Unlike the situation in the male, the functional morphology of the female gonad has been very little studied in wild species, and there is a paucity of information on the seasonal changes in ovarian histology. Most of our data have been derived almost exclusively from investigations of the domestic hen.

In seasonal breeding species, the ovary undergoes the same great variation in size that is evident in the testes, and in the European Starling (*Sturnus vulgaris*), for example, ovarian weight can fluctuate from 8 mg at the time of sexual regression to a weight of over 1400 mg at the height of the breeding season. In the quiescent state, the ovary is a compact, often triangular, flattened organ with small follicles giving the surface a granular appearance. But as it develops into the breeding condition, the follicles become distended and bulge conspicuously from the surface of the ovary, which now assumes the appearance of a bunch of grapes. The largest follicles ultimately become suspended from the surface by a narrow isthmus of tissue, the pedicle. The mature follicle is highly vascular and the surface has a conspicuous scarlike area, the stigma, which delineates the region where the follicle will eventually rupture during ovulation. The stigma appears macroscopically to be avascular, but it does, in fact, have small blood vessels traversing it. In mammals, the ejection of the ovum during ovulation causes a rupture of the follicular capillaries resulting in the blood accumulating in the follicular cavity, but in birds the constriction of spiral vessels prevents such hemorrhage. In domestic species, such as the fowl, the ovary remains more or less in this well-developed state throughout the year (except during molting), but in the seasonally breeding species, the ovary undergoes a postnuptial regression into its quiescent state.

Histologically, the ovary follows the basic vertebrate pattern in consisting of an outer cortex containing the developing follicles surrounding a highly vascular medulla forming the ovarian stroma and composed primarily of connective tissue. At the periphery of the cortex is the germinal epithelium, which in the embryonic stage, starts its gametogenetic activity and proliferates a vast number of primary oocytes. By the time of hatching, the gonad is already endowed with many thousands of oocytes, and no further oocyte formation occurs in adult life. Follicular development proceeds from the endowment of oocytes propagated in the embryonic stage. Of these, only a very small percentage eventually mature fully, and the majority undergo follicular atresia while still in an early stage of maturation. Atretic follicles are, therefore, a prominent feature of the avian ovary.

Unlike the testes of seasonal birds, which in their regressed winter condition are generally gametogenetically inactive and show complete uniformity in histological condition, the ovary always contains large numbers of follicles in various stages of development, as well as a spectrum of corpora atretica in various stages of atresia. In its primary stage the follicle consists of a single layer of flattened granulosa cells ensheathing the oocyte, but as it matures, a multilayered thecal tissue forms around the outside and becomes highly vascularized. During this growth, the granulosa cells become separated by intercellular spaces filled with perivitelline substance, and villus-like elevations arise on the oocyte cell membrane to form the zona radiata. The oocyte completely fills the developing follicle, so that there is no antrum or other internal organization as in the mammalian follicle.

In the Rook (*Corvus frugilegus*), which to date is the only seasonally breeding wild species in which the ovarian cytology and histochemistry has been studied in any detail, stromal fibroblasts become incorporated into the theca interna of the developing follicle and become rounded and glandular as maturation proceeds. With the approach of the breeding season, a proportion of the follicle population progress toward full maturity and eventually enter a phase of very rapid growth that results in conspicuous bulging from the surface. This culminating period of accelerated development is brought about mainly by the deposition of yolk, and Wyburn et al. (1965) have suggested that the ultrastructural elevations of the zona radiata may facilitate the passage of yolk materials into the egg at this time. The final yolk deposition can be very rapid and can, in the pigeon, produce an increase in the maximum follicular diameter from 5.5 mm to 20 mm within 8 days. Sturkie (1954) reports a sixteenfold increase in ovum weight in chickens, in the same period before ovulation. After ovulation, the ruptured follicles rapidly regress, disintegrate, and become cleared by phagocytic action. No persistent corpora lutea of the mammalian pattern are formed, and in the chicken and Rook the postovulatory follicle becomes resorbed within a few days, but in Ring-necked Pheasants (*Phasianus colchicus*) and in the Mallard they persist for as long as 3 months. In these species in which there is a more persistent postovulatory structure, it is derived mainly from the hypertrophic thecal cells.

The numerous follicles that undergo preovulatory degeneration may do so at different stages of maturity. The first observable histochemical indication of this is usually a lipoidal zone surrounding the oocyte, which subsequently spreads so that the whole interior of the follicle becomes occluded with a densely sudanophilic mass of cholesterol-positive material. In the Rook, Marshall and Coombs (1957) report that this fatty atresia is accompanied by a proliferation of fibroblasts that invade the follicle and, for a time, remain as a distinguishable cluster of cells in the ovarian stroma. This phenomenon can also be observed in the Tree Sparrow. These cellular masses subsequently lose their integrity with the disintegration of the follicle and become scattered singly, or in groups, in the stromal tissue. Marshall and Coombs (1957) use the term "exfollicular gland cells" to distinguish these from the stromal interstitial cells that have developed from the ordinary connective tissue cells of the ovarian stroma and are regarded as being homologous with the interstitial Leydig cells of the testis. As in the latter, the ovarian interstitial cells possess the general ultrastructural characteristics of steroidogenic tissue and react positively to histochemical tests for 3β-HSDH, cholesterol, and sudanophilia. Chieffi and Botte (1970) also distinguish a pattern of atresia in which the yolk becomes extruded through the granulosa layer into the surrounding stromal tissue, where it becomes resorbed by phagocytic action. Once yolk resorption is completed only a few vacuolized cells, some capillaries, and strands of connective tissue remain to mark the site of the resorbed oocyte. In other cases, the oocyte may be invaded by cells derived from a hyperplastic activity of the granulosa tissue.

Histophysiology

General histophysiology

Literature on the cyclic changes of the avian ovary is sparse, and very little is known beyond the follicular activity of domestic species. Data on the histophysiological changes in the ovary of the seasonally breeding bird throughout the year are almost nonexistent and have been reported, in any sort of detail, only in one species, the Rook (*Corvus frugilegus*). In this bird, the follicles of the December ovary are too small to be observed on the surface macroscopically, but histologically, sectioned material shows some oocyte development to be taking place internally, and follicles up to 1 mm diameter occur in association with numerous smaller ones. Marshall and Coombs report that the largest follicles at this time of year have distended, sudanophilic, glandular cells incorporated in the theca interna, and that the granulosa layer of some of the biggest follicles disappear as such, as a proliferation of this tissue fills the structure with large lipid-free cells, which later on apparently become secretory. There is relatively little lipoidal atresia at this time of year.

In temperate-zone species, the advent of the spring weather is, as in the male, generally accompanied by a recrudescence of gonadal activity, and developing follicles become increasingly obvious, macroscopically, over the ovarian surface. In *Corvus frugilegus*, follicular recrudescence begins in late January and early February, and from then until the first ovulation in March the whole organ becomes studded with cholesterol-positive lipoidal structures, as increasing numbers of the smaller follicles start developing and become atretic. The stromal interstitial cells also become heavily charged with lipids at this stage and react positively to histochemical tests, both for cholesterol and 3β-HSDH. In the Rook, by March some follicles have enlarged enormously and a few enter the final vitellogenic stage and become ovulated. In a species such as the Band-tailed Pigeon (*Columba fasciata*), on the other hand, no more than one such enlarged follicle is found in the breeding ovary, and this can be correlated with the fact that such species lay only a single egg per clutch. Generally, a species that lays several eggs per clutch will have an ovary containing several large follicles filled with yellow yolk.

After extrusion of the egg, the follicular wall collapses and becomes spotted with lipoidal material. In the hen, the granulosa cells become inflated with lipids and remain conspicuous for about 72 hours after ovulation, but luteinization does not occur and no corpus luteum develops. In seasonal breeders the completion of the last clutch is succeeded by a rapid regression of the gonad, and individual follicles become increasingly difficult to distinguish by the naked eye, though internally great numbers of follicles continue to undergo a slow development followed by lipoidal atresia. According to Marshall and Coombs (1957), a differentiation of a new generation of interstitial tissue, a phenomenon analogous to the testicular interstitial cell rehabilitation during the regeneration phase, takes place in the Rook ovary during the autumn.

There is unequivocal evidence that the avian ovary, like its mammalian counterpart, is a source of androgenic, estrogenic, and also progestogenic secretions, even though a true corpus luteum is lacking. Chromatographic studies have clearly established that this organ, both in its embryonic stage and in the adult bird, has the capacity to produce these hormones, and androgenic, estrogenic, and progestational steroids have all been extracted from the plasma of the domestic fowl. The respective intraovarian sites of biosynthesis of these various steroids, however, is still somewhat speculative. We have already noted that the thecal and granulosa tissue of the preovulatory follicle, the postovulatory follicle, the atretic follicle, the interstitial cells of stromal origin, and also the exfollicular gland cells of Marshall and Coombs, at some stage all contain cholesterol-rich lipoidal inclusions and also have other glandular characteristics that might suggest every one of them to be possible loci of endocrine function, and this is confirmed by the fact that a positive reaction for 3β-HSDH activity has been reported for all these tissues. Furthermore, electron microscopy has shown that all these tissues at some stage possess the fine structure normally attributed to steroidogenic tissue.

Interstitial cells

Because of their apparent homology with the testicular Leydig cells, the ovarian interstitial cells arising from the stromal tissue are generally regarded to be the most likely source of androgenic secretion within the female gonad. Techniques for displaying steroid dehydrogenase activity have indicated that in the hen the steroidogenic potential in this tissue develops early in the embryogenesis of the organ and not just in the post-hatching period, as had previously been supposed, and this has also been shown to be the case in *Coturnix*. In a seasonally breeding bird, such as the Rook, the cyclic development of these cells closely parallels the development of the sexual displays, becoming increasingly prolific and active as sexual activity heightens in the spring and autumn. In the hen, too, the 3β-HSDH activity in these cells increases in intensity as the animal enters its egg-laying period, and reduces in intensity as the follicular phase wanes. The work of Taber (1951) also strongly suggests that a stromal cell is responsible for androgen production. The comb of the domestic fowl is in both sexes under androgenic control and is sometimes used as a bioassay for androgenic analysis. Taber recorded a reappearance of foamy lipoidal interstitial cells concomitant with the decline of androgen production (comb regression) in hens after the cessation of gonadotropin stimulation. It seems likely that in the absence of gonadotropic stimulation and decline in androgen synthesis, the rapid accumulation of cholesterol-rich lipoidal precursor material in these cells resulted in them becoming more prominent. The more recent methods for the visualization of androgenic steroids by means of fluorescent antibody techniques confirm that the interstitial tissue is the primary locus of androgen secretion, though the thecal and granulosa layers also produce a weak reaction.

Boucek and Savard (1970) have recently investigated steroid formation in the hen's gonad at different stages of the ovarian cycle, by a chromatographic analysis of the bioconversion of [^{14}C] acetate into progesterone, androstenedione, testosterone, and estradiol-17β in an *in vitro* incubation. A parallel analysis of the distribution and intensity of 3β-HSDH and 17β-HSDH activity was also carried out to provide data on the intraovarian location of the sites of activity. It was discovered that whereas the same spectrum of radioactive steroids was manufactured, the relative ratios of one steroid to the other varied significantly with the state of the ovarian cycle. Thus, a relatively large proportion of androgenic steroids was synthesized by the gonad of the molting hen, whereas the steroid profile produced by the ovarian tissue of the laying hen showed equivalent amounts of progesterone, androgens, and estrogenic hormones. In the broody hen, chiefly progesterone and estrogen were synthesized with very little androgen. Significantly, the histochemical studies demonstrated that the high androgenic production by the molting hen coincided with an intense 17β-HSDH activity in the stromal cells, and also the thecal tissue, whereas only trace reactions were produced by these cells during the laying stage.

Developing follicle

In mammalian ovarian histophysiology, current opinion generally credits the thecal cells of the developing follicle with the secretion of estrogens, and the granulosa cells with the secretion of progesterone after they metamorphose into the granulosa lutein tissue of the corpus luteum. Support for this latter observation is also provided by the data showing that granulosa cells grown in tissue culture manufacture progesterone. In birds, the evidence is more conflicting, but the tremendous increase in the size of the oviduct which is coincident with sexual maturity is known to be an estrogen-dependent effect and indicates that estrogen secretion appears to be a consistent consequence of follicular development, as is the case in mammals. In a seasonally breeding species, the marked annual fluctuations in oviduct development closely parallel the activity of the ovary. Thus, in *Corvus frugilegus* the oviducal epithelial height in the sexually inactive winter condition is 8 μm, but cellular proliferation begins in early February as the vernal acceleration of follicular development starts, and the epithelial layer rapidly expands to a maximum height of about 40 μm by the end of the month. When follicular development diminishes, the oviducts become constricted again. Estrone, estradiol-17β, and estriol have all been found in avian ovarian tissue, and the former two substances have also been identified in the plasma of the laying hen. In addition to their vitally influential role in the seasonal preparation of the oviduct, estrogens have many effects on a wide variety of somatic and physiological processes in birds.

The seasonal histochemical changes described by Marshall and Coombs (1957) in *Corvus frugilegus* seem to support the hypothesis that the avian thecal cells may similarly be the site of estrogen synthesis as they apparently are in the mammal. Thus, an accumulation of cholesterol-positive lipoidal material builds up in the glandular cells of the theca interna in the largest follicles during the winter months, and then rapidly becomes depleted as estrogen titers are reaching their peak and the mature follicles are undergoing their culminating vitellogenic phase prior to ovulation. By the time yellow yolk appears in the oocytes, only slight traces of sudanophilic material remain in the cytoplasm of the greatly distended thecal cells. Vitellogenesis, and the consequent deposition of large quantities of yolk within the developing oocyte, is an estrogen-dependent phenomenon, and the rapid depletion of the thecal cholesterol and lipids at this stage of the seasonal cycle may be indicative of a rapid utilization of precursor material at a time of high steroid synthesis, as has been suggested to be the case when a similar depletion occurs in the testicular Leydig cells. Certainly, the fine structure of the thecal cells and their positive response to tests for steroid dehydrogenase activity suggest a possible site of steroid synthesis, and the more recent demonstration that isolated thecal tissue from growing follicles of the hen's ovary has the capacity to convert cholesterol to estrogens provides strong support for the suggestion that this tissue may be the locus of estrogen production in the avian ovary. Sayler et al. (1970) have investigated the effects of photostimulation on the distribution of 3β-HSDH in the ovary of young Japanese Quail and have attempted to express changes in steroidogenesis by expressing the intensity of the 3β-HSDH histochemical reaction by means of an arbitrary scale based on the density of the formazan granulation deposited by the different steroidogenic tissues. In *Coturnix* that were raised from hatch to maturity under a stimulatory (16L:8D) light cycle, 3β-HSDH activity increased in all the ovarian steroid producing tissues, but from day 35 onward, the index of 3β-HSDH activity in the thecal tissue increased very rapidly from an average value of 1.3 to 4.3, whereas the enzyme activity in the granulosa tissue was less intense, rising from an average of 1.0 to 2.9 in the same period. Interestingly, the enzyme activity in the thecal tissue of birds kept under a short day length (8L: 16D) was also much higher than that of the granulosa tissue.

According to Chieffi (1967) the granulosa cells are the probable site of estrogen biosynthesis, since 17β-HSDH is limited to this tissue. This enzyme catalyzes the transformation of testosterone into androstenedione and of estradiol into estrone. Arvy and Hadjiisky (1970) have also recorded a greater dehydrogenase activity in the granulosa cells both in the hen and in the quail. However, Marshall and Coombs (1957) point out that the granulosa cells in the seasonally breeding Rook are the only prominent ovarian cells that do not contain cholesterol during follicular development, and electron microscopy studies of the hen ovary have shown that granulosa cells start developing abundant agranular endoplasmic reticulum, mitochondria with tubular cristae, cholesterol, and sudanophilic granules, only at a time when the follicles are ready to ovulate, thus suggesting that their steroidogenic activity has been relatively slight up to this time.

Boucek and Savard (1970) have confirmed Chieffi's earlier observations of the localization of 17β-HSDH in the granulosa cells of the laying hen, but have also shown that in the molting bird the reverse situation develops and 17β-HSDH activity now becomes intense in thecal tissue but absent from granulosa cells, even though *in vitro* incubation and chromatographic analysis of the ovarian tissue show that there is a considerable bioconversion of radioactive acetate into estradiol-17β during this time. The same investigation has also shown that there was little or no progesterone production in these molting hens. In view of the absence of 17β-HSDH in granulosa tissue in these birds, and also the very marked reduction in the intensity of the 3β-HSDH as well, it seems not impossible that the granulosa cells might be involved in the synthesis of progesterone. The ovarian steroid profiles of the laying and broody hens both show considerable progesterone biosynthesis and, in both groups, steroid dehydrogenase activity occurs in the granulosa tissue, being particularly intense in the laying hen. These observations are also supported by the data of Furr (1969a), which show that progesterone occurs in the blood of laying and broody hens, but is not detectable in the blood or ovaries of molting birds. Furthermore, the highest levels of progesterone are found in the follicles and only negligible amounts occur in the ovarian stroma.

Postovulatory follicle

The postovulatory follicle has sometimes been suggested as the source of ovarian progestogenic steroids, and the ruptured follicles have been shown to contain progesterone and have the capacity of synthesizing steroids *in vitro*. Furthermore, the experimental removal of recently ruptured follicles influences the retention time of eggs in the oviduct. Conner and Fraps (1954) have established that there appears to be a quantitative relationship between the amount of postovulatory tissue and oviposition, so that the greater the proportion removed, the greater the retardation in oviposition. It has also been established that when the next-to-last ruptured follicle is removed it has no effect, suggesting that whatever endocrine activity the postovulatory follicle possesses, it is rather short-lived, and this is also indicated by the rapid degeneration that generally occurs in most avian species.

The very transient endocrine phase of postovulatory follicles outlined by the experiments of Fraps and his colleagues is also supported by the electron microscopic studies of Wyburn and his co-workers, who have recorded an extensively developed agranular endoplasmic reticulum in the granulosa cells of the recently ovulated follicle, together with an increase in sudanophilia and cholesterol content. There is also a very intense reaction to tests for 3β-HSDH, but this rapidly diminishes with the degeneration of the postovulatory structures. It seems likely that this very abbreviated postovulatory steroidogenic capacity in these cells is probably a hangover from the activity of the granulosa tissue during the period of high progesterone production immediately preceding the ovulatory process.

Atretic follicle

Cholesterol is abundant in the thecal tissue of the atretic follicles, and during the initial stages of atresia in the granulosa cells. In both layers, a positive 3β-HSDH activity has been reported. The latter investigators also report 17β-HSDH in the granulosa layer. Marshall and Coombs (1957) have suggested that the atretic follicles may be involved in progesterone production, since great numbers of follicles undergo such lipoidal atresia, and, in a seasonal breeder, this phenomenon builds up a considerable reservoir of cholesterol by the time of the seasonal ovulation period. However, Sayler et al. (1970) have established that when young *Coturnix* are reared to maturity under nonstimulatory light conditions (8L:16D), there is an increase in the numbers of atretic follicles, and when the 3β-HSDH reactions in the ovary are examined, sparse granulation is seen in the granulosa cells at the beginning of follicular regression. But in the later stages, this disappeared entirely, suggesting a loss of steroidogenic activity in developing follicles once atresis sets in. It is perhaps premature, however, to conclude that the atretia structures play no significant part in the endocrinology of the avian ovary, and it is evident that there is a need for much more information before the precise locus of any of the ovarian steroids can be established with any certainty.

Vitellogenesis

Attention has already been drawn to the fact that, in the days immediately preceding its ovulation, the avian follicle undergoes a very rapid expansion in size due to the storage of large quantities of yolk in the maturing oocyte (i.e., vitellogenesis). For example, in the chicken, the mass of yolk can increase from 100 mg to as much as 20 gm in the 7 days before ovulation. Although experiments using radioisotopes leave no doubt that protein synthesis can, and does, occur within the growing oocyte, this accounts only for a minor portion of the yolk materials synthesized during this short time interval, and current evidence indicates that the greatest portion of the yolk lipids and proteins laid down in vitellogenesis arise elsewhere in the organism. Basically, the process can be divided into three successive stages, all of which are causally related. These are (1) the synthesis of the yolk proteins and lipids, (2) their transportation to the ovary and uptake by it, and (3) the incorporation of the serum proteins into the yolk elements.

In the avian oocyte, the yolk contains relatively more lipid (33% wet weight) in relation to protein (16%) than is the case in many other vertebrate eggs. In fine structure, it is seen to consist of two major components, namely, the yellow and white yolk granules which appear to be freely suspended in the second component, the yolk fluid. The granules often have discrete limiting membranes and, internally, may contain osmophilic subgranules that are larger in the white granules than they are in the yellow ones. The yellow and white granules are laid down in alternating concentric layers that reflect a diurnal variation in carotenol deposition. Very few mitochondria are observable within the mature yolk mass, except during the early stages of yolk-granule formation, nor is there much evidence of an endothelial reticulum. Much of the protein, together with some of the lipid, largely occurs in the granules, whereas the yolk fluid contains large quantities of lipid globules and a smaller proportion of proteins. Although there are slight differences in their detailed chemistry, the major yolk proteins consist, as in all other groups of oviparous vertebrates, of lipovitellins and phosvitin.

Evidence for the extraoocyte origin of most of these constituents stems from the earlier observations that phosphoprotein appears in the plasma of the laying hen. In view of its chemical similarity with the yolk protein vitellin, it was suggested that plasma phosphoprotein might be a precursor of the latter, which eventually became translocated into the egg. The role of blood proteins in yolk formation was further emphasized by the demonstration, by immunological techniques, that even foreign proteins after being injected into the circulation could be detected in the yolk. Other changes in the constitution of the blood that have been observed to occur as a prelude to vitellogenesis are a very sharp elevation in the level of plasma calcium and also of lipids.

More recently, Schjeide and co-workers (1963) have been able to show that, at the onset of laying, two new serum proteins appear in the hen which are chemically and immunologically similar to the three major egg yolk proteins—phosvitin and α- and β-lipovitellin. The same investigators also confirmed the transference of serum lipovitellin into the egg yolk by means of radioactive tracers. In this experiment serum lipovitellin was first extracted and labeled with ^{14}C, then injected into a laying hen. Within 10 hours, much of the radioactive protein was lodged in the yolk proteins, indicating its transference from the circulation across the egg cell barriers into the yolk. The mechanism of the transmission of such blood proteins into the oocyte is far from clear, but it may be by micropinocytotic activity at the oocyte surface. The follicular epithelium certainly seems to play a significant part, and Patterson et al. (1961) have clearly demonstrated a preferential transfer of serum γ-globulins by the follicular tissue to developing ova and have found that its concentration in the yolk is many times that in the serum. Gonadotropins appear to be important for stimulating the mechanism of this final transference of blood proteins into the oocyte, and in some of the lower vertebrate groups, an increase in serum protein transmission and yolk deposition has been shown to result from FSH treatment, possibly by a stimulation of the process of micropinocytosis.

The liver appears to be the extraoocyte locus of the synthesis of yolk protein and lipid, and hepatectomy has been shown to prevent the typical plasma changes associated with the onset of the vitellogenic phenomenon. Furthermore, radioactive phosphate injected into the wing vein of laying hens has been shown to become incorporated first into a protein-bound component in the liver which then becomes circulated in the plasma as a phosphoprotein, before finally being transferred into the oocyte and localized in the egg yolk. Confirmation is also provided by the demonstration that the *in vitro* incubation of liver slices from laying hens can synthesize phosvitin, which is similar to the yolk phosphoprotein, and Hawkins and Heald (1966) have similarly demonstrated the *in vitro* synthesis of triglycerides and established that it is much greater in laying hens than in the liver of immature pullets. These triglycerides are mainly transported to the oocyte as β-lipoproteins in the plasma, and eventually become concentrated within the fluid yolk compartment of the egg as lipid globules. Although β-lipoproteins occur in the plasma of immature as well as mature birds, there is a pronounced lipemia at the onset of vitellogenesis in the latter.

Control of Ovarian Function

Effects of exogenous hormones

The exogenous administration of different gonadotropic and gonadal hormones and their effects on the female gonad have been less extensively studied than in males, and the little information that is available is generally based on observed changes at the morphological level, without any detailed records of the cytological and histochemical events that might have occurred. Most of our knowledge has been derived from studies on the domestic hen, with particular reference to the effects that such hormone treatment have on ovulation and oviposition. Information about the follicular phase is far less precise.

Androgen

The evidence for ovarian androgen synthesis, and the probable site of production, has already been discussed. The role of androgenic steroids in the regulation of secondary sexual structures and behaviour is also well established, but there is no good evidence that indicates that endogenous ovarian androgens have any significant role in the regulation of ovarian function. The available data are contradictory.

Chu and You (1946) have reported a stimulatory effect on the ovary of both immature and hypophysectomized adult pigeons after androgen administration, and testosterone stimulation of the ovary has also been recorded in *Passer domesticus*. In both these species, androgen injections greatly enhance follicular growth, and Ringoen has noted that the follicular epithelium of treated sparrows becomes stratified. Similarly, stratification of the follicular epithelium and a stimulation of follicular growth have also been noted in the hen ovary after testosterone propionate treatment. An increase in the vacuolation of the cells and a greater ovarian sudanophilia was also noted in the latter investigation. On the basis of observing the effects of a variety of dose levels (0.1-100 μg) of testosterone propionate, Breneman (1955, 1956) has reported that androgen administration increases the height of the follicular epithelium, but a subsequent statistical analysis of his data has led van Tienhoven (1961) to conclude that no dose-response relationship existed and that the differences between controls and androgen-treated pullets could be accounted for by a sampling error. In the American Sparrow Hawk (*Falco sparverius*), large doses of testosterone propionate (140 mg over 30 days) fail to produce any gonadal changes, neither in the left nor right ovary.

Estrogen

Exogenous estrogens are known to delay ovulation in Ring Doves and in domestic hens, and have been reported to inhibit the seasonal development of ova in House Sparrows. Injections into hypophysectomized pigeons failed to induce any significant follicular maturation, and in 30-day-old pullets, estradiol administration did not cause any significant change in ovarian weight, although Breneman (1955, 1956) noted a general increase in sudanophilia and cholesterol content. In contrast to these observations, Phillips (1959) has reported a 32% increase in ovarian weight in 6-week-old pullets injected with diethylstilbestrol, and an even greater ovarian stimulation after similar injections into adult nonbreeding Black Ducks (*Anas rubripes*). The same steroid can also stimulate ovarian development in wild Mallards and is said to act synergistically with gonadotropins to stimulate follicular development. It may be that such contradictory results reflect varying capacities of different steroids to inhibit gonadotropic secretion.

Although to date there is insufficient evidence to attribute any direct effect of estrogenic hormones on gonadal function (i.e., with concomitant pituitary influence), there are ample data to show that such steroids have a vital indirect role by influencing vitellogenesis in the mature ovary. It is now reasonably well established that estrogen stimulates the liver to produce the protein and lipid precursors of yolk. Estrogen injections into female birds lead to a very rapid rise in the plasma levels of fatty acids and lipoproteins, and Roos and Meyer (1961) report that estrogens also boost the serum level of lipoprotein lipase in pheasants. Phosphoproteins also appear in the circulation as a consequence of estrogen administration, and, significantly, the same treatment can even induce similar changes involving the appearance of lipovitellin and phosvitin in the plasma of immature pullets and cockerels.

As might be expected in view of the above observations, estrogen injections also cause parallel changes to take place in the liver, and Schjeide et al. (1963) report that a very large increase in liver volume is produced in domestic fowl as a consequence of estrogen administration. In estrogen-treated birds, the liver tissue undergoes a hyperplasia, and the cells become greatly enlarged with a highly developed endoplasmic reticulum and an increased ribosomal and cytoplasmic protein content. The nuclear RNA content is also doubled in comparison with the average liver cell of nonestrogenized birds. Schjeide (1967) has established that lipid metabolism, and the release of such lipids into the plasma, is greatly stimulated in such birds. Although exogenous estrogen administration brings about a mobilization of yolk precursors, it does not necessarily result in an increase in the number of yolky follicles in the ovary, since gonadotropins are necessary for the transference of these substances from the plasma into the oocyte. Thus, in nonbreeding ducks injected with diethylstilbestrol, only those that were additionally treated with anterior pituitary extract developed large follicles with yellow yolk, whereas treatment with either estrogen or gonadotropin alone failed to stimulate large yolky follicles. The same is true also of wild Mallards.

Progesterone

In domestic fowl, progesterone administration inhibits egg laying if given early in the 24 hour ovulatory cycle, or if given at any time in large doses, and results in an increase in follicular atresia. A similar blockage of ovulation, together with a decrease in ovarian weight and follicular diameter, also occurs in progesterone-injected California Quail when given early in the daily ovulatory cycle. In this latter species the oviduct weight of progesterone-treated females was also significantly smaller than in control birds, suggesting a decline in estrogen secretion presumably caused by an inhibition of FSH release. When, however, progesterone is implanted into immature pullets, it hastens follicular maturation and precipitates precocious egg production, and its administration late in the egg-laying cycle of the adult causes premature ovulation, presumably by stimulating LH release. The experiments of Ralph and Fraps (1960), which demonstrated the ovulation inducing effects of small quantities of progesterone stereotactically injected into the hypothalamus, but not when injected into the pituitary, indicate that this ovulatory influence is probably mediated via the hypothalamo-hypophyseal axis. The investigations of Kappauf and van Tienhoven (1972) on the progesterone concentrations in peripheral plasma of laying hens have demonstrated that there is a significant rise in plasma concentrations at about 6 hours prior to ovulation.

Prolactin

Although prolactin was the first anterior pituitary hormone to be isolated in pure form, data on its effects on the ovary are much fewer than on the avian male gonad. As in the male, exogenous prolactin injections have sometimes been reported to have an antigonadal effect also in the female, and in the hen, pigeon, and White-crowned Sparrow, treatment with this hormone inhibits follicular development with resultant atresia. In pigeons, hens, House Sparrow, and Bank Swallow (*Riparia riparia*), the ovary has been reported to be regressed during the incubation period, and, in view of the good evidence showing that prolactin release is probably high during this time, such observations are consistent with the experimental evidence showing an antigonadal effect with this hormone. However, in the Rook, the ovaries of incubating females possess many large follicles, and gonadal regression does not appear to take place until after the young have hatched.

Furthermore, Jones (1969a) has reported that prolactin only interrupts, but does not stop egg laying in *Lophortyx californicus* and does not effect follicular diameters or ovarian and oviductal weights. Similarly, Juhn and Harris (1956) found that prolactin blocked the antigonadal effect of progesterone in domestic fowl and only temporarily interrupted egg laying when given alone. Thus, in females, too, there is at present insufficient evidence to attribute a general ovarian regulatory role to prolactin.

Pituitary-gonad axis

Ablation of the anterior pituitary gland produces a regression in the female gonad and extensive follicular atresia. This atresia can be delayed for about 5 days by injections of mammalian gonadotropins and for a longer period by avian pituitary material. Furthermore, ovarian activity and a recovery of ovary-dependent organs can be induced in hypophysectomized hens in which postoperative regression has been allowed to develop, by injection of a crude avian pituitary preparation. In this investigation, Mitchell (1967a) allowed a 10 day

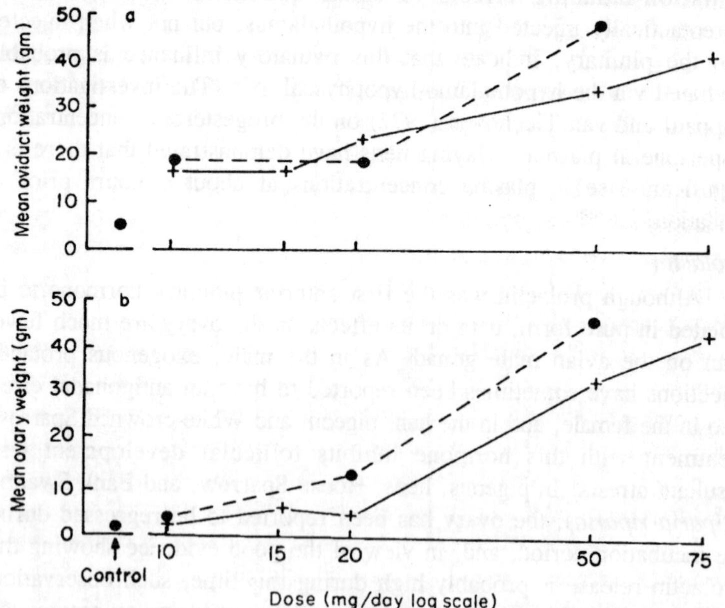

Fig. 7.5. Effects of chicken anterior pituitary extract on (a) oviduct weight and (b) ovarian weight of hypophysectomized hens, in which both organs had been allowed to become completely regressed before commencement of treatment; (●) treatment for 12 days; (+) treatment for 8 days.

postoperative period for complete ovarian and oviduct regression; he then injected different groups of hypophysectomized birds with varying doses of powdered acetone-dried material suspended in isotonic saline.

There is a very close correlation between ovarian function and gonadotropic concentration of the pituitary gland, both in domestic species and in a seasonally breeding bird such as the White-crowned Sparrow, and a similar relationship between pituitary gonadotropin content and ovarian function has also been shown in *Coturnix*, where gonadotropin potency and ovarian development undergo a parallel increase when birds are held under long daily photoperiods from the time of hatch.

Although female birds, like males, show a gonadotropic response to photostimulation, the ovary in most species can rarely be stimulated to full breeding maturity under experimental conditions. A block to growth occurs at the onset of vitellogenesis, presumably because other proximate factors are required. Captivity itself is adequate to stop development, and captive female White-crowned Sparrows have pituitary glands whose gonadotropic potency is only about one-fourth that of reproductively active females in their natural habitat. Similarly, a lack of gonadotropic potency has also been recorded in the pituitaries of captive wild Pintails (*Anas acuta*), which failed to reproduce. An exception appears to be domesticated strains of *Coturnix*, in which young females under long daily photoperiods can apparently develop to full reproductive condition, a fact that enables the species to be utilized for commercial egg production.

As is also the case in male birds, practically all our knowledge on pituitary-gonad interrelationships is derived from studying the effects of exogenous purified mammalian gonadotropins or, as outlined above, by determining endogenous gonadotropic potencies of the anterior pituitary by assay methods that are not specific but generally measure the combined effects of FSH and LH. Investigations using purified avian FSH and LH have still to be done, and, therefore, caution must be exercised in extrapolating conclusions based on mammalian gonadotropins, particularly since there appear to be pharmacological and physiological differences between the avian and mammalian hormones. For example, follicular development cannot be stimulated in the ovaries of immature chicks by mammalian gonadotropins, yet crude avian pituitary extracts produce a marked stimulation. The lack of response to mammalian gonadotropins, however, alters with age, and in pullets nearing sexual maturity, exogenous hormone administration induces follicular development. Dessicated avian pituitary material is also more effective than purified mammalian gonadotropins in causing continued follicle development and ovulation in starved pullets and in stimulating precocious follicular development in the immature ovary.

The differences between gonadotropins of avian origin and mammalian origin is further underlined by the fact that the partially purified avian hormones react poorly in mammalian assays. Interestingly, Licht and Stockell-Hartree (1971) have also tested avian FSH and LH in the lizard *Anolis carolinensis* and find that the relative potency is significantly greater than that estimated from standard rodent bioassays.

Bearing in mind the above reservations, there is good evidence that mammalian gonadotropins have considerable activity when injected into female birds, and ovarian stimulation as a consequence of such treatment has been demonstrated in a number of wild and domestic species. Furthermore, replacement therapy with mammalian gonadotropins is apparently effective in hypophysectomized pigeons. The evidence indicates that, as in mammals, FSH stimulates follicular development and estrogen secretion (as expressed by an increase in oviduct development) in a number of species. For example, the effect of daily injections of 0.1 mg/kg of mammalian FSH for 2 weeks into sexually regressed Green-winged Teal (*Anas crecca*) caught on their wintering grounds in Hong Kong. It can be seen that follicular development has been stimulated and there is a marked hypertrophy of the oviduct, indicating a stimulation of estrogenic secretion. So far, the effects of injecting purified avian FSH have not been recorded, but a partially purified preparation injected into hypophysectomized hens maintained the ovary and oviduct weights at the control level when therapy commenced within an hour after the operation. There was also evidence that yolk deposition also continued after hypophysectomy in these birds. When hormone injections were delayed until 10-15 days after hypophysectomy, however, a resurgence of follicular development could not be induced at the dose level used (0.33, 0.5, and 0.75 mg/day). When avian FSH was tested on lizards, it stimulated a rapid development of the follicles and a significant increase in oviduct weight.

Nalbandov (1959) has pointed out that in birds, unlike most other vertebrate groups in which groups of follicles mature together and are destined to ovulate together, the growth and ovulation of follicles is continuous within the breeding season, and there are mechanisms that must prevent the maturation and ovulation of more than one egg at one time. Thus, the avian controlling mechanisms provide for the existence of a hierarchy of follicles of graded sizes, only one of which becomes ovulated at any one time and usually on any one day. There is considerable evidence that ovulation depends upon a cyclic release of LH, but ovulation will be induced only if the follicle has reached sufficient maturity to respond. Nalbandov (1961) has pointed out that the ovarian follicle is extremely resistant to the ovulatory action of exogenous LH until shortly before the expected release of endogenous LH, and in the domestic hen and also in the Japanese Quail, there is good evidence that plasma LH levels display a significant peak some 6 to 8 hours prior to ovulation. The magnitude of the rise in plasma LH is much smaller than that found in mammals prior to ovulation, and may reflect an avian physiological adaptation, since a massive burst of LH might release more than one ovum from the ovary. Between clutches the level of LH declines and becomes almost undetectable.

The release of LH from the anterior pituitary gland is under nervous control, and lesions placed in the ventral preoptic region of the hypothalamus block the ovulatory process. Such lesions will also prevent progesterone-induced ovulations, indicating that this steroid stimulates LH release via a hypothalamic route, and this is also indicated by the observation that electric stimulation of the preoptic area will induce a premature ovulation

Accessory Sexual Organs

The accessory sexual organs are those parts of the reproductive system exclusive of the testis and ovary that are also vital to reproduction. These comprise the sperm ducts in the male and oviduct in the female, together with any special cloacal differentiation. The reproductive process may also be assisted by other differences between the sexes. For example, plumage colour or pattern may differ, special plumes or wattles may become elaborated in one sex, bill colour may change, or there can be voice differences. Such characters whereby the sexes differ are termed secondary sexual characters, the primary sexual characters (the gonads) obviously being excluded from this definition.

Two pairs of ducts, derived from primitive kidney ducts, develop in both sexes, these being the Mullerian and Wolffian ducts and each pair runs from the gonad to the urogenital sinus. In the male, the Wolffian ducts become the gonoducts draining the testes, and subsequently become the vasa deferentia and epididymides in the adult bird, while in the female they usually remain vestigial. In the female, the left Mullerian duct becomes the oviduct, while the right, and both these ducts in the male, remain vestigial; occasionally the ducts do not atrophy as appropriate. The occasional persistence of the right Mullerian duct in the female appears to be genetically controlled, since a high proportion of certain inbred strains of fowl exhibit two ducts instead of one.

Male

Mature spermatozoa are released into the lumina of the spermatic tubules, which form the highly convoluted network of the testis. From here they are passed by a small number of tubules, the rete tubules, to the highly convoluted vasa efferentia. Outside the breeding season, the rete tubules are inconspicuous, as they are sited in the tunica albuginea of the testis, but they do become enlarged and distinct as two to four tubes extending from the testis with the onset of reproductive activity. In view of the fact that spermatozoa are rarely observed in the rete tubules. Bailey (1953) has concluded that their passage through this region is rapid and probably periodic. The spermatozoa next pass to thin tubules, the vasa efferentia, which also show some seasonal enlargement and which may become secretory. These leave the tunica albuginea and join to form a long coiled tube, the epididymis. This comprises a compact structure closely bound to the testis and lying adjacent to the kidney. With the onset of sexual activity, the epididymis hypertrophies to become a prominent white knob on the side of the

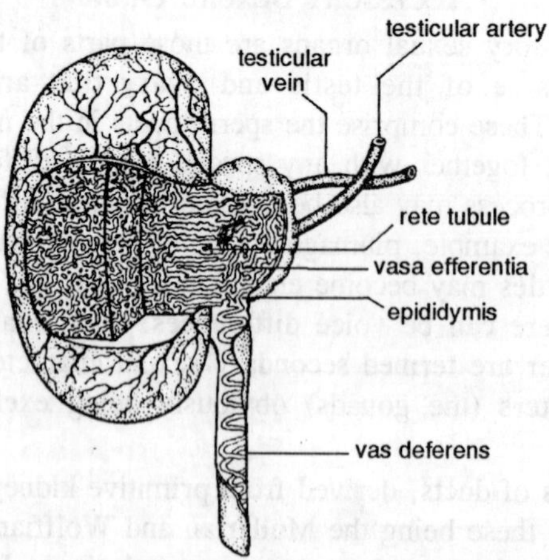

Fig. 7.6. Diagram of the internal network of the testis of the domestic fowl and its relationship with the efferent ducts.

testis. The cells of its epithelium are columnar and ciliated, and secrete a seminal fluid. Sperm are not stored in this organ but progress into a deferent duct, the vas deferens; two muscular tubes one from each of the paired epididymides therefore lead to the urodeum of the cloaca. These vasa deferentia become enlarged, more convoluted, and distinct from the kidney during the breeding season, and before entering the cloaca are expanded to form a seminal sac. This increases thirty-to fortyfold in weight during the breeding season, becoming heavily charged with spermatozoa. A muscular sheath forces the posterior wall of the sperm sacs to expand into the cloaca as erectile papillae and doubtless facilitates the extrusion of sperm during copulation. However, these papillae are minute in most species, and during copulation the cloaca of each sex is everted and the papillae are brought into contact with the opening of the oviduct.

In some species part of the cloaca is modified as a penis-like structure. In waterfowl the penis is a vascularized sac which can be protruded by a muscle and retracted by a ligament. The ligament is bound spirally round the sac and gives the penis a twisted appearance. Sperm move along a spiral groove that passes from the cloacal papillae at the organ's base to its tip. Hohn (1960) has shown that the weight of the penis of the Mallard (*Anas platyrhynchos*) increases sixfold with the onset of the breeding season and that castration prevents such development. Benoit (1936) had earlier noted that at puberty the phallus became enlarged under the influence of testicular hormones, but outside the breeding season it did not regress again to its juvenile dimensions but instead remained fairly constant in size. Females of those species in which the male possess a "penis" have a homologous but reduced structure termed the "clitoris." Their cloacal opening is slightly modified to receive the intromissive organ of the male.

Penis-like organs occur in tinamous (Tinamidae), curassows (Cracidae), the ratite birds, and the ducks and geese (Anseriformes). Alone among the Passeriformes, the Black Buffalo Weaver (*Bubalornis albirostris*) possesses an imperforate penis-like external organ. The groups possessing these modified genitalia are diverse, and the species appear to share no common feature that would indicate an adaptive function for their special attributes. However, it may be significant that all the species concerned, except the ducks and geese, habitually practice polygamy or polyandry. Thus, in tinamous, as in nearly all other ratites, the male usually incubates the eggs and cares for the young unaided by the female. In the Slaty-breasted Tinamou (*Crypturellus boucardi*), the female lays an egg for one male which incubates it and then, in turn, she moves on to lay an egg for another male. Conceivably, each male must be ready to fertilize the female with a reduced opportunity for precopulatory displays that would help synchronize the process, but this view is speculative. The Ostrich (*Struthio camelus*). Greater Rhea (*Rhea americana*), and other ratites, and such tinamous as *Nothocercus bonapartei* perform cooperative laying whereby several females lay their eggs in the same nest that is then attended solely by the male; in some cases the females may repeat the performance for another male. Again selection may be favouring more effective copulatory mechanisms in species that do not have the benefit of a long period of intrapair courtship during which time the endocrine state of the male can become adjusted. The ducks are also unusual in that pairing and copulation usually occur on the water.

As already mentioned, seasonal changes in the size and degree of development of the vasa deferentia, the epididymis, and the penis, if present, parallel the seasonal changes observed in the primary sex organs in those species showing cyclical activity; these structures remain permanently developed in continuous breeding species such as the domestic fowl. This seasonal development apparently depends on androgenic hormones in the case of the vasa deferentia, epididymides, and the seminal sacs, and ablation of the testes will prevent the seasonal development of these organs in seasonally breeding species, whereas injections of testosterone propionate will induce their development into a fully breeding condition. Some reservation is needed in that various pioneer experiments have not been repeated since purified hormones became available, and it is possible that early extracts were contaminated with gonadotropins; this reservation is occasioned by recent experiments into the factors inducing the black bill pigmentation of House Sparrows. Moreover, in the case of the female canary, it seems that estrogenic hormones will induce oviduct development only in subjects kept on stimulatory photoperiods and not birds treated in winter. This suggests that the synergistic action of a gonadotropin is necessary, and it is likely that a parallel situation may apply in the case of the male. Critical experiments with hypophysectomized birds have not been performed.

Maximum development of the epididymides is noted at the beginning of the preincubation behaviour cycle of the male pigeon, and the height of the columnar epithelial cells decreases up to the time of egg laying. If androgen-dependent, this change would indicate a declining androgen titer over the period leading up to egg laying, and for this there is other evidence.

Female

The right oviduct, like the right ovary, remains vestigial except in Raptores, and it is the left that becomes highly differentiated to satisfy the processes involved in producing the complex shelled and pigmented eggs. Even in species in which the incidence of two functionally developed ovaries is high, however, only one oviduct may develop. The developed oviduct is a tube attached to the body wall by dorsal and ventral ligaments and double folds of the peritoneum in which are situated a network of blood vessels; the oviduct is covered by the peritoneum. Its walls embody circular and longitudinal muscle fibers variously developed in different sections, and similarly, there is a variable development of a lining mucous membrane. In places, the membrane comprises longitudinal folds covered with a glandular and ciliated epithelium. Five anatomically distinguishable regions are discernible, namely, the infundibulum, magnum, isthmus, shell gland, and vagina. The extruded ovum is received from the ovary via the open anterior funnel-like infundibulum, which so surrounds the follicle that when the latter ruptures the ovum does not lay free in the body cavity. The ovum is also probably guided into the funnel by the ventral ligament. It is fertilized in the upper thin-walled neck of the oviduct, where the mucous membrane is many-folded and before albumin is added to the yolk. In domestic fowls, fertile eggs can be laid 10-12 days following a single copulation, emphasizing the length of time that sperm can remain viable. However, fertility declines after about 5 days, and it is a fact that wild birds normally copulate many times during the laying period. Polyspermy occurs as several spermatozoids enter the blastodics (12-25 in the pigeon), though only one unites with the female pronucleus. By the time the egg is laid, cell division causes the segmentation area or blastoderm to attain a diameter of nearly 5 mm in the domestic fowl.

The middle portion of the oviduct is the highly glandular magnum, where the white albumin is formed; this is deposited round the ovum in four layers to make a complex structure. In the hen, this process takes from 3 to 4 hours to complete. The histological development of the oviduct has been studied in fowls and in a careful investigation of the domestic canary. While the ovaries comprise up to 1.5% of body weight of the canary, there is an approximately linear relationship between oviduct and ovary weights (both represented as percentages of total body weight), as might be suspected since oviduct development depends partly on ovarian hormones. The formation of albumin coincides with the enlargement of one ovarian follicle more than the rest, and this also correlates with the completion of defeathering in the course of brood-patch development and intensive nest-building behaviour. Relatively little oviduct development is apparent until nest-building activity begins; indeed only the division of epithelial cells of the magnum is to be noted prior to the collection of the first nest material. However, within the period of active nest-building, the oviduct increases markedly in length (decreasing again after egg laying), the epithelium divides and invaginates to form tubular glands, and these become distended and their cytoplasm charged with albumin granules; at the same time, the mucosa becomes very folded. The epithelium differentiates during this time into ciliated columnar cells and goblet cells. The histology of the magnum apparently remains uniform along its length at all stages of development.

After albumin has been deposited, the egg next passes into a less glandular and more muscular isthmus of the oviduct where the two shell membranes are secreted, these being comprised of felted protein fibers cemented together with albumin. The outermost membrane has a rough texture and so provides a key for the calcarious shell. This last, together with pigment, is added in the wider and expandable uterus, or shell gland. The shell is secreted in the form of calcium salts (about 97%) supported on an organic matrix of protein fibers. Any ground colour is deposited during the final stages of shell formation, while blotches, streaks, or other surface markings are acquired after the shell has been completed. Because the egg is rotated on its long axis surface markings are often given a spiral configuration. The rotation is achieved by a powerful sphincter muscle, which when relaxed, allows the egg to pass into the cloaca and to be laid pointed end first. In the domestic fowl, about one-third of the eggs become reversed; the vagina bulges beyond the sphincter thereby turning the egg over, whereupon it is eventually laid blunt end first; whether this happens in other birds remains uncertain. It appears that this eventual oviposition is effected by a muscular contraction of the shell gland (uterus) regulated by the secretion of posterior pituitary hormones. Thus, a sudden decrease in the content of vasotocin in the neurohypophysis takes place in domestic hens and is correlated with a rise in plasma levels, coincident with oviposition. Furthermore, a premature expulsion of the egg can be induced by exogenous oxytocin and vasopressin and also vasotocin, the latter being the most effective.

The seasonal hypertrophy of the oviduct that occurs with sexual recrudescence (gonadotropins are implicated) is dependent on estrogen secretion by the ovary and estrogenic hormones have been shown to cause marked morphological changes in immature chick oviducts and to stimulate the differentiation of mature cell types. But other hormones are involved in the full development of the tract, for progesterone and prolactin augment the increase in weight of the oviduct following estrogen injection in Ring Doves and canaries. In domestic fowls, estrogen alone will cause growth of the albumin-secreting glands and the secretion of ovalbumin. Daily injections (5 mg) of diethylstilbestrol into 4-day-old chicks causes a 300-fold increase in ovalbumin production within 15 days, and an increase from 5-10 mg to 1.5-3 gm in the oviduct weight is recorded after 3 weeks of such estrogen treatment. Progesterone administration, on the other hand, appears to stimulate only the synthesis of the egg-white protein avidin, and this has been shown to be effective, not only *in vivo*, but also *in vitro* when minced chicken oviduct tissue is incubated with progesterone. At the cellular level, O'Malley and his colleagues have demonstrated that estrogen treatment stimulates a large increase in nuclear transfer RNA and a definite but smaller increase in cytoplasmic transfer RNA.

Secondary Sexual Characters

Darwin appreciated that the physical differences between sexes in plumage and voice or their possession of various adornments were associated with courtship and that sexual selection depends on the success of certain individuals over those of the same sex in terms of producing surviving progeny. Because females normally choose their partner, and frequently rely on visual cues, isolating signal characters have evolved in many male birds to prevent confusion with, and hence hybridization between, closely related species. The avoidance of such interspecific competition for mates is to be distinguished from intraspecific competition for a mate, which, as Huxley (1938) recognized, leads to the maximum elaboration of display plumage in polygamous species.

Where a permanent plumage difference between the sexes imposes no other disadvantage, sexual dimorphism may have a genetic basis, so that hormones are unable to change the inherited sex type. Examples are provided by the House Sparrow (*Passer domesticus*) and European Bullfinch (*Pyrrhula pyrrhula*).

Witschi (1961) has provided examples of the endocrine basis of various secondary sexual plumage changes and acquisition of other nuptial adornments. Several of these variations have served for the bioassay of hormones. The development of the comb and wattles of newly hatched chicks, irrespective of sex, depends on androgen and can be stimulated by injections of exogenous testosterone. The pigmentation of the bill of birds is of two main kinds: (1) that produced by deposition of brown and black melanins in melanophores that become injected into those epidermal cells moving out from the area of proliferation and (2) yellow, red, and orange carotenoids that are not synthesized by the bird but are absorbed from the diet. The European Starling has a black bill during the contranuptial season, while both sexes acquire a yellow bill with onset of the breeding season. Injections of androgens induce the yellow colour, but estrogen and progesterone do not; castration causes the loss of the yellow colour. In *Quelea quelea*, the bills of both sexes are red during the contranuptial season, but in females this colour changes to a straw colour during the breeding season and is an estrogen-dependent response.

The bill of the House Sparrow (*Passer domesticus*) turns black during the breeding season under the supposed influence of androgen; castration results in a loss of pigmentation in subjects with black bills. Testosterone propionate, however, will not induce a black pigmentation in sparrows captured in winter and held under nonstimulatory day lengths, suggesting the implication of gonadotropins in the response. Exogenous LH plus FSH or androgen plus FSH will induce a black bill pigmentation in subjects kept on short day lengths so that endogenous gonadotropin secretion is inhibited. Colour change in the beak of the Paradise Whydah (*Vidua paradisaea*) apparently depends on LH and not gonadal hormones.

The assumption of secondary sexual plumage seems in many cases to depend on steroid hormones; for instance, castrated Ruffs fail to don the pectoral display plumes described above. Similarly, castrated male Black-headed Gulls (*Larus ridibundus*) fail to acquire the brown head typical of the breeding season. In phalaropes the female dons the more colourful breeding plumage, and in both Wilson's Phalarope (*Phalaropus tricolor*) and the Red-necked Phalarope (*Phalaropus lobatus*), it has been shown that the assumption of this plumage is dependent upon androgenic steroids. Thus, administration of testosterone propionate (but not estradiol or prolactin) to birds of either sex will stimulate the growth of feathers of the nuptial type in plucked areas. The reason that this colourful plumage is normally shown by the female and not the male can be attributed to the fact that there is a higher androgen secretion in the female, the reverse of the situation in nearly all other species. In contrast to the mechanism in the preceding species, in which the acquisition of the nuptial plumage is apparently determined by the elaboration of gonadal androgens, the male-type plumage of the weaver *Euplectes orix* has long been known to depend on a gonadotropin, supposedly LH, and the species has long served as the bioassay animal for this hormone. In this, and other weaver birds, a castrated male continues to develop a nuptial plumage. Females do not normally develop a gaudy plumage, but will do so after ovariectomy. In the natural condition ovarian estrogens inhibit potentiality of LH for stimulating nuptial plumage, and this is confirmed by the injection of estrogen into male weavers, which inhibits their assumption of the nuptial dress. Exogenous estrogens have a similar effect when injected into male fowls. Many of these experiments, however, were performed before purified hormones became available. Furthermore, because subjects were often held under photostimulatory day lengths, some reservation is needed in accepting previous conclusions about many of the supposed hormonal mechanisms governing secondary sexual characters.

Incubation Patches

In many avian species (but not all) the ventral apterium becomes defeathered and highly vascular and edematous before or during egg laying. Hyperplasia of the epidermis also occurs, producing up to a sixfold increase in thickness. These so-called incubation patches, or brood patches, are in close contact with the eggs during incubation and supply the necessary warmth for development of the embryos. In addition, such a structure provides a sensitive surface in contact with the nest and eggs, which in the domestic canary at least, is a source of tactile stimuli influencing nest-building behaviour. The sensitivity increases as the patch develops.

Incubation patches usually develop in the females only (e.g., House Sparrow, White-crowned Sparrow), but in some species they may occur in both sexes (e.g., starlings and swallows). In the Red-necked Phalarope and Wilson's Phalarope, they develop only in the males, which in these species incubate the eggs, the female taking no part in this activity. Generally, there seems to be little correlation between the incubation habits of male birds and the possession of a brood patch, since the incubating males of some species, such as the House Sparrow and Bushtit (*Psaltriparus minimus*), lack it, whereas the nonincubating cocks of others, such as the American genus of flycatchers (*Empidonax*), possess it.

The chronology of the morphological changes that constitute the formation of an incubation patch appears to differ in various species. In the passerine species, defeathering occurs before egg laying, but the timing of patch development in galliformes appears to be more retarded, so that defeathering becomes marked in Ring-necked Pheasants only after the laying of the fourth egg, and in California Quail after the tenth egg. In *Zonotrichia leucophrys*, Bailey (1952) reports that defeathering from the ventral apterium is completed several days before oviposition, and increased vascularization and edema is not apparent until completion of this stage, but in the Red-winged Blackbird (*Agelaius phoeniceus*), Bank Swallow, canary, and House Sparrow, epidermal hyperplasia, increased vascularization, and mild edema occur before defeathering is complete and, in some birds, begins weeks before egg laying. Thus, in both the European Starling and House Sparrow an appreciable increase in the numbers of dermal blood vessels is noticeable a month before oviposition. The increase in vessel diameter, however, occurs mainly during the egg-laying period.

The development of the incubation patch is under endocrine control and involves hormones of both gonadal and adenohypophyseal origin. Generally, it is formed under the influence of estrogen and prolactin, which are synergistic in their effects. In the passerines, exogenous estrogen administration has been shown to induce increased vascularization and defeathering in both intact and ovariectomized birds, but similar treatment of hypophysectomized birds produces vascularization only. When estradiol in combination with prolactin is given to hypophysectomized specimens, the complete patch is developed; in hypophysectomized *Zonotrichia* prolactin alone has no effect. Since the effect of estrogen on defeathering is augmented by prolactin, and the combination of these two hormones is necessary to produce full development in hypophysectomized birds, it seems likely that, in female passerines at least, this phenomenon is normally dependent upon the endogenous secretion of both these hormones. In California Quail, however, prolactin appears to play a more significant role than in the passerine mechanism, since Jones (1969b) reports that in this species injection of estradiol into reproductively active females fails to have any significant effect. Prolactin alone, on the other hand, stimulates epidermal hyperplasia, and when given in combination with estrogen produces full incubation patch development (i.e., defeathering, edema, vascularization, and dermal thickening). In this species, and also in the Bobwhite Quail (*Colinus virginianus*), prolactin-induced hyperplasia of the incubation patch can be stimulated not only *in vivo* but *in vitro* as well. It seems that only the ventral skin is responsive to hormonal effects, since Jones et al. (1970) have shown that estrogen plus prolactin induces the ventral abdominal skin of the chicken to undergo hypervascularization and epidermal hyperplasia only if the skin is *in situ*. Ventral skin transplanted to the back exhibited epidermal hyperplasia but not hypervascularization. Dorsal skin was not responsive *in situ* nor when transplanted to the ventrurn. Thus, hypervascularization of the chicken incubation patch is site-specific, but the epidermal hyperplasia is not. Furthermore, the data show that epidermal hyperplasia is not dependent on hypervascularization.

Progesterone has been found to augment the effects of estrogen in inducing defeathering in the female canary but is not essential for this process, since it can be induced by estrogen alone in ovariectomized birds. When progesterone is given alone it has no effect on vascularization in canaries, House Sparrows, or California Quail, and in the last species it is also less effective as a synergist with estrogen. In the canary, this hormone has been found to be necessary for the characteristic increase in sensitivity of the ventral skin during development of the incubation patch.

In *Phalaropus lobatus*, *P. tricolor*, and *Lophortyx californicus*, in which the males develop incubation patches, it has been shown that exogenous testosterone plus prolactin can induce incubation-patch development. In the natural situation it seems likely that in such species, vascularization occurs when circulating androgen levels are high, and before defeathering of the patches is completed, the latter event being delayed until circulating prolactin levels increase, probably at the time of testicular regression. In the passerine species in which males usually do not incubate, exogenous androgen plus prolactin do not cause such feather loss in the cock, and in the parasitic Brown-headed Cowbird (*Molothrus ater*), a species that does not incubate its own eggs or develop a brood patch, the ventral skin of either sex is insensitive to estrogen-progesterone injections.

In the Red-winged Blackbird, it has been shown that the experimentally induced patch (by exogenous estrogen or estrogen plus prolactin) differs histologically from those of wild incubating birds, and in the House Sparrow, Selander and Yang (1966) have shown that the normal patch is more edematous than the hormonally induced ones, and it may be that the sensitivity to tactile stimulation that has been demonstrated by Hinde and his colleagues may be necessary for the complete development of the patch.

Breeding Behaviour (Endocrinological Basis)

As already made clear in this chapter, gametogenesis depends on a changing pattern of hormone production culminating in the animal entering the phase of overt reproductive activity. At this stage, various behaviours become manifest to assist the process of breeding. These include the acquisition of a territory and its advertisement by song, the attraction and pairing with a mate, and the initiation of displays that lead to egg laying and nestling production. Increasing titers of androgenic or estrogenic hormones have long been assumed to be responsible for many sexual behaviours in male and female, respectively, but it is becoming increasingly apparent that the endocrine basis of reproductive behaviour is complex and often highly specific. Research has been concentrated on various pigeon species and the canary, and while these are possibly reasonably representative of the class Aves, cognizance must be made of the possibility that other species have evolved special endocrine adaptations to suit their own peculiar ecological requirements.

Unpaired, yet sexually mature, male feral pigeons will usually advance with the bowing display toward other members of the same species. Sometimes, an otherwise responsive male will fail to react to a receptive female but respond to other females, and it appears that the birds can recognize particular plumage morphs with which they are reluctant to pair. In free-living populations, this results in certain pairings of like individuals occurring with less than expected frequency, and the aversion appears to function as an outbreeding mechanism preventing the production of homozygous birds that have a reduced fertility; the gene concerned improves fertility and lengthens the breeding season if present in the heterozygous condition. In Ring Doves, much variation in the range of displays exhibited by males during the initial courtship phase could be attributed to their previous experience; an experienced bird might reduce or omit certain of the more aggressive postures. Previous breeding experience by either male or female contributes to and improves the reproductive performance of Ring Doves under captive conditions. Fundamental endocrine-behavioural interactions of this kind doubtless underlie the general fact that under field conditions older and more experienced birds reproduce more successfully.

There is good evidence that androgens mediate bowing behaviour in both male Ring Doves and also feral pigeons, and they restore male-type behaviour to castrates, increase social status and aggressiveness, and induce malelike behaviour in females. In the phalaropes, in which there appears to be a reversal of the normal situation and females display more aggressive patterns than the male, Hohn and Cheng (1967) have shown an unusually high ovarian testosterone content. Both Goodwin (1967) and Davies (1970) have shown that the bowing display differs from species to species, exhibiting a typical intensity in each. Davies (1970) who also examined F_1 hybrids of *Streptopelia* doves showed that these displayed with a characteristic bowing pattern that depended on inherited components of the display. In particular, the rate of bowing proved to be species-specific, while components of the bowing vocalization were similarly inherited in the same manner as morphological characters.

Bowing appears to function as a ritualized appeasement display; it combines crouched submissive postures (at termination of downward bow) with tail-fanning and head-lowering reminiscent of and probably derived from braking movements. However, the upright and advancing components of the display sequence embodying an almost "goose-stepping" gait represent aggressive components. The ambivalent nature of the display is indicated by the fact that if a bowing male catches up with the subject of his attention, he may (1) peck aggressively, (2) mount and attempt to copulate, or (3) turn and flee. Thus, the display involves causal factors for attacking, fleeing, and mating, and it is significant that different exogenous hormones injected into males at the beginning of their preincubation behaviour cycle will alter the relative importance of these various display components. Exogenous FSH increases the incidence of driving, an aggressive display, and facilitates the appearance of sexual-aggressive components generally, while exogenous androgen facilitates a more natural sequence with bowing progressing to the more sexual manifestations of mounting and copulation. Hutchison's data (1970b) indicate that the chasing component of the display persisted in castrates, whereas bowing was eliminated, but the extent to which exogenous FSH will evoke displays in castrates has yet to be determined. It does seem probable that it may act synergistically with androgen. Moreover, on the basis of behavioural evidence, it is feasible that FSH is involved in the release of androgen. Exogenous LH had little effect on behavioural sequences in this species, but it did increase the probability that attacking would follow bowing. In *Quelea quelea*, although androgen injections would not increase the social status of low rank birds kept in small groups, exogenous LH would. Similarly, LH and not androgen seems to determine social status in *Sturnus vulgaris*.

Natural selection will favour males who can fertilize a potentially good mother, while females must ensure they have adequate protection from a mate and territory before accepting a male's advances. Initially the male's behaviour is aggressive and centers on depositing his sperm in the female before another male can do so; hence copulation occurs early in the behaviour sequence when aggressive behaviour is paramount. Females will usually squat for a male only in a territory in which they also have become accepted, but this does not prevent males making sexual advances to all potential mates when on the feeding grounds and away from a territory. So, selection tends to advance the time of copulation for the male, retard it for the female, and results in a compromise balance. Having fertilized a female, a male must now ensure adequate protection for the forthcoming eggs and young; and to a large degree, the remaining needs of his courtship sequence are, so to speak, to mollify his partner and provide her with suitable nesting facilities. But it appears that the male is initially endowed with high titers of plasma androgen (and probably also of FSH) which would be detrimental to the progression of the cycle. In fact, it appears that feedback mechanisms become initiated in consequence of courtship and lead to a reduction in androgen levels.

It is known that implanted testosterone propionate in the preoptic region of the hypothalamus reestablishes copulatory behaviour in capons so long as the necessary external stimuli are provided, and such work is beginning to define the hormone-sensitive regions of the central nervous system which must underlie sexual behaviour. This led Hutchison (1970b) to examine whether other sexual patterns of courtship preceding copulation might be linked to androgen-sensitive areas of the central nervous system, especially since there had been a suggestion that aggressive and copulatory activity might depend on separate androgen-dependent mechanisms in the central nervous system. Hutchison, therefore, investigated the effects of intracerebral implants of crystalline testosterone propionate on castrated Ring Doves; he was careful to segregate his experimental subjects according to their experience and preparedness to perform the nest-demonstration display as a sequel to the bowing display. The type of courtship evoked and whether fragmentary or the complete sequence of chasing, bowing, and nest soliciting (nest demonstration) depended on the anatomical position of the implant. Only those males with anterior hypothalamic and preoptic implants showed the complete reestablishment elbowing, while implants in the area basalis and hypothalamicus posterior medialis resulted in maximum nest soliciting. Hutchison's data (1970b) are consistent with the hypothesis that the stimulatory effects resulting from posterior hypothalamic implants depended on the diffusion in low concentration of testosterone to an androgen-sensitive area in the preoptic and anterior hypothalamic area; i.e., mechanisms underlying chasing and bowing require higher hypothalamic concentrations of androgen than mechanisms underlying nest soliciting. This evidence does not support the view that different androgen centers control different behaviours. Nonetheless, separate sites are possibly involved if a wider range of behaviour is considered. Electrical stimulation of the preoptic and anterior hypothalamic areas of the forebrain of pigeons induced an intensive bowing with all normal components, and this display could then become transformed into nest demonstrations via displacement preening. Defense and escape behaviour in the same species was induced by electrical stimulation of di- and telencephalic structures in the forebrain.

Following castration, bowing and chasing are the first displays to be eliminated while nest demonstration persists for longer, suggesting that it depends on a lower threshold of androgen stimulation. Indeed, Hutchison (1970b), using different sized hypothalamic testosterone implants, found that high-diffusion implants, delivering high concentrations, resulted in longer durations of bowing than smaller medium-diffusion implants, while low-diffusion implants induced nest soliciting in the virtual absence of chasing and complete absence of bowing. This at first suggests a simple and plausible explanation for the mechanism underlying the natural display sequence. But Murton et al. (1969b) also found that nest demonstration became fully established, albeit after a delay, in subjects maintained on high levels of exogenous testosterone treatment, so that these high dosages did not inhibit the onset of nest demonstration. Instead, these data suggest that a new mechanism becomes activated that can nullify the effects of exogenous androgen. The exogenous administration of estradiol monobenzoate immediately and consistently caused the appearance of nest-demonstration displays in intact feral pigeons and also in Ring Doves, while hypothalamic implants similarly induced long durations of nest demonstration in castrated doves. We do not know whether estrogen is involved in the normal cycle of the male, but the capacity to synthesize estrogenic steroids by avian testicular tissue is well established.

There is other evidence that androgen levels may decrease during the preincubation display phase of the feral pigeon. Lofts et al. (1973b)

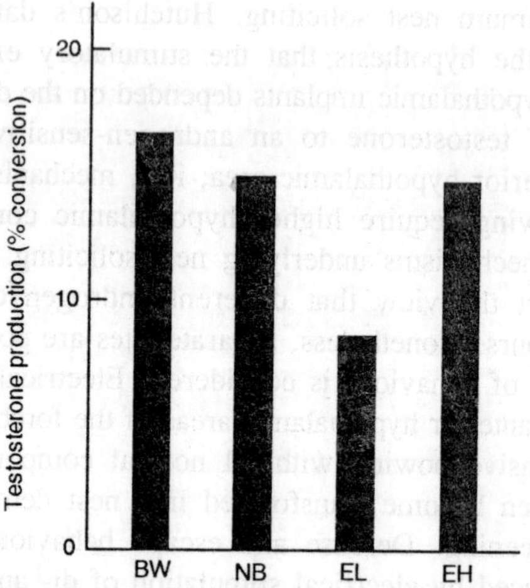

Fig. 7.7. The in vitro production of testosterone from radioactive pregnenolone by the pigeon testis during the bowing (BW), nest building (NB), egg laying (EL) and hatching (EH) periods of the reproductive behaviour cycle.

incubated testicular and ovarian material taken from feral pigeons at different stages of the reproductive cycle with tritiated pregnenolone as the steroid precursor. The synthesizing ability of the gonads at different stages was then measured using thin-layer chromatography to separate the different steroids produced. During the initial period of bowing display, there was a peak in androgen synthesis (i.e., during the first 3 days of the cycle), but approximately 10 days later and just prior to egg laying, androgen synthesis was markedly reduced. An increase in progesterone levels was also noted at this time. The capacity of the Leydig cell to synthesize a substance need not imply that the same substance is released from the cell, but it is almost certainly significant that quite separate evidence points to the existence of a high plasma titer of progesterone prior to egg laying in Ring Doves. Thus, males of this species that have previous breeding experience can be induced to incubate eggs by injections of progesterone or, in a smaller percentage of cases, following prolactin administration. All the same, when progesterone-injected and prolactin-injected doves were tested in bisexual pairs, more incubated than when tested alone, so the presence and behaviour of a mate significantly influences hormone-induced incubation responses.

Progesterone crystals chronically implanted into the brains of reproductively experienced Ring Doves induced incubation behaviour if placed in the preoptic nuclei and lateral forebrain, while the sexual

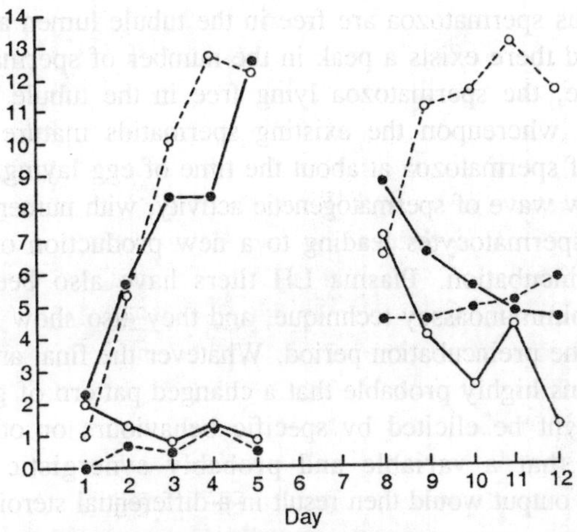

Fig. 7.8. Daily incidence of nest demonstration by paired male pigeon subjected to different hormone treatments.

and aggressive components of male courtship were suppressed. Male castrates, caused to exhibit courtship displays by treatment with testosterone, were selectively inhibited from bowing and induced to display nest soliciting alone when simultaneously injected with progesterone. So it seems conceivable that rising progesterone levels could inhibit androgen effects leading to the suppression of sexual-aggressive displays and instead allow the emergence of pair-cementing displays, such as mutual allopreening (caressing) with the onset of nest building and egg laying. But we must still not neglect the possibility that an estrogenic phase occurs in the male pigeon during the middle part of the preincubation cycle, and certainly exogenous estrogen is more effective than exogenous progesterone in causing nest demonstration.

A changed endocrine state during the preincubation cycle must depend either on exhaustion of a mechanism already engaged, which seems unlikely, or on the specific stimulation of neural-hormonal pathways elicited by specific behavioural events in the cycle. This second proposition is more likely in allowing for changes consequent to the natural or artificial disruption of the cycle. The preferential production of progesterone, for example, must depend initially on neural or hormone mechanisms initiated at higher levels of integration. We do not yet have the means to measure plasma FSH levels accurately and can only suspect that FSH titers decline during the preincubation behaviour phase. The spermatogenetic sequence in feral pigeons shows that numerous spermatozoa are free in the tubule lumen at the start of the cycle and there exists a peak in the number of spermatids. By the middle phase, the spermatozoa lying free in the tubule lumina have disappeared, whereupon the existing spermatids mature into a new generation of spermatozoa at about the time of egg laying. Thereafter, there is a new wave of spermatogenetic activity, with numerous divisions of primary spermatocytes leading to a new production of spermatids during late incubation. Plasma LH titers have also been measured using a radioimmunoassay technique, and they also show a fluctuating titer during the preincubation period. Whatever the final answer proves to be, it seems highly probable that a changed pattern of gonadotropin secretion might be elicited by specific behaviours or other external stimuli and that a variable and probably synergistic balance of gonadotropin output would then result in a differential steroid production by the gonads.

Craig (1911) was the first to demonstrate that the caressing display of the male pigeon could facilitate ovulation in the female. Matthews (1939) later showed, by separating males and females with a glass plate, that the stimulus causing ovulation was visual and not tactile. Subsequently, Erickson and Lehrman (1964) found that castrated males separated from females by a glass plate were far less effective in stimulating oviduct development in the female than were intact birds. The elegant work of Lehrman and his colleagues has now well established that stimuli arising from association with a mate cause the birds to become interested in nest building, and that once in this state the stimuli produced by the presence of nest material facilitate the birds becoming interested in sitting on eggs.

The behavioural acts involved in nest building by the male stimulate gonadotropin secretion (presumably FSH) in the female, causing gonadal hormone secretion from the ovary, which in turn leads to oviduct development. Thus, in both pigeons and canaries, nest-building activity is temporally associated with rapid oviduct development and ovulation followed a few days later by oviposition and the onset of incubation. In both doves and canaries, estrogen causes a moderate increase in oviduct weight, and whereas progesterone alone has no effect, it potentiates the effects of estrogen to produce maximum oviduct development. Prolactin can also augment the estrogen response, possibly by causing progesterone release. But while exogenous estrogens readily cause oviduct development, very high doses are necessary to initiate actual nest building in doves and canaries. Paired and photostimulated canaries in winter show nest-building behaviour, while birds injected with PMS (pregnant mare serum) can lay without nest building. These observations support the likely situation that gonadotropin secretions potentiate nest-building behaviour, which exogenous progesterone or prolactin certainly do not.

In the female canary, incubation-patch development occurs coincidentally with nest-building and oviduct development. Lehrman's studies make it clear for doves that exogenous prolactin, unlike progesterone, will not initiate incubation behaviour but that prolactin can maintain such behaviour once initiated. Moreover, judging from the development of the crop gland in pigeons, prolactin secretion is stimulated by the specific act of incubating eggs, though prolactin secretion can be induced in the male if he can see the incubating female and if he has previously associated freely with her during earlier courtship. The appearance (artificially) of squabs once the female has laid eggs can also accelerate crop growth and, presumably, prolactin secretion. The pituitary glands of domestic hens contain prolactin only so long as they are allowed to sit on eggs.

To summarize, the preincubation reproductive cycle of the female appears to involve a period of increasing estrogen production, which it is tempting to attribute to increases in FSH production. Next follows a phase during which progesterone or prolactin secretion, or both, become involved, and the important endocrine problem needing to be resolved is whether increasing estrogen levels give rise to progesterone production which in turn stimulates prolactin secretion, or whether prolactin secretion is first stimulated leading in turn to progesterone production. Since Meites and Turner (1947) found no increase in pituitary prolactin level following progesterone injection, and since Lehrman's studies (1963) reveal no crop sac development in pigeons following progesterone administration, we may prefer the second alternative or acknowledge the possibility that nonmeasurable levels of prolactin were involved.

In the male feral pigeon, administration of exogenous estrogen causes (1) a regression of the testis tubules with only spermatogonia remaining present; (2) the encapsulation of the old Leydig cell generation, which becomes densely lipoidal; and (3) the appearance of a new Leydig cell generation, also heavily lipoidal. Coincidentally, plasma LH titers are markedly elevated, and it seems likely that the histological manifestations depend on the known elevation of LH coincident with a presumed suppression of FSH activity. In the male feral pigeon, LH titers increase during the end phase of the prelaying cycle and after the emergence of nest-soliciting displays. While it has yet to be resolved whether estrogen is liberated during the male's prelaying cycle (though it is known that the testis can produce estrogen), it does seem reasonable to anticipate that rising estrogen production in the female suppresses FSH activity and stimulates LH production (the Holweg effect) and that such a change in gonadotropin balance provides the necessary conditions for progesterone (prolactin) secretion. It is of course known that there is a boost of LH production associated with ovulation in the domestic hen. Analogous endocrine changes apparently occur in both male and female, particularly at the hypothalamic level, though the male cycle is advanced compared with that of the female. In the male, an initial FSH androgenic phase gives way to an LH and progesterone phase leading to incubation. In the female an FSH-estrogen phase is stimulated in consequence of the males' courtship, and this leads to oviduct development relatively late in the cycle; it is probably significant that in feral pigeons the female assumes the more aggressive and dominant role coincident with the nest-building phase, and she may even drive the male away at this stage of the cycle. Significant developments in this field can be expected in the next few years.

8

MALE SEX HORMONES

From the remotest times, perhaps since prehistoric days, it is known that castrated domestic animals show great differences when compared with normal members of the same species. A bull, castrated as a young animal becomes a tame ox, which can be handled easily, and a castrated cockerel develops only very small head appendages; moreover, its behaviour differs largely from that of a normal cock; the quiet capon does not crow and the pugnacity of the normal animal is not present.

Successful transplantations of testes were performed as long ago as in about 1770 by John Hunter, who in addition, in 1792, was the first to describe postcastrate changes in the accessory sex organs of mammals.

In Hunter's time and also during the first half of the nineteenth century it was still generally accepted that the functional relation between the gonads and the so-called secondary sexual characters was exclusively mediated through the nervous system; thus, it was thought that the changes occurring in a castrated animal were caused by the severance of the nerves between the testes and the rest of the body.

Experiments of Berthold (1849), however, were not in favour of this view; this investigator removed the testes of cocks from the normal site but left them in the body cavity between the intestines; moreover, he implanted the male gonads of another cock into the body cavity of a capon. If the concept of the nervous connection had been right, his experimental animals should have shown castration phenomena. This, however, was not so; the animals remained completely normal. At autopsy Berthold found that the testes had not regressed and had attached themselves to the wall of the body cavity, where they had become richly vascularized. Implantation experiments with testes into other parts of the body had the same result.

Although Berthold did not conclude that the testes secrete substances into the blood which influence other parts of the body, he was the first to assume that the influence exercised by the testes on the rest of the body is not mediated through the nervous system but should be ascribed to changes in the blood that circulates through the testes. Later Claude Bernard and his pupil Brown-Sequard expressed similar ideas.

Brown-Sequard's rejuvenation experiments (1889) in particular led to the concept that active substances which regulate certain processes in the body are present in the testes. It took, however, several decades before David et al. (1935) in Laqueur's laboratory succeeded to obtain a pure, active extract from bull testicles. But since Brown-Sequard attempted to prepare testis extracts for the first time, he may be regarded as the founder of modern sex hormone research.

Several important papers relating to the internal secretion of the testis were published during the end of the nineteenth and the beginning of the twentieth centuries.

In 1894 Steinach reported that in castrated male frogs the copulation pads and the strong musculature of the forelegs regress so that amplexus is no longer performed. Moreover, he found in castrated juvenile male rats that such accessory sex glands as the prostate and vesicular glands remain very small and that neither erection nor copulation occur in castrated adults.

Bouin and Ancel concluded in 1903 that the mammalian testis hormone is formed in the interstitial or Leydig cells. Some years later the castration experiments with fowls of Pezard (1911, 1918) and the masculinizing and feminizing experiments of Steinach (1912, 1913) and of his associate Lipschutz (1919) attracted much attention. When after about 1935 chemically pure sex hormones became available, these investigations stimulated much other work, first mainly concerned with morphological and later with functional aspects.

Physiology of Androgens

In vertebrate embryos the Mullerian as well as the Wolffian ducts with their accessory organs are generally present in both sexes. In the male embryo the Wolffian ducts and their derivatives develop further, whereas the Mullerian ducts and their derivatives remain rudimentary. The reverse applies to the female embryo.

In young or adult animals the accessory sex organs react readily to castration and to heterologous gonad transplantations or to administration of heterologous sex hormones. Thus, the sex hormones do not induce new characteristics but (like other hormones) only modify the development of pre-existing physical or mental characters.

It should also be stressed that frequently the results of hypophysectomy resemble those of castration: hypophysectomy is followed by regression of the gonads and when the latter is very pronounced castration effects develop. A good example of this has been given recently by Dodd et al. (1960), who described the results of hypophysectomy in the male lamprey, *Lampetra fluviatilis*.

Experiments relating to castration, transplantation of gonads and administration of sex hormones, have been done in representatives of nearly all vertebrate classes.

Secondary Sexual and Ambisexual Characters

In male vertebrates the testis hormone causes the phenomena related to male copulatory behaviour and develops and maintains the so-called secondary sexual characters. The latter can be defined as the characters, morphological, physiological, or mental, by which the male and the female of a certain species differ from each other. The gonads are defined as the primary sexual characters. According to this definition the accessory sexual glands and organs arc considered secondary sexual characters.

Scientific castration experiments have shown that some sex characters change after castration, and are consequently influenced by the sex hormones (dependent sex characters) but that others are not affected by gonadectomy and they are therefore called *independent sex characters*. This may be elucidated by an example derived from investigations on birds.

When a male of one, of the common breeds of poultry, for instance a single-combed Brown Leghorn, is castrated, either as a chicken or as a cockerel or cock, the following differences from control animals appear after some time: the head appendages (comb, wattles, and ear lobes) are small and pale; the carriage of the bird is not erect, and the male sexual instinct is absent. Capons do not crow and fight each other, and they no longer take interest in the normal hens of the pen. The weight of the castrated animal increases, owing to deposition of fat. Castration, however, does not influence the development of the spurs in chickens; on the contrary, in capons the spurs generally develop even more strongly than in normal birds. Neither is the plumage affected

Fig. 8.1. Brown leghorns; (a) a castrated male or female bird; (b) a normal hen, and (c) a normal cock.

by castration: thus a capon possesses the same beautifully coloured long feathers on neck, shoulders, rump, and tail as the normal cock. Further, the size of the bird is not affected by castration. Of the internal organs the ductus deferens remains thin and straight in a capon, castrated as a chicken.

It may therefore be concluded that in cocks of the Leghorn breed the following sex characters are dependent on the testis hormone: the carriage of the cock, the marked development of the head appendages, the male sexual instinct and behaviour, and the convoluted deferent ducts. Independent sex characters, i.e., sex characters which are not affected by castration, are the plumage, the size of the bird, and the development of large spurs. With the exception of the size, which is a genetic characteristic in both sexes, a young ovariectomized female chicken develops into a castrate with the same characteristics as a capon of the same breed. For instance, the plumage of the cock or capon develops also in the ovariectomized hen. It is clear, therefore, that in the female the development of the cocky plumage is inhibited by the ovary. It can thus be said that after castration the male as well as the female chicken are turned into a "neutral" animal.

In some species in which the male and female possess the same external appearance, characters may be present which are dependent on the gonads in both sexes. In addition to the plumages of the hen-feathered poultry breeds a good example of such "ambisexual" characters is the yellow bill of the male and female starling (*Sturnus vulgaris*) during the reproductive period (outside this period or after castration the bill becomes dark in both sexes). The almost black head, and the crimson bill and feet of the male and female. European black-headed gull (*Larus ridibundus*) and the North American laughing gull (*L. atricilla*) during the breeding period provide other examples. Van Oordt and Junge (1933) have demonstrated that in the black-headed gull, castrated in the nonbreeding season, the dark head and the crimson bill and feet do not develop during the next reproductive period and Noble and Wurm (1940) have observed the same phenomenon in the castrated male as well as in the ovariectomized female laughing gull.

Action of Androgens in Fishes

The lampreys (Petromyzontidae) possess well-developed sex characters in both sexes, but castration experiments have not as yet been performed.

Knowles has shown (1939) that testosterone and estrogen injected into immature lampreys induce changes associated with sexual maturity. Following injection of testosterone or estrogen the cloacal labia of immature male or female adults became swollen; ammocoetes reacted only slightly.

In elasmobranchs some investigations have shown the existence of sex characters, dependent on the male hormone. Hisaw and Abramowitz (1939), Goddard and Dodd (cf. Dodd, 1955), and Thiebold (1954) have found some growth of the claspers after the administration of testosterone to embryonic and young sharks or skates.

Teleosts, which usually have pronounced sex characteristics, have often been used for castration experiments. Castrated male sticklebacks

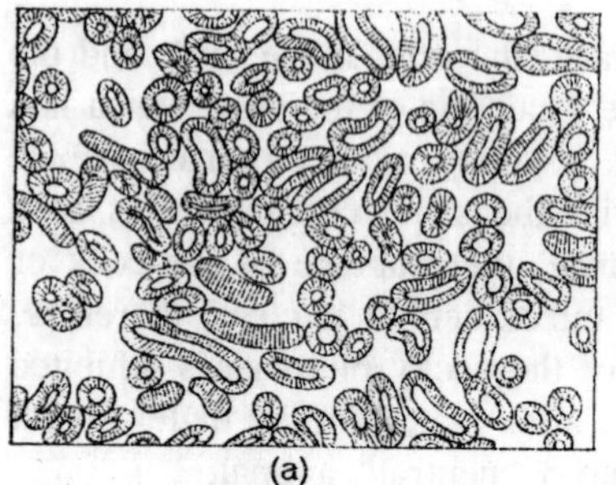

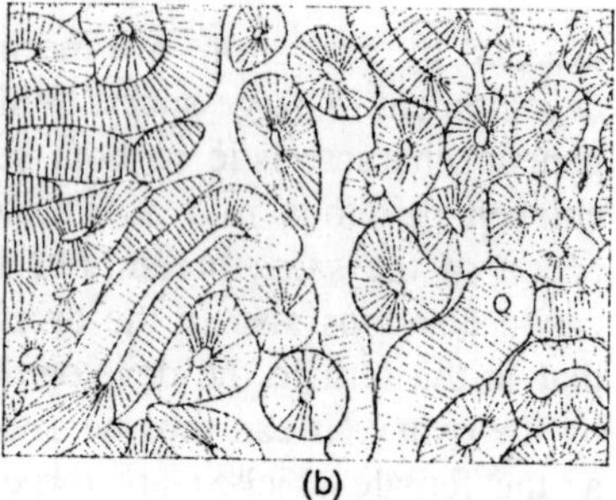

Fig. 8.2. Section through kidneys of stickleback, Gasterosteus pungitius; (a) during the autumn and (b) during the reproductive period.

(*Gasterosteus* sp.) lose their nest-building instinct. As in the Japanese bitterling, *Acheilognathus*, the beautiful nuptial colouration (caused in *Gasterosteus aculeatus* mainly by pigment dispersal in the erythrophores) does not develop after castration. Moreover, the kidney tubules, which in the male increase markedly in diameter during the reproductive season and secrete the mucus with which particles of the nest are attached to each other, return to the nonbreeding condition after total castration and can again be increased in size by administration of testosterone. The same phenomenon could be established by Ogura (1958) by injecting testosterone into female *Gasterosteus*. According to Tavolga (1955) the seminal vesicles of *Bathygobius* are also dependent on the male sex hormone.

In many other teleosts seasonal assumption of nuptial colouration as well as permanent sexual differences in pigmentation have been obtained by administration of testosterone. That the gonads influence the sex characters follows also from the fact that in functional sex reversal many sex characters are also changed.

In the cyprinodonts, a teleost family in which most males have very distinct sex characters, it has been found that many of these characters (e.g., the gonopodium, a transformed anal fin or the "sword," the elongated tail fin and the pointed pectoral fins of the swordtail, *Xiphophorus helleri*, are dependent on the testis hormone).

If testosterone propionate is administered to female guppies (*Lebistes reticulatus*) male sex characters are formed in the same order of appearance as during the normal development of the male. In castrated male guppies, in which the gonopodia have been removed, femalelike anal fins are regenerated. Ishii and Egami (1957) have pointed out that under the influence of testosterone the anterior rays of the

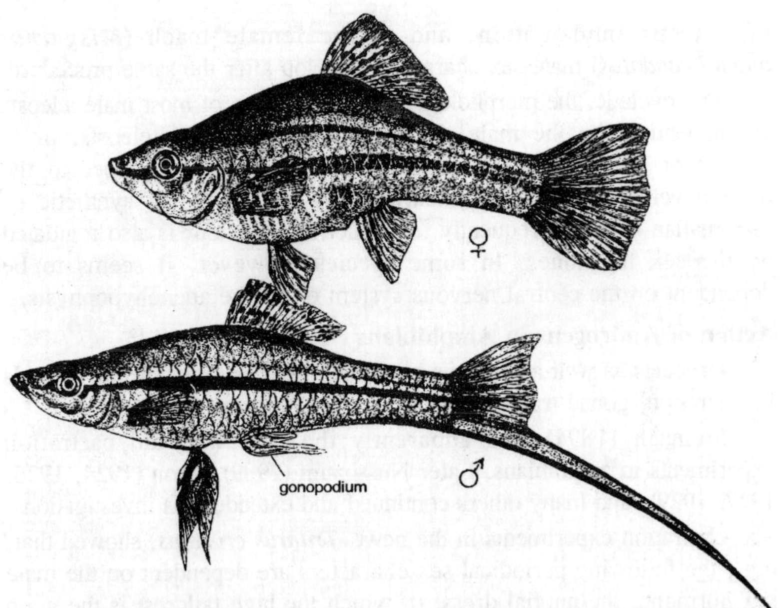

Fig. 8.3. Adult female and male swordtail (Xiphophorus helleri).

dorsal fin of juvenile male as well as female filefish (*Monacanthus cirrhifer*) resemble the adult male.

According to Ishii (1960), testosterone pellets implanted in pregnant females of *Ditrema temmincki* cause the death of the embryos and regression of the ovary walls.

During the breeding season the adult male of the minnow (*Hyborhynchus*) is characterized by several tubercles on the head, whereas the female possesses a large anal papilla. By injecting testosterone propionate Ramaswami and Hasler (1955) induced the growth of tubercles in males outside the breeding season as well as in females; in the latter, however, a paradoxical effect was obtained in addition: the anal papilla increased enormously. Ramaswami and Hasler think therefore that in female *Hyborhynchus* factors for tubercle development are inherently present, and that under normal conditions the tubercle growth in the female is suppressed by the ovary. Similarly, in *Amia* the caudal ocellus, present in males only, develops in females after ovariectomy. Paper on the effect of testis transplantation or the administration of androgens to normal or ovariectomized female fishes are numerous. Okada and Yamashita (1944), for instance, succeeded in producing masculinization in female ovariectomized *Oryzias latipes* after testis implantation, and in the female loach (*Misgurnus anguillicaudatus*) male sex characters develop after the same procedure.

To conclude, the morphological sex characters of most male teleosts are dependent on the male gonad. In castrated male teleosts, or in normal or ovariectomized females the sex characters proper to the male develop after the administration of androgens (of synthetic or mammalian origin). Frequently reproductive behaviour is also regulated by the sex hormones. In some species, however, it seems to be dependent on the central nervous system or on the adenohypophysis.

Action of Androgens in Amphibians

Urodeles as well as anurans have been used as experimental animals in castration, gonad transplantation, and sex hormone experiments.

Steinach (1894) was apparently the first who did castration experiments in amphibians. Later Nussbaum (1909), Aron (1924, 1926, 1927, 1929), and many others continued and extended his investigations.

Castration experiments in the newt, *Triturus cristatus*, showed that, e.g., the following periodical sex characters are dependent on the male sex hormone: the nuptial dress, of which the high tailcrest is the main feature, and the pronounced development of the cloacal glands.

In male anurans belonging to the genera *Rana* and *Bufo*, it was found that the marked development of the "thumb-pads" or copulation-callosities and their glands, of the strong muscles of the forelegs, and of the epithelium of the seminal vesicles are caused by the male sex hormone. Conversely, testosterone produced precocious thumb-pads in prepuberal male toads (*Bufo arenarum*). Steinach (1894, 1910) found that in male frogs, castrated some time earlier, the clasping reflex appears in the next breeding season, but is subsequently reduced or even lost. Similarly, Burgos (1950) found that the clasping reflex disappeared gradually in castrated toads.

Since all sexual characters are potentially present in both sexes, it is understandable that testes, grafted into ovariectomized females, cause the development of male sex characters. Welti (1925, 1928), for example, in ovariectomized female *Bufo bufo* was able to induce thumbpads, sex call, and mating behaviour by grafting testes.

Action of Androgens in Reptiles

This vertebrate class has been scantily investigated.

Sauria

In lizards the skin colouration, dorsal crests and spines, femoral and pre-anal pores, the epididymis, and the urinary sex segment are distinct male sex characters. Mathey (1929), Padoa (1933) and Neeser (1940) showed in male lizards that after castration the femoral glands and the skin pigmentation become identical (femoral glands) or almost identical (skin pigmentation) with those of the female. Herlant (1933) and Regamey (1935) found that in the male lizard the sex segment of the kidney is dependent on testosterone. Takewaki and Fukuda (1935) observed that testis implantation in castrated male *Takydromus* causes regeneration of the kidneys and that the regressed epididymis becomes again secretory. According to Kehl and Combescot (1955), the female urinary sex segment of *Uromastix* after the injection of androsterone is changed into a segment similar to that of the male; testosterone has also a masculinizing effect on the female ducts of this lizard.

Ophidia

No experiments on the regulation of the sexual characters by the reproductive glands have been performed in snakes.

Crocodilia

Forbes (1938, 1939) pointed out that in young alligators (*Alligator mississipiensis*) the genital tubercles of both sexes hypertrophy after administration of testosterone propionate. In the female the oviducts were paradoxically stimulated, in young male alligators the epididymis and ductus deferens hypertrophied after intra-abdominal implantation of testosterone pellets.

Chelonia

Risley (1941) has reported that testosterone propionate, injected into 2-month-old female *Malaclemmys*, masculinized the sex ducts and copulatory organ; the same applied to young specimens of *Emys leprosa*. Moreover, Evans (1951a, 1951b, 1952) found that in male *Pseudemys scripta* the tail and the claws of the forelegs (and in male *Terrapene carolina* the claws of the rear legs), which play an important role during mating, are dependent on the male sex hormone.

Action of Androgens in Birds

The dependence of the plumage on the gonads varies widely in avian species. In the house sparrow (*Passer domesticus*), for instance, plumage of both males and females is independent. According to Zawadowsky (1926b) and Hachlow (1927), this is so also in the chaffinch (*Fringilla coelebs*) and in the bullfinch (*Pyrrhula pyrrhula*). Nowikow (1939) even found that in the ptarmigan (*Lagopus mutus*), in which the male has four and the female three different successive types of plumage in the course of a year, the same types of plumage succeed each other

Fig. 8.4. Castrated ruff (Philomachus pugnax); plumage resembles that of a reeve during the summer.

after castration. In the male ostrich the long feathers are an independent sex character. In the ruff (*Philomachus pugnax*) the beautiful male breeding plumage, which is only present during the reproductive period, is a dependent character. It is also accepted that the female plumage, worn by the reeve during the whole year, is not dependent on the ovaries.

Keck (1934) has stated that the bill colour of the male sparrow (*Passer domesticus*) is a dependent character: in spring, under the influence of testosterone, black pigment is deposited in the horn-coloured bill. In the female red-billed weaverbird (*Quelea quelea*), however, during the breeding season the ovary inhibits the deposition of red pigment in the bill which turns bright yellow. In ovariectomized birds and in birds outside the breeding season the bill is red in both sexes.

Fig. 8.5. Castrated ruff (Philomachus pugnax) with a regenerated testis; plumage resembles that of a normal ruff during the reproductive period.

Many experiments have also been done in which testis hormones have been injected or testes have been implanted into intact or ovariectomized adult female birds. No unexpected results have been obtained; generally, sex characters dependent on the testis hormone develop in the experimental animals. In intact female birds characters appropriate to both sexes can be obtained.

It follows from the castration experiments reported that the sex hormones have stimulating as well as inhibitory activities. Neither the male nor the female hormone are stored in the body; indications of gonadal insufficiency appear therefore very soon after castration.

It has also been established—first by Pezard (1922)—that there are no intermediate stable states between the castrate and the normal male with fully developed sex characters. Pezard stressed the fact that, as a rule, the development of the sex characters follows an "all-or-none" law, i.e., the sex characters develop as the hormonal threshold is reached, and if so, they develop fully. This threshold differs for different organs. However, in some instances this law does not appear to be valid.

The results of Rowan (1926, 1938) in Canada, who exposed juncos (*Junco hyemalis*) in the autumn to ordinary electric light and obtained male birds in full spring condition in midwinter (i.e., singing birds with large, fertile gonads) suggested to him that the sex hormones play a prominent role in starting the northward migration in spring. Since, however, castrated birds also migrate the role which the gonads play in migration cannot be as important as Rowan thought. From more recent investigations, especially those of Wolfson who found (1945) that photo-stimulation leading to gonad activation is followed by fat deposition in migratory, but not in nonmigratory birds, it may be assumed that this fat deposition is induced by the sex hormone.

Investigations on a breed of birds in which the sexes strongly differ from each other when the regulation of the sex characters by the gonads was discussed. In the "hen-feathered" breeds, however, the cock and the hen possess the same "henny" plumage. In the Sebright Bantam for example, the hen-feathered cocks are distinguishable from the hens by their size, the strongly developed head appendages, the spurs, the sexual behaviour, etc., but not by the feathers, which are "henny," i.e., short in both sexes. When the cocks are castrated, they develop during the next molt a plumage which differs completely from the typical plumage of the breed, but which is present in the cocks of most other breeds of fowls, i.e., "long" neck-, shoulder-, rump-, and tail feathers are grown. Ovariectomized hens of this breed wear the same long feathers as the Sebright Bantam capon. Hence the neutral plumage of the hen-feathered breeds is different from the cock's as well as from the hen's plumage; the development of the "cocky" plumage is inhibited by the testes. In the normal breeds the cock's plumage represents the neutral one.

Finally, the interesting results of work on sex reversal in hens should briefly be mentioned. True functional sex reversal has been obtained in addition to the changes caused in the hen after ablation of the left ovary. It may occur that the right, rudimentary gonad of such ovariectomized hens regenerates some time after the operation (after several months) and is transformed into a hormone-secreting and even sperm-producing testis. Such birds begin to crow and the small head appendages of the ovariectomized hen develop into large ones, indistinguishable from those of a normal cock. At autopsy convoluted deferent ducts may be found. The hormone secreted by the right gonad

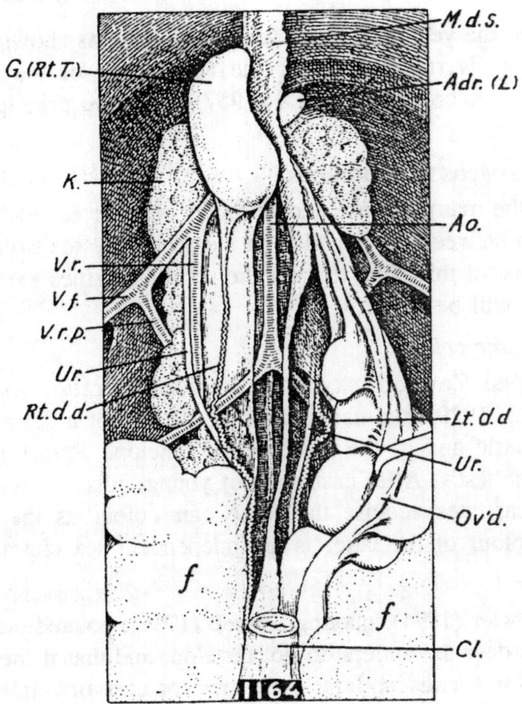

Fig. 8.6. Internal reproductive organs of an ovariectomized hen. The right gonad has developed into a testis.

is thought to be testosterone or a closely related compound. Hewitt (1947) has claimed that the presence of intact adrenals is essential to the transformation of the right gonad into a testis.

Summing up, we can say that the testis hormone has a very pronounced effect on many avian sex characters. However, the development of at least some sex characters may also be regulated by other hormones.

In weaver finches of the genera *Euplectes*, *Vidua*, and *Pyromelana*, the anterior pituitary and gonadal hormones cooperate in regulating the plumage. According to this author the male birds, which during the period of sexual quiescence wear the same modest plumage as the females, grow their brilliant plumage under the influence of the pituitary; if a female bird is ovariectomized it also acquires this plumage in the breeding season. In the intact female, however, the ovary impedes this development, and in the male the testis hormones have no influence on it; castrated birds show the same sequence of plumage during the year as normal birds. Witschi has shown that the anterior pituitary is responsible for the appearance of the beautiful nuptial plumage. According to Segal (1957) the active principle is the luteinizing factor.

Action of Androgens in Mammals

Although the mammals are exceptionally suitable for the study of the relationship between the gonads and the accessory sex organs, it is beyond the scope of this book to treat the results obtained extensively. Examples only will be given.

Regulation of hair colour

According to Zawadowsky who studied castration effects in ruminants (1922, 1929), the hair colour of the bull of a certain breed of Ukrainian cattle and of the male of the antelope *Portax pictus* is dependent on the testis. After castration of young bulls or male *Portax* antelopes the castrates acquire the same hair colour as the female; thus the hair colour of the latter is an independent sex character.

Antlers of deer

Already Fowler (1894), quoting Russell (1755), pointed out that in young castrated deer the antlers do not develop, and that if the deer is castrated when it carries antlers, they are not cast off. It has also been known for a long time that in old male roe deer (*Capreolus capreolus*), in which the male sex hormone is apparently no longer secreted, the velvet of the antlers is not shed and that the antlers persist after the midwinter period of reproduction. Since the velvet continues to grow steadily, it assumes the form of a wig ("wig-antlers").

In the white-tailed deer (*Odocoileus virginianus*) in which the growth of the visible antler starts in April, and in which the velvet is shed in September, Wislocki et al. (1947) and Waldo and Wislocki (1951), in castration experiments, have found that the male sex hormone is responsible for the shedding of the velvet, turning living skin into dead skin. When the testes regress, the casting off of the antlers follows. The maximal regression of the testes and of the male accessory glands occurs at the time when the antlers begin to grow. When the animal is castrated after the velvet has been shed, the calcified antlers are cast off; in the next year they are renewed and then retained. The hypothesis of Waldo and Wislocki (1951) that antler growth is mediated by the adenohypophysis and that the hormones of the adenohypophysis and of the testes supplement each other in the regulation of the

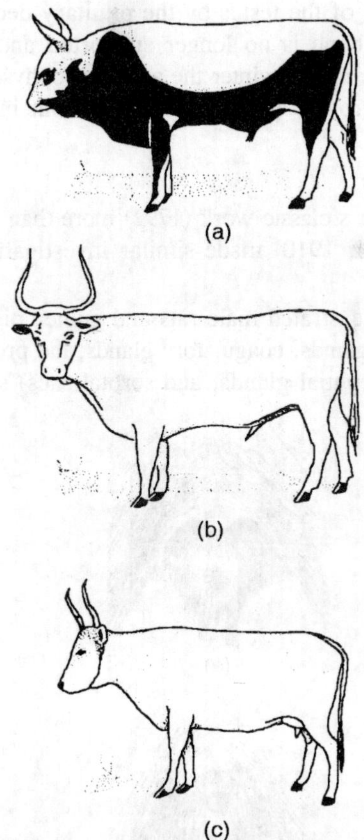

Fig. 8.7. A normal bull (a), an ox (b), and a normal cow (c) of the same Ukrainian breed.

periodicity of antler growth was proved by Hall et al. (1960) who succeeded in hypophysectomizing a young male white-tailed deer, and observed that the antlers did not grow at all. From these results and from those of Waldo and Wislocki, Hall et al. concluded that antler growth in late winter is initiated by the adenohypophysis, probably in response to the lengthening days. Later (in July), when the secretion of testicular androgen, influenced by the pituitary, follows, antler growth is almost complete. It is thought that the "antler-growth" adenohypophyseal hormone is inhibited by androgen secretion and that as a result the velvet dries out and is cast off in August or September. The testicular androgens appear to maintain the stability of the union of the dead antler tissue with the live frontal bone. When the days are short, the stimulation of the testes by the pituitary declines, the union between antlers and bone is no longer supported and the antlers are shed (December). During the winter the adenohypophyseal and testicular function and antler growth are all at low ebb, but in late winter the cycle is resumed.

Accessory sex organs

After John Hunter's classic work (1792) more than a century passed before Steinach (1894, 1910) made similar investigations on a larger scale.

He found that in castrated male rats and guinea pigs the accessory sex organs (vesicular glands, coagulatory glands, the prostate, Cowper's glands, the penis, preputial glands, and scrotal sacs) showed not only

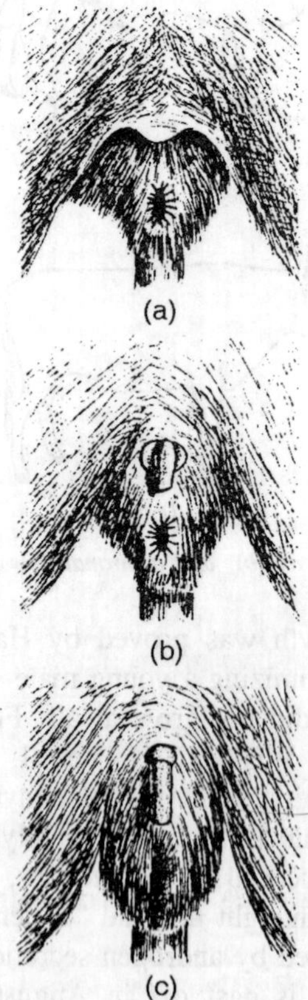

Fig. 8.8. External genital organs of (a) an ovariectomized female rat, (b) a masculinized female rat with markedly hypertrophied cliotris, and (c) a normal male rat.

a marked decrease in size but also in function. After implantation of testes into castrated animals the organs recovered. He showed also that after transplantation of gonads of the other sex into castrated or spayed animals the accessory organs developed under the influence of the hormone secreted by the transplanted gland. One of the most conspicuous examples of this has been described by Steinach's co-worker Lipschutz (1918a) and by Sand (1926) who noted that after a testis had been transplanted into a spayed guinea pig or rat, the clitoris developed into a large penislike structure.

The submandibular salivary glands of the mouse are also dependent on the male sex hormone. Lacassagne (1940a) found that these glands are larger in the male than in the female, and that after extirpation of the testes the serous tubules of the submandibular gland regress in size. The glands assume female characteristics but revert to the male type after an injection of testosterone. The salivary glands in the female mouse remain almost unchanged after ovariectomy. The enzyme content of the saliva secreted by the submandibulary glands is also dependent on the male sex hormone; the enzymatic activity is greater in males than in females. It decreases in males after castration, and increases in females after the injection of testosterone.

Since thyroidectomy produces this atrophy of the serous tubules of the submandibulary gland and atrophy is more pronounced after hypophysectomy than after castration, it is thought that the thyroid gland is also necessary for the maintenance of the serous tubules.

When testosterone is administered to hypophysectomized mice the amylase content of the atrophic gland increases. It is assumed therefore that testosterone affects the amylase content directly.

Spermatogenesis

The results of investigations on the influence of testosterone on mammalian spermatogenesis have often been contradictory, but it is usually accepted that testosterone has a stimulating effect on this process. It has been stated by Walsh et al. (1934) and confirmed by Nelson (1940) and Wells (1942) that in hypophysectomized rats the degeneration of the seminal epithelium can be prevented and spermatogenesis can be maintained by normal doses of injected testosterone. According to Gaarenstroom and de Jongh (1946), this depends on a lengthening of the duration of life of the sperm cells (maintenance effect). Moreover, testosterone supports the growth of the seminiferous tubules under the influence of the pituitary follicle stimulating hormone (FSH). Gaarenstroom and de Jongh believe, therefore, that for normal spermatogenesis in mammals two factors are necessary: FSH and a certain amount of testosterone within the testis.

On the other hand, testosterone, injected in large quantities into normal mammals, seems to depress the gonadotropic action of the anterior pituitary and this is followed by a decrease of the testosterone concentration in the testis and a regression of spermatogenesis. If, however, testosterone is withdrawn, the gonadotropic hormone is secreted in increased quantities and spermatogenesis may even be enhanced.

In addition it must be mentioned that Riisfeldt (1949) found in rats that the amount of hyaluronidase in the testes increases under the influence of androgens.

In amphibians Blair (1946), Penhos (1953, 1956), and Iwasawa (1957) have shown that after administration of testosterone spermatogenesis is slightly stimulated or that the testicular weight increases. However, these authors did not state whether the whole spermatogenetic process or only part of it is affected by testosterone. Cei et al. (1955) and P.G.W.J. van Oordt and Basu (1960) investigating *Leptodactylus chaquensis* and *Rana temporaria* obtained the following almost identical results. In *Leptodactylus* large doses of testosterone caused degeneration of the secondary spermatogonia and primary spermatocytes, and the mitotic capacity of the primary spermatogonia was impeded; in *Rana temporaria* the implantation of testosterone pellets was followed by impaired formation and development of secondary spermatogonia but the mitotic capacity of the primary spermatogonia was not affected. The primary and secondary spermatocytes and the spermatids were slightly stimulated. Testosterone was also found to impair the formation and development of the secondary spermatogonia in *Rana esculenta*. Since in almost all amphibians the period of the most intense spermatogenetic activity does not correspond with the secretory activity of the interstitial cells, P.G.W.J. van Oordt concluded in his survey (1960) that normally the male sex hormone does not stimulate the development of spermatocytes and spermatids and that in amphibians with a discontinuous spermatogenetic cycle the male sex hormone might be one of the factors preventing the onset of spermatogenesis during the resting and spermiation periods.

It is generally accepted that the normal influence of testosterone upon spermatogenesis is mainly an indirect one, i.e., that the hormone acts by way of the gonadotropic activity of the pituitary gland.

Mammalian semen

Testosterone not only influences the vesicular glands, the prostate, and other accessory organs of the male mammal, but also the secretory function in the accessory glands, viz., the formation of fructose and citric acid. In castrated mammals these accessory glands are very small and both secretions are absent. If, however, androgens are administered the glands regenerate and fructose and citric acid reappear in the semen.

Thermoregulation of the scrotum

Andrews (1940) has described an interesting action of testosterone on the scrotum. In castrated mammals the capacity of the scrotum to relax and to contract in response to changes in environmental temperature is lost; thus, in the normal male the thermoregulatory function of the scrotum is controlled by the testes.

Action of androgens in the female

The ovaries of many vertebrates are known to secrete androgens also. Suitable animals (e.g., castrated and juvenile females, or females in the quiescent season, when the reproductive organs are small) will therefore react to injections of testosterone. For instance, testosterone appears to have a slight effect on the endometrium and on the opening of the vagina in prepubertal female rats and mice.

A direct influence of testosterone on mammalian ovaries has also been described. According to Gaarenstroom and de Jongh (1946), the antra in the follicles of the rat ovary are formed under the influence of testosterone, which is presumably secreted by the interstitial cells of the ovary. These workers assume also that ovarian testosterone causes one or more blood vessels (which have grown out from the theca interna into the cumulus oophorus of the follicle) to burst at the distal ends; consequently the pressure in the follicle increases and ovulation follows. There is also some evidence, summarized by Folley and Malpress, that testosterone causes mammary growth, especially when given with an adequate dose of estrogen.

The testicular hormone has also distinct actions on other endocrine glands:

Adenohypophysis

Although androgens are less effective than estrogens in inhibiting the gonadotropic function of the anterior pituitary, the effects of castration upon this gland are very distinct. This follows from the following observations: (a) After castration, the anterior pituitary gains

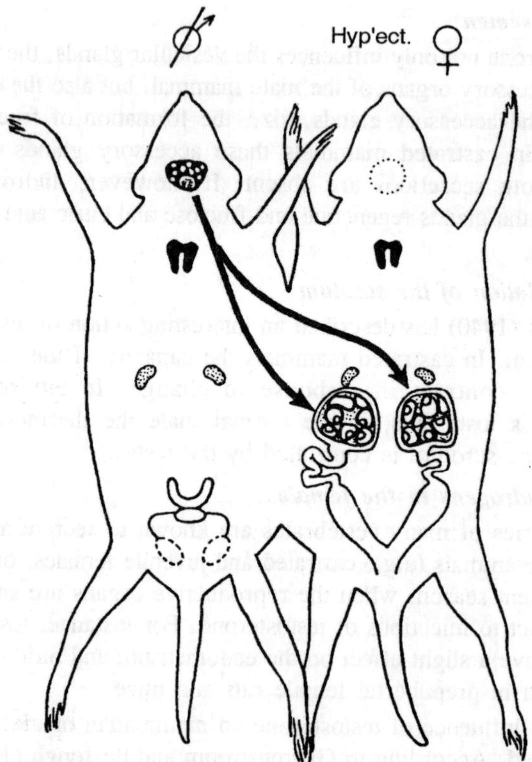

Fig. 8.9. Parabiosis of a male castrated and female hypophysectomized rat. The gonads and reproductive ducts of the hypophysectomized female have markdely increased in size.

in weight and secretes a larger quantity of gonadotropic hormones; this can be proved by implantation or by injection of an extract of such a gland into young female mice or by parabiosis. If in the latter case a castrated male or an ovariectomized female rat is united with a hypophysectomized female rat, the anterior hypophysis of the castrated or ovariectomized animal secretes such a large quantity of gonadotropic hormone that the gonads and, as a result of this, also the accessory organs of the hypophysectomized female, increase markedly in size. (b) Distinct cytological effects of castration are detectable in the adenohypophysis: the so-called "castration cells" appear, i.e., strongly secreting chromophil cells, to which the formation of gonadotropic hormone is attributed. (c) The hypersecretion of the anterior lobe in a castrated male can be prevented by injection of testosterone or implantation of testes.

Adrenal cortex

Not only the hypophyseal secretion of the gonadotropins but also that of adrenocorticotropin and of thyrotropin appears also to be controlled by the gonads. In particular, the influence of the sex hormones on the adrenocorticotropin secretion should be stressed. Corticotropin secretion is stimulated by small quantities of estrogens but inhibited by androgens. Corticotropin is, therefore, secreted in larger quantities by the female than by the male. After castration of a male mammal, the adrenal cortex increases slightly in weight, but decreases in weight in an ovariectomized female. In most mammals the cortex in the female is much more developed than in the male especially in the deeper layers. The sexual differences in the adrenal cortex must thus be considered as a true sex character.

In the embryonic adrenals of many mammals, and especially in the adrenals of newborn mice a "X-zone" is interposed between the cortical zonula reticularis and the medulla. This zone is particularly well developed in the young female mouse and adds much to the sexual differences in size between the adrenal cortices of young female and male animals. In young male mice the X-zone regresses and is transformed into a connective tissue capsule. In female mice the X-zone persists until the first pregnancy. It also persists in young castrated male mice or it redevelops. After testosterone administration to normal young male mice the X-zone disappears precociously but in castrated young male mice injected with estrogen it is still present. It is probable, therefore, that the X-zone of young mice disappears due to the influence of the male sex hormone. Sakiz (1958) found that the X-zone persists throughout life in ovariectomized mice, and that it reappears after castration of 34-day-old males.

The function of the X-zone is unknown. Grollman (1936) claimed that it produces androgens but this has been denied by Chester Jones (1955). Since, however, ACTH can stimulate the regression of the X-zone in both normal and ovariectomized virgin female mice, Deanesly (1958) has recently suggested that the mouse adrenal cortex itself produces androgens.

Metabolism

Information about the influence of androgens on the metabolism in the lower vertebrates is very limited. In fishes, Raffy and Fontaine (1930) noted an increase in oxygen consumption when male guppies (*Lebistes*) mature sexually, and in male goldfish when injected with extracts of carp pituitaries. Since this does not happen in castrated goldfish, this action must be mediated by the testes. An increase in oxygen consumption in young salmon after injection of methyltestosterone has also been reported.

In the immature fowl, increased nitrogen retention has been found after androgen administration. In doves and fowl the male has a higher basic metabolic rate than the female. In castrated fowl and mammals the basal metabolic rate is subnormal. Whether this is mediated by the thyroid gland is not certain.

As long ago as in 1936 and 1937 Kochakian found that androgens cause nitrogen retention in the castrated dog. Since then, many investigations on the anabolic effect of androgen on proteins in mammals have been published. Body weight is increased and muscle growth stimulated.

Testosterone causes hypertrophy of the temporal muscles of prepubertally castrated male guinea pigs and of normal and spayed female guinea pigs. In the rat, however, only the perineal musculature reacts to testosterone. The respiration of the epithelium, but not that of the musculature of the vesicular glands, is also conditioned by the male sex hormone. The fact that androgens promote protein anabolism in the muscles serves as a basis for the Hershberger test in the rat. The weight of the levator ani in androgen-treated castrated rats is used as the index for the anabolic, and the weight of the glandula vesicularis as the index for the androgenic activity.

The anabolic action of testosterone in mammals is so strong that testosterone has been regarded as a growth hormone. Its action differs, however, from that of the somatotropic hormone.

Other regulatory effects of the male sex hormone

The thymus increases after castration and regresses under the influence of androgens (and estrogens). Gonadal hormones appear also to have a delaying action on body growth. This is very obvious in the human male, for after prepuberal castration the male develops long limbs because the epiphyseal fissures close later than normally. The muscles are weak. In high dosages testosterone propionate produces inhibition of hair growth in normal, castrated, and hypophysectomized rats. Androgens have also a "renotropic" influence.

Finally it seems that androgens (like estrogens) enhance mitoses in certain organs as, for instance, in the epidermis of mouse ears. The increase in size of the sebaceous glands of mice after testosterone administration may also be attributed to this mitogenic influence.

Psychic Effect of Androgens

Sexual behaviour is affected by the sex hormones. In fishes, there are distinct alterations of sexual behaviour after ablation of the gonads. In castrated male sticklebacks (*Gasterosteus aculeatus*) nest-building activities are absent. Tavolga observed (1955) on the other hand that, after castration, males of the gobiid fish *Bathygobius soporator* show the same courtship behaviour as intact males, and Aronson (1959) reported that in males of the cichlid fish *Aquidens latifrons* all elements of the mating pattern were still present up to 1$^1/_2$ months after castration, whereas certain nest-building activities had noticeably declined. That in some fishes behavioural phenomena are not controlled by the testes has also been found by Noble and Kumpf (1936, 1941) in *Hemichromis* and in *Betta* and by Aronson (1951) in *Tilapia*.

It seems possible therefore that in some species the reproductive behaviour is mainly regulated by the central nervous system or perhaps directly by the adenohypophysis. It has been claimed that the migratory behaviour of some fishes is caused by the gonadal hormones. Since, however, sterile *Colisa labiosa* × *lalia* hybrids showed this behaviour very markedly, this assumption cannot be true.

In castrated male frogs, the clasping reflex is absent if castration is performed well before the breeding period. Immature toads, on the other hand, treated with androgens will develop mating responses.

In reptiles, castrated male *Anolis carolinensis* show full sexual activity after the injection of testosterone.

As a rule, the sexual activities of castrated birds are reduced, notably when they are operated upon as young birds. Carpenter (1933), however, found that some of his completely castrated pigeons copulated after the operation, but later mating activities gradually declined.

Hamilton (1938) induced precocious aggressiveness and crowing in 10-day-old male Leghorn chicks by injecting testosterone propionate, and Noble and Zitrin (1942) evoked the complete sexual behavioural pattern of the adult cock in male chicks: crowing started on the fourth, and mating on the fifteenth day of age. In female canaries Leonard (1939), Shoemaker (1939), and Frederiks (1941) induced singing by the administration of testosterone.

The regulation of the behavioural pattern of the laughing gull (*Larus atricilla*) is particularly interesting. As mentioned before, the morphologically ambisexual characters of this species are regulated by testosterone in both sexes; this also applies to calls and postures, common to male and female, as well as to those of the male during the breeding period. Purely female behaviour, however, is caused by estrogen.

In conclusion some findings may be mentioned related to a phenomenon called "peek-order" by Allee (1942). It is well known that in a group of hens, confined together, some individuals dominate others by pecking but are not pecked in return. Thus a kind of social hierarchy exists in such a group, and if the position of one individual has been established it is usually permanent. It is thought that this behaviour is based on recognition. A similar hierarchy is found in many animals, which live in groups.

The hierarchy in a group of hens can be changed by the injection of testosterone; if hens of lower rank are repeatedly injected with male sex hormone they improve their rank in the brood, and may even rise to the highest rank, dominating all others. It is, therefore, assumed that the rank in a brood of hens derives from the quantity of testosterone secreted by the ovaries of each bird. A similar rise in hierarchy was induced with testosterone by Noble and Borne (1940) in teleosts, i.e., in swordtails (*Xiphophorus helleri*), by Evans (1946) in a group of male reptiles (*Sceloporus*), and by Shoemaker (1939) in female canaries.

Although castrated male mammals may continue for long periods to show copulatory behaviour to a diminished degree, it is generally accepted that androgens have an important but not an exclusive influence on sexual behaviour and mating reactions; usually injections of testosterone restore normal behaviour very soon. On the other hand, precocious sexual behaviour can be easily induced by treatment with testosterone propionate in young male rats, but not in guinea pigs. Beach (1947) has pointed out that in primates the sexual behaviour in both males and females, castrated prepuberally, is much more pronounced than in castrated lower mammals.

It is thought that the stimulating effect of testosterone in the immature rat is mainly due to an increased tactile sensitivity of the glans penis, in which the cornified, genital papillae disappear in castrates. However, their number increases markedly under the influence of androgens.

In female mammals estrus is chiefly regulated by estrogen; it has, however, been found that injection of large quantities of androgen can cause an estruslike behaviour in prepuberally castrated female rats or mating behaviour in normal female rats. Male rats treated with testosterone may also show female mating behaviour.

In his book "Hormones and Behaviour" (1948) Beach has suggested that the sex hormones operate by sensitizing preorganized mechanisms within the central nervous system. Very important in this respect are the character and intensity of external stimuli offered by the presence of a receptive female; untreated male rats display the male copulatory behaviour in the presence of estrous females.

Antagonism and Synergism of Androgens and Estrogens

Generally speaking the antagonism between the effects of the male and female hormones on the sex organs is an indirect one. This conclusion is based on the fact that androgens as well as estrogens inhibit the gonadotropic action of the adenohypophysis, or, according to recent investigations, impede the secretion by the hypothalamus of the gonadotropic hormone releaser. When, therefore, high doses of testosterone are injected into a normal female mammal, the gonadotropic activity of the anterior pituitary decreases and ovarial function is reduced followed by a regression of the female sex characters. Conversely, a regression of the male sex characters occurs after the injection of estrogen into normal male animals. Estrogens arc more efficacious than androgens in this respect.

A direct effect of testosterone, when injected into an ovary of a guinea pig, has also been postulated. In hypophysectomized rats, Payne et al. (1956) found an inhibitory action of testosterone on the ovarial gain of weight induced by stilbestrol.

Moreover, there appears to be a direct antagonism between the effects of male and female hormones on the accessory sex organs. For instance, testosterone diminishes the local effect of estrogen on the cornification of the vaginal epithelium, and the effect of androsterone on the capon's comb can be impeded by injecting estrogens simultaneously. This effect on the capon's comb is held to be a direct one. Bruzzone and Lipschutz (1953) found that the growth of the clitoris, produced by administering androgen to castrated female guinea pigs, can be considerably retarded by a simultaneous injection of small quantities of estradiol benzoate.

After pregnant mare serum had been given to juvenile female sparrows Pfeiffer and Kirschbaum (1941) found that their ovaries secreted an androgen so that their bill became jet black. When, however, estrogen was injected simultaneously, the bill remained yellow.

A summation of the effects of androgens and estrogens given in certain concentrations has been observed in some instances. For example, the normal structure of some accessory sex organs appears to be the result of the action of both androgen and estrogen. Korenchevsky and Dennison (1936) demonstrated that testosterone is capable of enlarging the uterus and of increasing the gain of weight produced by estrogen. The same applied also to young spayed rats when both hormones were administered in high doses. Leathem and Wolf showed (1955) that in the immature rat a combination of testosterone propionate and estradiol benzoate produces larger vesicular glands than did either steroid alone. Again, androgens which normally inhibit the effect of estrogens can, when administered to young spayed rats in high doses, increase the uterine growth produced by large quantities of estrogens.

Action of Testicular Estrogen and Ovarian Testosterone

It is well known that the testis produces not only the male hormone but also an estrogenic hormone, and that the ovary secretes not only estrogens but also androgens. The testicular estrogen has, in addition to testosterone, an important action on the development of certain male accessory organs.

In the early 1930's de Jongh stressed this "paradoxical" influence of estrogen on such accessory male glands as the vesicular glands (or "seminal vesicles") and the prostate of some mammals (mouse, rat, guinea pig). When, on the other hand, castrated rats with rudimentary accessory glands were injected with testosterone, only the epithelium of the glands reacted. To induce the development of the fibromuscular parts of the vesicular glands and of the coagulatory part of the prostate, estrogen had to be injected as well.

Thus it became more and more probable that both testosterone and estrogen are necessary for the normal development of the accessory glands mentioned, and that their normal structure is the result of a synergistic action of both hormones.

The fact that both male and female sex hormones are present in the urine of the male and female of many mammalian species provides a further indication that male and female sex hormones are formed in both sexes. Beall (1940) extracted estrogens from horse testes, and Zondek as long ago as in 1934 pointed out that large quantities of estrogen are present in the urine of the stallion, but none in the urine of geldings. If compounds with sex hormone activity persist in the urine of castrated animals, they are assumed to derive from the adrenal cortex.

The production of androgens by the ovaries has been demonstrated as follows: Pfeiffer and Kirschbaum (1941) showed that the yellow bills of sexually inactive female sparrows become jet black—like those of the male during the reproductive period—after pregnant mare serum has been administered. The much enlarged ovaries secreted an androgen comparable to that of the testis in the reproduction period.

It has already been pointed out that the so-called ambisexual characters are dependent on the gonads. This is due to the influence of the male sex hormone, for in castrated male and female black-headed or laughing gulls (*Larus ridibundus* or *L. atricilla*) these characters do not appear. They develop in castrated male or female gulls after the administration of testosterone, but not of estrogen. Thus we see that in the female gull the ovary secretes not only estrogen but also testosterone. The same holds for the starling, in which the yellow bill of both sexes represents an ambisexual character during the breeding season, caused by testosterone secreted by the testes or the ovary.

Since the comb of the hen of most poultry breeds is larger and more vascular than that of the capon, and becomes very turgid at the time of laying, it is thought that the left ovary, in addition to estrogen secretes a combgrowth-promoting substance, probably related to or perhaps identical with testosterone.

The gonads of cock-feathered and hen-feathered poultry breeds offer another interesting problem. In the first place, the question had to be answered whether there is a hormonal difference between the testes of a cock-feathered and of a hen-feathered breed. This has been answered in the negative. For, when testes of a Leghorn (a cock-feathered breed) are exchanged for the testes of a hen-feathered breed the plumage of the experimental birds does not change.

Danforth and Foster (1929) and Danforth (1930) concluded from these experiments that the feather follicles of these breeds reacted differently to the female sex hormone, and Danforth formulated the hypothesis that the feather follicles of the hen-feathered cock are so sensitive to the estrogen secreted by the testis, that they react only to estrogen, or react better to estrogen than to the testosterone which is secreted by the testis. By means of skin transplantations in chicks of "normal" and "henny" feathered breeds Danforth and Foster (1929) produced convincing evidence for this hypothesis. In summary, their results were as follows: (a) Skin of a Leghorn hen (a "normal" breed) grafted on a Leghorn cock: development of cocky feathers in the transplant. (b) Skin of a Campine hen (a "henny" breed) grafted on a Leghorn cock: development of henny feathers. (c) Campine cock skin grafted on a Leghorn cock: development of henny feathers. (d) Leghorn cock skin grafted on a Campine cock: development of cocky feathers.

From these results and from the fact that no changes occur following cross-transplantations of Leghorn and Campine hen skin, it can be deduced that the testes secrete estrogen in addition to testosterone. According to Parkes and Emmens (1944), however, it is probable that the capacity of the testis to feminize the plumage in such a breed as the Sebright Bantam is due to the fact that the plumage reacts to testosterone in the same way as to estrogens.

The secretion of androgens by the ovaries of mammals has also been convincingly demonstrated. When, for instance, extracts of mammalian ovaries or ovarian transplants from mammals are brought into contact with sensitive receptors, the capon's comb or the vesicular

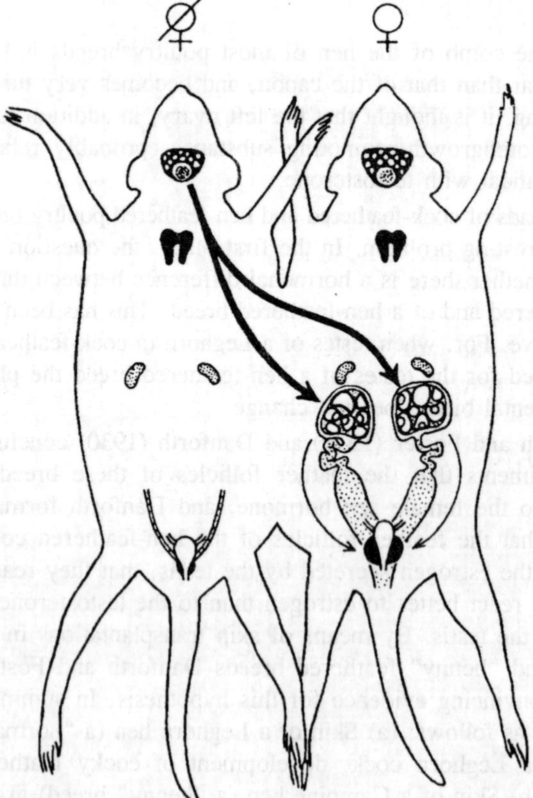

Fig. 8.10. Parabiosis of an ovariectomized female rat and a normal female rat.

gland of castrated adult rats, the changes in the receptors indicate that androgens have been secreted by these ovaries.

This is particularly obvious when the ovaries have been stimulated by gonadotropins. Pfeiffer and Hooker (1942), for instance, transplanted ovaries to the ears of castrated male mice and later administered pregnant mare serum with the result that the vesicular glands were manifestly stimulated. Johnson (1958) connected a castrated male rat parabiotically with another castrated male in which an ovary was implanted. Because of the large quantity of gonadotropic hormone formed in the anterior hypophysis of the castrated male, the ovary of the other rat produced such large amounts of androgen that the ventral prostate of the male increased at least fivefold. When a female rat was parabiotically connected to a castrated female, the gonadotropin also stimulated the development of the residual prostatic tissue in the female by way of the enlarged ovaries.

In conclusion it may be said that testosterone is neither a specific male nor estrogen a specific female hormone. Both sex hormones—though in different quantities—are necessary for the development of most, if not all, sex characters.

Source of Sex Hormones in the Male

It is by now generally accepted that the mammalian testis produces only one male hormone, i.e., testosterone. The existence of a second testis hormone, "inhibin," was formerly postulated; this hormone was supposed to inhibit the production of gonadotropins by the anterior pituitary. This is no longer accepted since the inhibiting action of the testis on the anterior hypophysis can be ascribed to testosterone or to estrogen, which as already mentioned is also secreted by the testes of mammals and birds. Testicular estrogen has also an important influence on certain sexual characters. Hence the entire physiological role of the testis derives from the secretion of testosterone and estrogens.

Source of Testicular Androgen

The first biologists interested in the source of the male sex hormone were Bouin and Ancel, who claimed in 1903 that it is formed in the Leydig or interstitial cells, present in the interstitial tissue between the tubules. They concluded this from observations on cryptorchid pigs, in which the generative part of the testes was almost totally degenerated, whereas intact interstitial cells were present in normal quantities indicating that the sex characters were not affected. The same observations were made on animals with tied ductus deferentes.

This concept was later adopted by many other workers. Steinach and Lipschutz even went so far as to call the sum of the interstitial cells the "puberty gland." Other workers however believed that the male sex hormone is produced in the generative part of the testis, i.e., in the testis tubules. The most prominent supporter of this theory was Stieve (1921).

In addition to the investigations on cryptorchid testes, a number of other results have been obtained which also demonstrate that the male sex hormone is secreted by the interstitial cells. For example, after irradiation with X-rays in doses which cause atrophy of the seminal epithelium without injury to the interstitial tissue, androgen continues to be secreted. On having damaged the interstitial cells of the testes of rats by feeding a diet deficient in vitamin B complex, Moore and Samuels (1931) found that the seminal epithelium remained normal, but that the accessory sex glands like the prostate and the vesicular glands underwent atrophy.

It is now generally accepted that in the homeothermic amniotes the male sex hormone is secreted in the interstitial cells. This assumption rests on the following evidence: First, the appearance of secretory phenomena in interstitial cells is histologically demonstrable. Second, after the injection of interstitial cell-stimulating pituitary hormone into young mammals and birds or into birds during the nonreproductive period, the nonfunctional interstitial cells change into cells with a distinctly secretory appearance. Third, these alterations are correlated with certain changes in and developments of sex characters such as those occurring during the reproductive period; and last, during this period the level of 17-ketosteroids in the urine rises.

Although the presence of true Leydig cells in mammals and birds is well established, it is still a matter of dispute in which part of the testis the male hormone is secreted in fishes, amphibians, and reptiles.

Courrier in 1922 described interstitial cells in the testes of the three-spined stickleback, *Gasterosteus aculeatus*, and correlated their presence with the development of sex characters (nuptial colouration, development of secretory kidney tubules). Van Oordt (1924) found that in the ten-spined stickleback maximal numbers of interstitial cells are present after the spermatogenetic period, whereas in the testes of *Xiphophorus helleri* interstitial cells are lacking when the sex characters ("sword," gonopod, and transformed pectoral fins) are developing. According to Craig-Bennett (1931), however, the maximal development of interstitial cells in *Gasterosteus aculeatus* coincides with full male nuptial colouration, and Follenius (1953) noted in guppies (*Lebistes reticulatus*) irradiated with X-rays that the tubules of the testis became sterile, though the interstitial tissue was not affected and the development of the sex characters proceeded normally.

A well-developed interstitial cell cycle has also been described in some amphibian species, and a correlation with changes in sex characters has been established. Although definite proof that the male sex hormone is formed in the interstitial cells of the amphibian testis is still lacking, there is some evidence for assuming this; the histochemical investigations of Ashbel et al. (1951) and of Burgos (1955) in *Rana pipiens*, of Niwelinski (1954) in *Xenopus laevis*, and of de Kort et al. in *Rana esculenta* indicate that the interstitial testis cells of these amphibians produce ketosteroids. Moreover, Burgos and Ladman (1957) have found that low doses of purified luteinizing hormone (LH) stimulate the interstitial cells in *Rana pipiens*.

According to Kehl and Combescot (1955) the interstitial cells of the saurians change seasonally in size, number, and structure, and alterations in accessory sexual characters are associated with changes in the interstitial cells; therefore, a correlation between these two facts seems entirely possible. In the lizard *Xantusia vigilis*, however, neither quantitative nor qualitative changes occur in the interstitial cells of the testis at the time the sex characters develop.

The available data on the histology and cytology of the interstitial cells and their correlation to the development or maintenance of sex characters in poikilothermic vertebrates are so scarce that detailed investigations with up-to-date methods seem highly desirable.

Source of Testicular Estrogen and of Ovarian Androgen

It has already been mentioned that the testes of mammals and birds secrete estrogen. It is not known, however, where this estrogen is formed. Some workers assume that its source is the Sertoli cells, since it has been observed that in animals with Sertoli cell tumors the quantity of secreted estrogen increases enormously. Others, like Maddock, claim that estrogen is formed in the Leydig cells.

It has also been postulated that the androgen, found in the ovaries of birds and mammals, has its source in the ovarial interstitial cells, but definite information is lacking.

Many workers have claimed that the mammalian adrenal cortex produces corticosteroids with a masculinizing effect. For instance, Katsh et al. (1948) found that adrenal cortical tissue, transplanted to the vesicular gland in adult adrenalectomized and castrated rats, almost prevented the regression of this gland, which usually occurs after castration. The investigations of Deanesly (1958) make it likely that the X-zone of the mouse adrenal cortex produces androgens.

It is, however, still doubtful whether the adrenal cortex secretes androgens under normal conditions. According to Chester Jones (1955), "there is evidence that the adrenal of young rats can at best secrete minute amounts of androgens transiently." But under clinical conditions (in the adrenogenital syndrome) there is "an unequivocal and copious adrenal secretion of sex hormones." On the other hand, Dorfman (1955) claims that four adrenal steroids produced by the adrenals may be designated as proandrogens, since they are in part metabolically converted to androgenic steroids.

It is also accepted that small amounts of androgen are formed by the placenta. For instance, Stark and Voss (1957) found an androgen in human placenta that activated the capon's comb.

To sum up, one can say that androgens are produced by the testis, ovary, adrenal cortex, and perhaps by the placenta. Testosterone is a product of the testis and perhaps also of the ovary but probably not of the adrenal cortex.

Factors Regulating Testicular Functions

Testicular functions are regulated by interoceptive or internal and by exteroceptive or environmental factors.

One of the most important interoceptive factors is the gonadotropic influence of the adenohypophysis. Another interoceptive factor which is particularly significant for the spermatogenesis of most mammals is the temperature-regulating function of the scrotum.

While the majority of the mammals possess scrotal sacs, some animals, viz., rhinoceroses, seals, elephants, and whales, are testicond, i.e., the testes remain in the abdominal cavity throughout life. Some Insectivores and Chiroptera have no typical scrotal sacs, but during the breeding season the thin abdominal wall is distended by the testes and thus forms a, pseudo-scrotal sac. In most mammalian species there is, however, a true "descensus testiculorum." In many mammals the testes descend only during the reproductive period. In man and in domesticated mammals, the testes occasionally do not descend through the inguinal canal to the scrotal sac, but are retained in the abdominal cavity. Such cryptorchid testes are sterile, because sperm is not formed.

It was found, notably by Moore and his co-workers that testes of rats, guinea pigs, and rabbits when replaced through the wide inguinal canals into the abdominal cavity, became sterile after some weeks (experimental cryptorchism). In further experiments it was proved that this sterility is due to the fact that the temperature in the abdominal cavity is about 3°C higher than in the scrotal sac.

It follows from these findings that in mammals with a descensus testiculorum normal sperm function is only possible at temperatures which are some degrees lower than that in the abdominal cavity. The interstitial cells are not affected by the high temperature; Bouin and Ancel (1903) have shown that cryptorchic mammals still possess their sex characters. This observation led these authors to assume that the male sex hormone is produced by the interstitial cells.

In many vertebrates which inhabit the temperate zones and also in animals exposed to the dry and rainy periods of the tropics, a very pronounced seasonal testicular (and ovarial) periodicity is present and gametogenesis is strictly a seasonal event. This periodicity is often influenced by exteroceptive factors, the most important of which are temperature and light.

In some fishes a rise in the environmental temperature appears to play a very important role in stimulating the testes and, indirectly, the male sex characters. In other fishes the influence of light is much more important than that of temperature.

In amphibians light does not seem to play a very important role. High temperatures, however, can strongly stimulate spermatogenesis.

In frog and toad tadpoles with undifferentiated gonads, Witschi (1914, 1929) found that the temperature is an effective factor in modifying the balance between the gonadal cortex and the medulla. High temperatures (25-30°C) favour the development of the medulla into a testis; low temperatures (10°C) inhibit the development of the medulla and favour that of the cortex. Thus male tadpoles dominate at high temperatures and female tadpoles at low ones.

The most extensive and valuable studies of the influence of light on testicular functions have been made in birds. Rowan (1926, 1929, 1932, 1938), working in Canada, discovered that light, but not temperature, affects the reproductive activities of certain species of birds in a very conspicuous manner. Because in spring the enlargement of the gonads of birds coincides with the increase in daylight, Rowan submitted juncos (*Junco hyemalis*, a North American species of bunting), in late autumn and in midwinter when their testes are minute, to increasing amounts of light by means of artificial illumination after sunset. Thus the length of day was increased and Rowan could induce a breeding condition in the male birds during the winter months. At the end of December the male juncos were singing all the time, despite the very low temperatures (–20° to –45°C) at which they were kept. Post-mortem investigations revealed that their testes were much enlarged, spermatogenesis being maximal. Since these birds showed male reproductive behaviour and characters, it was concluded that the interstitial cells were also stimulated.

Among seasonally breeding mammals Bissonnette (1932) found that the testes of ferrets, which were artificially illuminated outside the breeding season, matured precociously, and that female ferrets came into heat earlier under these circumstances. Species, such as the ground squirrels (*Citellus*), which hibernate underground and emerge in spring in full breeding condition failed to respond to light treatments in the laboratory.

It was concluded from these investigations that in temperate zones light, though not the only factor, plays a very important role in stimulating the gonads of male birds in spring.

In some instances it was found that it was not the longer duration of daylight in spring but the decrease in autumn which caused the gonads to develop. This applies to goats and sheep and also to deer since their rutting season occurs in autumn. The same phenomenon has been established in the brook trout (*Salvelinus fontinalis*), a species of fish which spawns in autumn.

By which pathway does the stimulus of light influence the testes of birds? Bissonnette (1936), Ringoen and Kirschbaum (1937), and others experimented with birds which were completely draped in light-proof material, or which were made to wear silk caps with or without eyeholes over their heads. Benoit and his co-workers exposed young ducks to high intensities of light after their optic nerves had been severed, or whose eyes had been blinded or completely removed, or which had been hypophysectomized. These experiments led to the conclusion that in normal birds light stimulates the anterior lobe of the pituitary by way of the eye and the central nervous system (the hypothalamus), and that the anterior lobe is activated to secrete gonadotropic hormones. These in turn act on the testis so that spermatogenesis starts and sperm cells are formed.

It is interesting that after the reproductive period has passed the testes do not react to light for a considerable time. This refractory period was first recognized in the starling by Bissonnette and Wadlund (1932), later emphasized by Riley (1936, 1937) and studied in detail by many other investigators, notably A.H. Miller (1948), Wolfson (1952, 1959), and Laws (1961).

This refractory period is very useful to birds, first because after the breeding season there is still enough light to maintain the enlargement of the gonads, and, second, because the migratory bird may during its southward flight reach low latitudes or even pass the equator (i.e., regions where the exposure to light increases). This would normally produce stimulation of the gonad but is prevented because the bird is temporarily refractory to light.

It is probable that the adenohypophysis and (or) the hypothalamus are insensitive to light during the refractory period. Thus, the testes do not react because they are not stimulated by injected gonadotropins.

In addition to the exteroceptive factors mentioned several other influences are known which affect reproductive processes. However, only darkness and confinement may be considered to influence reproduction or periodical reproductive phenomena unfavourably.

Biochemistry of Androgens

Isolation, Metabolism, and Biosynthesis

Although it is known that testicular extracts were prepared by Brown-Sequard (1889), and that McGee in 1927 obtained a preparation from bull testes which induced growth of capon's combs, it is now generally accepted that pure testicular hormone (testosterone) was first isolated in 1935 by David et al. in Laqueur's laboratory. Shortly afterward testosterone was synthesized by Butenandt and Hanisch (1935) and by Ruzicka (1935).

In mammals the group of androgens consists of several chemical substances, the most potent of which is testosterone. These compounds are synthesized in at least two endocrine glands (testis and ovary), and perhaps also in the adrenal cortex and in the placenta, as well as in peripheral tissues of which the liver is the most important.

The first studies on the metabolism of androgens were reported by Callow (1939) and by Dorfman et al. (1939), who found that testosterone was converted to the urinary 17-ketosteroids, androsterone and etiocholanolone. The introduction of new techniques (microchromatography, spectrophotography, and isotopic labeling) greatly increased our knowledge of the biosynthesis and metabolism of the androgens. Popjak (1958) has elucidated the biosynthetic pathway from acetate to the steroid molecule of cholesterol which is thought to be the basic material of all steroid hormones.

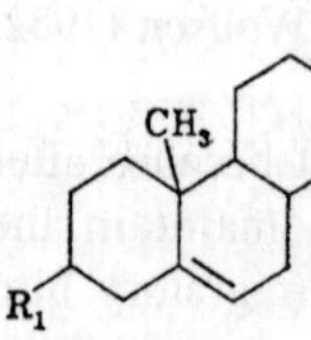

(*a*) Cholesterol $R_1: -OH$, $R_2: -C_8H_{17}$

(*b*) Dehydroepiandrosterone $R_1: -OH$, $R_2: =O$

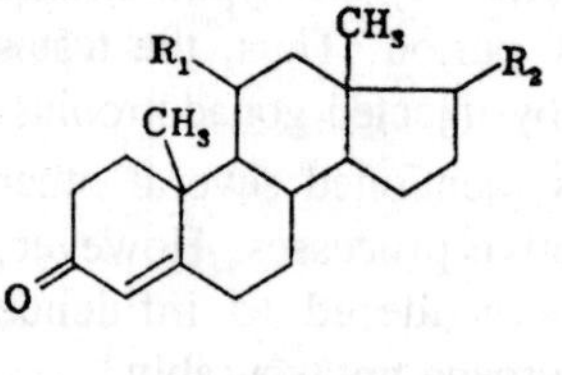

(*c*) Progesterone $R_1: -H$, $R_2: -CO{\cdot}CH_3$

(*d*) Androstenedione $R_1: -H$, $R_2: =O$

(*e*) Testosterone $R_1: -H$, $R_2: -OH$

(*f*) 11β-Hydroxy-androstenedione $R_1: -OH$, $R_2: =O$

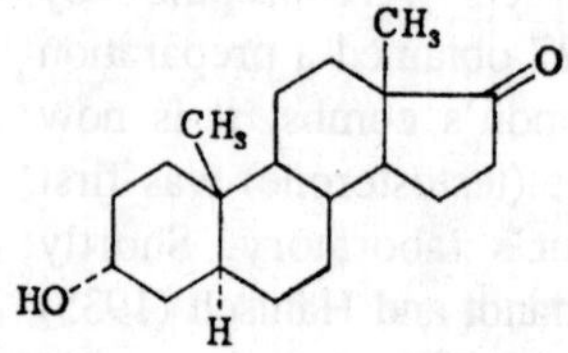

(*g*) Androsterone

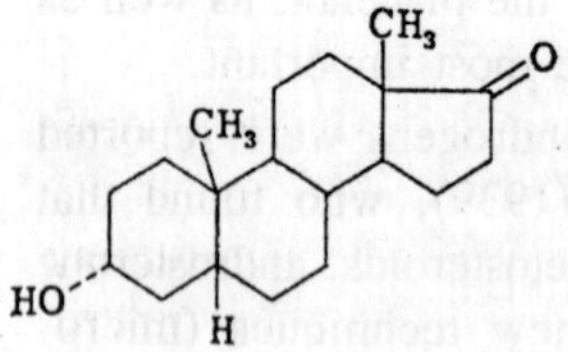

(*h*) Etiocholanolone

Fig. 8.11. Formulas of various steroid hormones.

At present two major pathways for the biosynthesis of androgens have been established. The first concerns the testis and ovary, and perhaps to a limited extent the adrenals and probably also the placenta. This pathway is as follows: acetate → cholesterol → progesterone →

androstenedione → testosterone. The second biosynthetic pathway applies exclusively to the adrenals and probably involves: acetate → cholesterol → dehydroepiandrosterone → androstenedione → 11β-hydroxy-androstenedione. In the liver, androgens may be formed from circulating corticosteroids by enzymatic removal of two carbon atoms.

The catabolic products of testosterone and other androgens are formed by oxidative and reductive changes in the molecule, as well as by hydroxylation; they appear in the urine as 17-ketosteroids. The *in vivo* conversion of testosterone to estrogenic steroid hormones has been established.

Steinach and Kun reported in 1937 that after testosterone administration to normal and castrated male rats, increased amounts of estrogens are excreted in the urine. That testosterone can be converted into estrogen in the organism was proved by Heard et al. (1955), who showed that in a pregnant mare testosterone labeled with C^{14} was converted to labeled estrogen.

It is thought that androgens (and estrogens) are inactivated by the liver. This certainly occurs *in vitro* in liver homogenates. Moreover, Burrill and Greene (1940) and Biskind (1940) have found that, after implantation of a testis into the spleen, castration phenomena cannot be prevented, which, of course, is the case, when the testis is implanted subcutaneously.

According to Freud et al. (1937) a substance can be extracted from testes and urine, which has no action by itself but which increases the action of testosterone and related compounds on the vesicular glands (but not on the capon's comb!). This unidentified substance (the X-substance) is thought to be a higher organic acid.

It has already been mentioned that many representatives of lower vertebrate classes react to androgens of mammalian origin or synthetic androgens. Therefore, it is sometimes assumed (but not proved) that testosterone is also the male hormone of these lower vertebrates.

From the investigation of Hazleton and Goodrich (1937), who, after the injection of extracts of salmon testes into capons, obtained comb growth, or of Potter and Hoar (1954), who stimulated comb growth in chicks with similar extracts, it has also been concluded that these extracts may have contained testosterone. However, biochemical investigations concerning the identity of the sex hormones in nonmammalian vertebrates are insufficient; more work on this problem is needed.

Methods of Biological Assay

Biological assay methods for androgens are numerous. The most important ones are:

Ejaculation test

In the guinea pig as well as in the rat, a copulation plug is formed after emission of the semen in the vagina of the female. This occurs because the male ejaculate hardens rapidly, due to the action of a prostatic enzyme on the secretory product of the seminal vesicle. By passing an alternating electrical current through the head of a male guinea pig (or rat) an ejaculation can also be obtained, but this is not possible in castrated animals, in which the seminal vesicles and prostate are atrophied. The function of these glands can be restored by the injection of testosterone (or other androgens) and the weight of the ejaculate may then serve as an index of the potency of the injected androgen.

Fructose test

Mann and Parsons (1947) found in many mammalian species, including man, that fructose (and citric acid) are absent in the accessory gland secretion after castration but reappear after the injection of androgens. The quantity of fructose in the semen may be used for assaying androgenic compounds.

Spermatozoon motility test

This assay method is based on the observation that androgenic substances can prolong the life of spermatozoa in the epididymis of castrated guinea pigs, and that spermatozoa in an isolated epididymis retain their viability longer under their influence. Further tests are based on the regression of the vesicular glands, the prostate, Cowper's glands, and the epithelium of the ductus deferens after castration and on their restitution after androgen treatment, or on the increase in weight of these organs after injection of the hormone into immature male rats.

Capon comb growth test

The capon's comb which responds to injected androgens or to direct inunction of the male hormone with an increase in size can also be used as an assay method.

Function of the Fetal Testis

Medullarin and Cortecin

There is some evidence that true sex hormones are formed in the fetal testis. However, it has also been shown that in earlier developmental

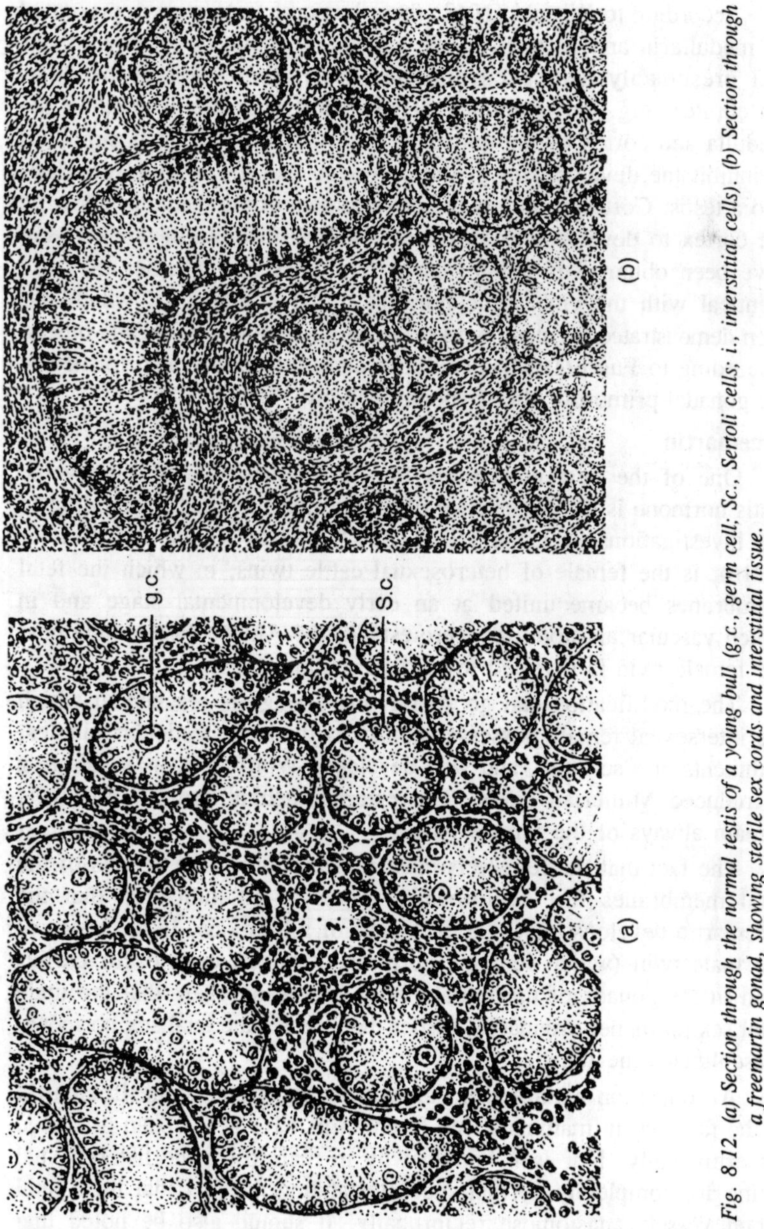

Fig. 8.12. (a) Section through the normal testis of a young bull (g.c., germ cell; Sc., Sertoli cells; i., interstitial cells). (b) Section through a freemartin gonad, showing sterile sex cords and interstitial tissue.

stages in addition to genetic factors, humoral factors play an essential role in the sexual differentiation of the gonads.

According to Witschi (1942) these humoral factors are represented by medullarin and cortecin in the gonads of young larval amphibians and presumably also occur in the "indifferent" gonads of other vertebrates, e.g., in mammals. These substances are formed in the medulla and cortex of the gonadal primordia. Medullarin is supposed to inhibit the development of the cortex so that the medulla develops into a testis. Cortecin inhibits the development of the medulla, allowing the cortex to develop into an ovary. Neither medullarin nor cortecin have been obtained in a pure state; it is uncertain whether they are identical with the embryonic hormones, the existence of which has been demonstrated by many workers as, e.g., by Jost, Wolff, and others. According to Fugo (1940) medullarin and cortecin are also formed in the gonadal primordia after hypophysectomy.

Freemartin

One of the best indications for the existence of an embryonic testis hormone is the occurrence of the freemartin which, according to the investigations of Lillie (1916, 1917) and of Keller and Tandler (1916), is the female of heterosexual cattle twins, in which the fetal membranes became united at an early developmental stage and in which vascular anastomoses were established. Under these conditions, the female twin is diverted toward masculinization.

The modification can go so far that a stage is reached in which the intersexual female possesses gonads with sterile testicular tubules, rudiments of a scrotum, and well-developed Wolffian ducts in addition to reduced Mullerian ducts. The external genital organs, however, remain always of the female type.

The fact that the female is masculinized only if the vessels of the fetal membranes fuse reciprocally serves as an indication that the freemartin develops under the influence of a hormone. In the testes of the male twin (which remains normal) interstitial cells occur earlier than in the gonads of the female twin, indicating again that the male fetal sex hormone, originating earlier than the female fetal sex hormone, masculinizes the female.

An objection against this concept of the origin of the freemartin is the fact that in mammals true freemartins have never been produced experimentally. Nor is it known why in marmosets the heterosexual twins are completely normal, notwithstanding the fact that their fetal blood vessels anastomose reciprocally. It should also be noted that with the exception of cows, freemartins are extremely rare in mammalian species.

In birds a spontaneous case has been described by Lutz and Lutz-Ostertag (1958) in a double-yoked hen's egg, the development of which could be followed to the eighteenth day of incubation. In the female embryo, which was connected by vascular anastomoses with the male, the posterior region of the Mullerian ducts was reduced; in the male the left testis contained some cortical nodules.

Wolff and Haffen (1952) succeeded in producing parabioses between primordial male and female duck gonads, cultured *in vitro*. Feminization of the male gonad by the secretions of the female gonad was observed, i.e., the male gonad developed a very distinct cortex.

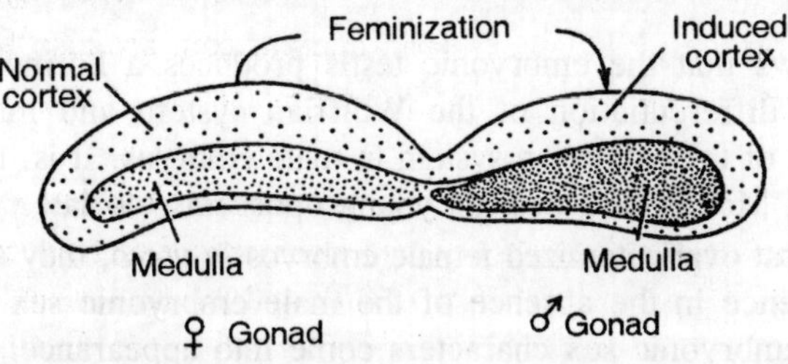

Fig. 8.13. Parabiosis between a potential male and female gonad cultivated in vitro; feminization of the male gonad.

Experimental freemartin-like conditions have also been obtained in mammals, birds, and amphibians: In mammals by administration of the heterologous sex hormones directly to embryos or in the circulation of the pregnant mother, in birds by injecting the embryo directly or by implantation of embryonic gonads, and in amphibians by the use of parabiotic larvae. In amphibians the female is usually masculinized, but feminization of the male may also occur.

Fetal Testis Hormone

To prove the existence of fetal sex hormones, castration experiments have to be carried out. The investigations of Moore (1941, 1943), however, who used the embryo-like pouch-young of the American opossum (*Didelphys virginiana*) as experimental animals, did not appear to be in favour of their existence. Though Moore succeeded in castrating the young opossums, changes in sex differentiation could not be detected. He, therefore, concluded that in opossums sex differentiation takes place in accordance with their genetic constitution and that embryonic sex hormones do not play a part. These investigations favoured the concept that during embryonic life the gonads of mammals do not secrete sex hormones, and that those sex characters which are macroscopically visible at birth do not develop under their influence.

More recent investigations, particularly those of Jost (1947, 1948, 1953, 1955, 1960) who succeeded in castrating rabbit embryos *in utero*, seem, however, to suggest that in placental mammals the development of certain sex characters takes place under the influence of the fetal gonad. When male rabbit embryos were castrated on or after the twenty-second day after conception (the duration of pregnancy is 32 days) development proceeded normally. When this operation was done before the twenty-second day, and preferably on the nineteenth, the existence of a male embryonic sex hormone was strongly indicated in that the male embryonic sex characters did not develop, and the embryo became femalelike.

It follows that the embryonic testis produces a hormone which favours the differentiation of the Wolffian system and inhibits the development of the Mullerian system in male embryos. It is, therefore, possible that Moore castrated his young opossums too late.

When Jost ovariectomized female embryos *in utero*, they developed normally; hence in the absence of the male embryonic sex hormone the female embryonic sex characters come into appearance.

Some of these results were confirmed by Raynaud and Frilley (1947) in mouse embryos after destruction of the gonadal regions by X-rays. Wells and Fralick (1951) and Wells et al. (1954), who castrated fetal rats *in utero*, showed that the development of some accessory reproductive organs of the male rat is influenced by the testis before birth. Wolff and Wolff castrated male and female duck embryos with X-rays. The Mullerian ducts persisted in both male and female castrated duck embryos. The bulla of the syrinx, as well as the genital tubercle (which are large in normal male and small in normal female ducks) were large in the embryos of both sexes. Thus in the female

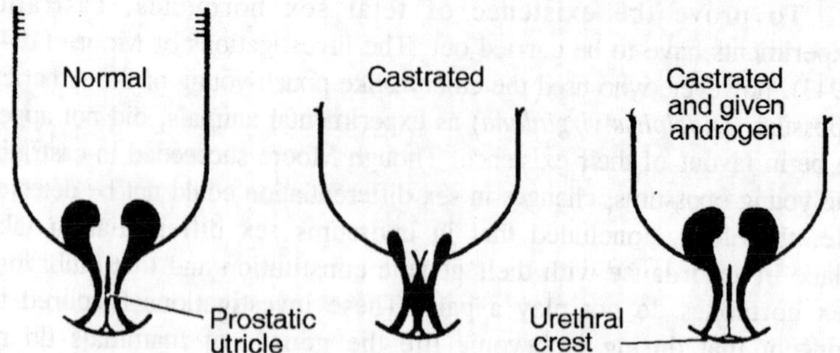

Fig. 8.14. Accessory sex organs of a normal and a castrated fetal rat and of a castrated fetal rat treated with androgens.

embryo the development of the syrinx and of the genital tubercle is impeded by the female embryonic hormone. In male embryos on the other hand, the development of the syrinx can be prevented by injected estrogen. The results of Price and Pannabecker (1956, 1959), who cultured the reproductive tract of male rats with fetal testes or ovaries, also favour the concept of the existence of embryonic sex hormones.

The chemical structure of these hormones is unknown; they may differ from the steroid hormones produced by the adult gonads.

9

Testicular Diseases

Hypogonadism

Since the two functions of the gonads are to produce hormones and to produce gametes, it would be logical to include, under the heading "*Hypogonadism*", insufficiency of either of these functions. In general, however, the term is confined to insufficiency of endocrine activity and in what follows attention will be restricted mainly to this aspect of the subject.

Two main forms of hypogonadism may be recognized, namely, *primary* where the main lesion resides in the gonad itself; and *secondary*, where gonadal changes are due to deficiency of pituitary gonadotrophic stimulation. Before puberty a condition of physiological hypogonadism exists because there is little or no production of gonadotrophins by the pituitary gland. When, at puberty, the pituitary gland becomes functionally effective, oversecretion of gonadotrophin arises if the gonad is unable to respond, so that in primary hypogonadal states an excessive urinary excretion of gonadotrophin is a diagnostic feature. In secondary hypogonadism, however, the excretion of gonadotrophin is very low or non-existent. It follows, therefore, that determination of the urinary gonadotrophin level is a key diagnostic test in differentiating these kinds of hypogonadism, but, since the existing methods of gonadotrophin assay are crude, it is not surprising that doubtful cases are sometimes. encountered.

Hypogonadism in Males

Hypogonadal males are called eunuchs when their hypogonadism is due to castration before puberty, and eunuchoids when they possess the physical and psychological characteristics of eunuchs, but have gonads which, however, remain infantile. These terms are less commonly applied to hypogonadal females.

PRIMARY HYPOGONADISM

This can be either congenital or acquired. A convenient grouping of congenital primary male hypogonadism, due to Sohval and Soffer, is as follows: (1) anorchism (testicular agenesis); (2) prenatal testicular atrophy; (3) germinal aplasia (testicular dysgenesis); (4) failure of normal differentiation of Leydig cells (testicular dysgenesis) due to: (a) long-standing cryptorchidism, (b) unknown causes; (5) as part of syndromes or diseases characterized by multiple congenital anoma including; (a) Laurence-Moon-Biedi syndrome (certain cases), (b) dystrophia myotonica, (c) the male counterpart of Turner's syndrome; (6) non-classifiable because of inadequate clinical, histologoical, or hormonal data.

Bilateral anorchism is exceedingly rare, though the unilateral variety is probably less uncommon. It is of considerable interest that the first authentic case of bilateral anorchism was recorded as long ago as 1564 in a soldier who was hanged for rape; post-mortem examination failed to disclose testes either in the scrotum or in the abdomen. It may be very difficult, or impossible, to differentiate prenatal testicular atrophy from true testicular agenesis.

Attention was first drawn to the form of testicular dysgenesis called germinal aplasia in 1947 by Engle and later in the same year by del Castillo, Trabucco, and de la Baize. In this condition, which is not an uncommon cause of azopermia, there is complete absence of cells of the spermatogenic series, though Sertoli and Leydig cells appear to be intact. The suggested cause of this condition is a failure of migration of the primordial germ-cells during embryonic life from the yolk-sac endoderm to the median-cell ridge. The gonadotrophin excretion in these patients may be entirely normal, though in some instances it appears to be elevated. The normal gonadotrophin excretion suggests that there is no deficiency of testicular hormones, particularly of that moiety which inhibits the overactivity of the pituitary gland. Since the Sertoli cells are intact in these patients, it has been argued that these cells are the source of the pituitary inhibiting X hormone. A condition in which the testicular biopsy is indistinguishable from the described above may arise where normal testes have been expected to irradiation of a degree sufficiently intense to destroy the germinal epithelium, but not so intense as to destroy the Sertoli or Leydig cells.

The other form of testicular dysgenesis is that in which the Leydig cells are not normally differentiated. McCullagh, Beck and Schaffenburg drew attention to "A syndrome of eunuchoidism with spermatogenesis, normal urinary FSH, and low or normal ICSH ("fertile eunuchs"). In two of their five patients the ICSH (or LH) excretion was normal, yet Leydig cells were apparently absent. In the other three patients of ICSH excretion appeared to be low, so that these patients should therefore be regarded as special examples of secondary hypogonadism. It is interesting to note that sperm counts in these patients ranged from very low figure (though not azoospermia) to 102 millions per ml. Landau has described a similar case of hypogonadism with spermatogenesis; this man had mind hypogonadism and unusually low seminal fructose. Androgen treatment led to a rise in the seminal fructose, the completion of masculinization and a temporary increase in the output of spermatozoa.

The hypogonadism of the Laurence-Moon-Biedi syndrome has generally been assumed to be secondary to pituitary gonadotrophic deficiency and Roth was the first to prove this by urinary gonadotrophin assays. More recently Francke has shown that in some of these patients gonadotrophin production is probably not deficient. In a 45-year-old patient who came to necropsy the testicular condition was indistinguishable from that found in the Klinefelter syndrome (a primary hypogonadal syndrome to be discussed later). It is therefore probable that, in some at least of the cases of Laurence-Moon-Biedl syndrome the hypogonadism is primary, the gonadal defect being just one more of the multiple congenital anomalies.

The association of gonadal deficiency with dystrophia myotonica has been known for a long time and several recent reports have described the testicular lesions in this condition. These consist of complete tubular sclerosis with normal Leydig cells and no interstitial fibrosis. The urinary 17-ketosteroids are usually low and the gonadotrophins usually, though not always, elevated.

It would seem that Turner's syndrome in the male forms a less well-defined entity than when found in the female. The combination of deficient stature, congenital anomalies of various kinds, testicular hypoplasia with androgen deficiency, and increased urinary gonadotrophins would qualify a male for inclusion in this category. Of the various cases described in the literature, no uniformity of testicular pathology has been found. This is in contrast to the condition in the female where the gonads consist merely of stromal elements with associated structures such as rete and medullary canals, but with total absence of the follicle apparatus. As will be seen later in the section on Turner's syndrome in the female, some of these so-called females appear to be of male chromosomal sex, their apparently female morphology being considered to arise as the result of the failure of development of a male-type gonad in early intrauterine life.

The above remarks on the heterogeneity of male Turner's syndrome apply also to a variety of miscellaneous cases of congenital primary hypogonadism in which mixed testicular lesions are found. Two such cases occurring in brothers were reported by Sohval and Soffer. These men had a moderate degree of androgen deficiency, small testes, aspermatogenesis, gynaecomastia, and increased urinary gonadotrophins. The testicular lesions differed from those of the Klinefelter syndrome; there was germinal aplasia together with a varying extent of completely hyalinized tubules. A further example of male hypogonadism with a mixed testicular lesion was described by Swyer and' Hughesdon. This patient also showed hypogonadism, slight gynaecomastia, azoospermia, and moderately increased gonadotrophins, but the testicular histology showed not only areas of germinal aplasia and of complete hyalinization but also some tubules with apparently normal spermatogenesis.

Acquired primary hypogonadism includes the following main conditions:

1. Surgical castration;
2. Functional prepuberal castration (some cases);
3. Klinefelter—Reinfenstein-Albright syndrome (sclerosing tubular degeneration);
4. Mumps orchitis and less commonly, tuberculous, syphilitic, or other infective conditions of the testes; Male climacteric.

The first of the above conditions will be considered further in the clinical section.. Functional prepuberal castration was first described by Heller, Nelson, and Roth. Some of their cases probably fell into the congenital group, but in others the lesion was very likely an acquired one. In this condition the testes are absent or completely atrophic, though characteristically the Wolffian duct derivatives descend into the scrotum. It is probable that many of these patients are mistakenly lebelled as cryptorchid. *Gynaecomastia* is common.

The syndrome of sclerosing tubular degeneration was first described by Klinefelter, Reinfenstein, and Albright in 1942 as "A syndrome characterized by gynecomastia, aspermatogenesis without aleydigism, and increased excretion of *follicle-stimulating hormone*'. It has since been the subject of numerous further reports. The condition has its onset at about the time of puberty, either earlier or later, and involves fibrosis and hyalinization of the basement membrane of the seminervous tubules, thereby leading to the cutting off of the blood supply to the tubular contents. There is in consequence a greater or lesser degree of destruction of the entire tubular contents-germinal epithelium and Sertoli cells. In addition the *Leydig cells* may be affected in varying degrees; in many cases they appear to be increased in numbers and, by the use of special staining techniques, it can be shown that they are morphologically abnormal.

Although Albright and his colleagues consider that the elevated gonadotrophin output in patients with the *Klinefelter syndrome* provides evidence that the Sertoli cells are the source of the pituitary-inhibiting X hormone, it is possible that an alternative hypothesis might more adequately explain the aberrant findings in this condition. According to this the primary lesion would lie in the inability of the Leydig cells adequately to convert the precursor substances into their normal hormonal products (that is, both testosterone and X hormone). Because of this the pituitary is uninhibited and secretes excessive gonadotrophin which in turn causes hyperplasia of the Leydig cells. The lack of normal Leydig-cell hormones is then held to be responsible for the tubular lesions and for the other clinical defects. This hypothesis, it will be noticed, corresponds with that explaining the nature of the adrenogenital syndrome due to adrenal hyperplasia.

According to Heller and Nelson, patients with the Klinefelter syndrome can be subdivided into three groups: (1) eunuchoidal,.(2) moderately eunuchoidal, and (3) non-eunuchoidal. This grouping depends upon the degree of failure of Leydig cell function. In general, the extent of gynaecomastia appears to be related inversely to the degree of Leydig cell failure, so that it is greatest in the least eunuchoidal types. There are, however, exceptions to this general rule.

Bradbury, Bunge, and Boccabella have reported the finding of female-type sex chromatin in the nuclei of buccal mucosal cells in all of five cases of the *Klinefelter syndrome*. (There have been several subsequent confirmatory reports). This indicates that some of these patients may represent a hitherto unrecognized form of female pseudohermaphroditism. Nelson has put forward reasons for supposing that there are two varieties of Klinefelter's syndrome, which he calls "true" and "false". The former are genetic females in whom the autosomal genetic male influence overrides the feminizing effect of the pair of X chromosomes so as to produce a male-type gonad and, in consequence, secondary sex characters. Mostly, the germ cells are sufficiently resistant to the male influence to prevent their functioning, but Segal and Nelson have described complete spermatogenesis in some tubules in a testicular biopsy taken from a case of "true" Klinefelter's syndrome with sex chromatin of female type, and Bunge and Bradbury had encountered similar findings in three of their patients. The "false" cases of Klinefelter's syndrome, according to Nelson, are genetic males in whom the sclerosing tubular degeneration is an acquired attribute. On this basis of differentiation, "true" Klinefelter's syndrome should be grouped with female pseudohermaphroditism.

Secondary Hypogonadism

As previously mentioned, the normal prepuberal child displays secondary *hypogonadism*, but requires no further notice here. In pituitary dwarfism there is presumably a failure of all the trophic functions of the anterior lobe of the pituitary gland; the testes along with the other target endocrine glands, remain in an unstimulated condition. The commonest variety of eunuchoidism is that due to idiopathic deficiency of follicle-stimulating hormone production by the pituitary gland. The 17-ketosteroid excretion ranges between levels such as would be found in prepuberal boys and low adult figures. The FSH excretion is, of course, low. The testicular histology is virtually indistinguishable from that found in the prepuberal state.

The presence of an organic lesion in or near the pituitary gland may lead to insufficiency of gonadotrophin production and so to secondary hypogonadism. The testicular lesion appears then to depend on the stage of testicular development attained prior to the onset of gonadotrophin suppression. When this happens before the completion of testicular maturity the general appearance is similar to that in idiopathic eunuchoidism, except that the tunica propria of the tubules is found to be thickened. When the lesion is of later onset more extensive degenerative change is usually found. Again, thickening of the tunica propria occurs but in these cases it is very marked, and is accompanied by peritubular fibrous tissue proliferation and Leydig cell degeneration. There is extensive impairment of germ-cell activity and the Sertoli cells contain lipoid-filled vacuoles.

Hypogonadism Without Endocrine Disturbance

In this group the main complaint is usually of infertility which may range from azoospermia to varying degrees of oligospermia. Azoospermia due to duct obstruction is, of course, excluded. The three main varieties are: (1) germinal aplasia, which has been mentioned already; (2) arrest of maturation; and (3) hypospermatogenesis. Little or nothing is known of the causes of these two conditions which, by the definition of this group, would appear not to be of endocrine origin.

Clinical Features of Hypogonadism in the Male

Genital organs

In eunuchoids the testes are always smaller than normal and frequently minute pea-like bodies such as might be found in an infant. However, in some of the hypogonadal syndromes mentioned in the previous action (e.g. germinal aplasia), the testes may be of apparently normal size. The penis is usually smaller than normal and may be infantile in size. The size of the penis is much more variable than that of the testes. The scrotum is usually, but not invariably, smaller than normal. In some patients the testes are undescended. The prostate is small and rarely palpable per rectum.

Hair

Pubic and axillary hair is nearly always scanty and the pubic hair does not ascend in triangular fashion to the umbilicus, but is limited horizontally as is typically the case in the female. This distribution probably merely reflects only the decreased extent of body hair generally. There is little or no facial hair and that on the rest of the body is usually very scanty, though axillary hair may sometimes be present in normal amounts. The scalp hair is typically luxuriant and baldness is most unusual.

Skin

The skin generally is thinner than that in normal adult males, and the skin of the face is soft in texture as in the female. Although the haemoglobin and red cells are normal, the face usually has a pallid appearance in strong contrast to the plethora which may be seen in Cushing's syndrome. The face and body are usually free from acne, although this may make its appearance in the course of treatment with testosterone.

Sexual behaviour

Libido is typicallly absent in the eunuchoidal patient and erections and emissions may never occur. In those hypogonadal syndromes where Leydig cell function is unimpaired, normal sexual behaviour may be found. Some eunuchoids may experience relatively infrequent erections and may be able to effect coitus, but their *sexual potency is usually of a low order and may frequently prove* unsatisfactory to their wives, should they marry. It is said that a high incidence of gonorrhoea has been found among Eastern eunuchs and it may therefore be concluded that some, at least, of these individuals are able to undertake sexual play. Postpuberal castrates, who have previously had normal libido and potency, may retain these in spite of the loss of their testes.

There is no evidence that homosexuality, either active or passive, is more frequently. encountered among eunuchoids than among other males.

Skeletal changes

Many, though by no means all, eunuchoids are tall, and some indeed (e.g. male cases of Turner's syndrome) may be below the average height. The tall eunuchoids may have been above average in height throughout life, or their tallness may have been due to continued growth after the age of 18 when it stops in normal people. This continued growth may proceed until the age of 26 or later, as a result of delayed apiphyseal union which is invariable among eunuchoids. The bone age in childhood or adolescence is younger than the chronological age. The younger bone age and delayed union of the epiphyses are also found in infantilism and is almost certainly due to lack of androgenic secretion by the testes.

In contrast sexual precocity is associated with advanced bone age, premature union of the epiphyses, and an ultimate height less than average. Because of the delayed epiphyseal union there is a relative overgrowth of the long bones, leading to the typical eunuchoidal proportions in which the span of the outstretched arms exceeds the height and the lower measurement (from the soles of the feet to the symphysis pubis) exceeds the upper measurement (from the symphysis pubis to the vertex).

The fingers are usually long and thin, but not invariably so. The general bodily build of eunuchoids varies in much the same way as does that of normal men, some being slender with rather thin bones, and others being more sturdy.

The pelvis tends to be gynaecoid, and is of greater breadth than the shoulders, which is the reverse of normal males, and similar to the skeletal proportions of females; or it may be intermediate between the male and female dimensions. The discrepancy between shoulders and pelvis is more marked in some than in others.

Not uncommonly eunuchoids show poor dentition, and the pattern of large central and small lateral incisors with blunt canines is said to be typical, though it is far from invariable. Unerupted permanent teeth may often be seen in X-ray films of the skull.

Gynaecomastia

Enlargement of the breasts, due to the development of mammary tissue, as opposed to the mere localized accumulation of fat, is encountered in certain groups of hypogonadal males. It is seen particularly in the *Klinefelter-Reifenstein-Albright syndrome*, though, as mentioned previously, it is then more likely to occur in the less eunuchoidal subjects. It is frequent in patients with functional prepuberal castration and has also been reported in miscellaneous cases, some of them having mixed testicular lesions.

No satisfactory explanation for the *gynaecomastia* of these patients is forthcoming, nor inded, is there any for the benign gynaecomastia not uncommonly encountered at the time of puberty in otherwise perfectly normal males. In particular there is no evidence that it is due to excessive oestrogen secretion. The possibility that it is due to excessive pituitary "*mammotrophic hormones*"-still uncertain in nature, perhaps prolactin and growth hormone-remain open.

Adiposity or leanness

Contrary to general belief, many eunuchoids are not fat, and, indeed, there may be a conspicuous absence of fat. In some of these patients, however, although their general appearance is not that of a fat person, there may be localized deposits of fat, for example, on the pubis, breast, abdomen, and buttocks. Thin eunuchoids may become obese in later life. One patient, for example, was tall and thin as a child; at the age of 21 he was still very thin and weighed only 112 lb., although his height was 70 in. In the next five years he grew 4 in. in height and put on 56 lb. in weight, becoming obese. In this particular instance a phase of pituitary activity manifested by growth was also apparently associated with the deposition of fat.

Cardiovascular system

The blood pressure is normal, or moderate hypotension may be found. The pulse rate tends to be slow. Radiographic studies may show subnormal heart measurements, findings which are consonant with the experimental fact that testosterone produces hypertrophy of cardiac muscle.

Muscular system

It is typical that eunuchoids have poorly developed and flabby muscles, so that they are unable to do heavy muscular work or to play games requiring skill or stamina. This state of affairs is directly due to the deficient secretion of androgens, since the muscular development and increased strength normally associated with puberty fail to occur. In contrast, sexual precocity is often associated with abnormal muscular development and strength.

Larynx

The larynx tends to remain small, and the voice highpitched. This may not always be obvious, except in emotional states, or on the telephone. On the other hand, it is important to realize that some otherwise entirely normal males retain a high-pitched voice. The possession of such a voice, in the absence of other stigmata of eunuchoidism, cannot therefore be used to diagnose testicular insufficiency.

Emotion and intellect

Intelligence is normal, and in some patients may be well above normal. The behaviour pattern, however, shows many interesting features. On the whole, eunuchoids are passive and accommodating. They are not necessarily afraid, but they have little pugnacity or external aggressiveness in their make-up. Their inertia may sometimes pass into somnolence. In contrast to their general placidity, they may exhibit phases of obstinacy, contrariness, sensitivity, or irritability; and may flare up like a *prima donna*. Such outbursts or tantrums are usually short-lived. Eunuchoids are sometimes introspective and secretive and may be given to intrigue. They may be depressed and have a sense of inferiority about their subnormal genital development. Commonly, though by no means invariably, they have little interest in the opposite sex. The range of variation in emotion and intellect among these cases is wide, and not a few eunuchoids have achieved outstanding intellectual and social success.

Familial incidence

There have been several reports of eunuchoidism among many members of different families. An excellent summary of familial hypogonadism has been given by Ferriman.

Incomplete eunuchoidism

This is probably more frequent than is generally recognized, the patient showing some undoubted stigmata of eunuchoidism, but being normal in other respects. One may also encounter examples of dissociation of androgenic effects. Thus, the genital, and libido may be normal, but the facial hair almost non-existent, or, as previously mentioned, the voice may remain high-pitched. It must be presumed in these cases that there is a failure of target-organ response rather than any insufficiency of male hormone production.

Diagnosis

The diagnosis of hypogonadism requires the demonstration of insufficiency of either androgen production of gametogenesis, or both. The most important criteria of androgen deficiency are the clinical features described in the previous section. The determination of androgen excretion in the urine is not a practical procedure for routine use and the measurement of 17-ketosteroid excretion is not always helpful because only a fraction of the urinary 17-ketosteroid in the normal male arise from the testes, the remainder being of output is usually low in eunuchoids, there may often be overlap with normal figures.

Insufficiency of gametogenic function can be determined by seminal analysis or by testicular biopsy. The latter is becoming increasingly used as a diagnostic measure, and although the evidence obtained from it seldom influences one's choice of therapy, there is no question that it greatly increases the precision both of diagnosis and of prognosis.

Testicular biopsy is essentially a simple procedure which can be carried out either under general anaesthesia or under local analgesia. It entails little or no discomfort for the patient and should not incapacitate him in any way. It is sometimes none too easy to perform when the testes are very small and soft, and occasionally it may be complicated by the development of a haematocele if unusually free bleeding occurs.

The important differentiation into primary and secondary hypogonadism can be made on the basis of urinary gonadotrophin estimation. In the first of these conditions, the output is increased for reasons which have already been explained, and in the second it is reduced or absent. Differentiation can also be made on the basis of the clinical response of the patient to treatment with chorionic gonadotrophin. The intramuscular injection of 1,000-3,000 I.U. of chorionic gonadotrophin twice weekly for six to ten weeks will be followed by distinct enlargement of the penis and increase in the growth of pubic hair in patients with secondary hypogonadism, but in primary hypogonadism this treatment will have no effect.

Treatment

The theoretically correct treatment for secondary hypogonadism is pituitary gonadotrophic hormone. Since commercial preparations of this are not available, recourse may be had to chorionic gonadotrophin and there have now been numerous reports that it may be wholly effective, not only in dealing with the eunuchoidal physical feature but also in promoting normal spermatogenesis. For this purpose large doses are necessary. For example, Maddock, Epstein, and Nelson used 5,000 I.U. of chorionic gonadotrophin thrice weekly for three months or more.

The great drawback of gonadotrophin therapy is the need for indefinitely prolonged and oft-repeated intramuscular injections. Since testosterone is fully effective in dealing with the eunuchoidal physical features and is the only available treatment for patients with primary hypogonadism, and since it can be administered far more conveniently than gonadotrophins, it is the treatment of choice in most cases. Moreover, it seems possible that in some patients with secondary hypogonadism, testosterone may also lead to spermatogenesis as well as to the development of normal secondary sex characters; Hurxthal, Bruns, and Musulin described four patients who showed such a response; one of these subsequently had two children and another also had a child (the other two patients were unmarried). A similar response to testosterone with the development of full spermatogenesis which persisted long after the last implantation was reported by Swyer. This was in a man who at the age of 27 still had genital development only of a degree to be expected in a boy of about 14. It is very doubtful if the argument sometimes raised, that the alleged results of treatment in these patients are, in fact, no more than the spontaneous development of late puberty, which occurred-or would have occurred-in spite of treatment; can apply to all the patients reported. For example, one of the patients of Hurxthal *et al.* was still eunuchoidal at the age of 43, only to become fertile after testosterone treatment.

Testosterone may be given by intramuscular injection of the propionate (50 mg. three times a week), the phenylpropionate (100 mg. weekly), or the oenanthate (125-250 mg. weekly or fortnightly); orally as methyltestosterone, effective when swallowed (25-50 mg. daily); or as testosterone linguets absorbed sublingually (25-50 mg. daily), the latter being a more cumbersome method; or by the subcutaneous implantation of pellets. Methyltestosterone has very rarely appeared to produce jaundice in adults but not in children, although the proof of the relationship is still *sub judice*. The implantation method is simple, and the implantation of six or eight 100 mg. pellets usually provides effective therapy for some six months.

The effect of treatment is usually dramatic. Libido and potency develop and may become supernormal. Orgasm occurs but detumescence may be delayed. Fluid emission is usually slight and sometimes absent. The penis and prostate grow in size but the testes remain unchanged.

There is a remarkable increase in strength, muscular development, and weight. The latter is largely due to muscular development, since there is marked nitrogen retention. The appetite usually increases remarkably and, strangely enough, though fat eunuchoids may lose some weight through increased muscular activity, the thin ones usually gain some fat.

Clear changes in psychology and personality usually follow treatment. The shyness and diffidence are lost and the patient may become extroverted, creative, energetic, and, occasionally, aggressive. Depression and apathy disappear, being replaced by initiative and the capability of assuming responsibility. Naturally these changes are more marked in some patients than in others.

Marked growth of hair on the pubis and in the axillae occurs, and to a lesser extent, hair growth on the rest of the body is also stimulated. It is, however, rare for the facial hair ever to grow at the normal rate, perhaps because of some initial failure of development of the hair follicles. The voice deepens.

The disadvantages of treatment which may be encountered are the development of acne in some patients, the occasional development of gynaecomastia, and the occurrence of penile erections at unwanted times. However, the advantages far more than outweigh the disadvantages.

If treatment ceases, some residual physical benefits usually remain, although libido and potency may disappear completely.

Post-Puberty Castration

Because of the varied accounts in textbooks, often based on other textbooks or literature, what follows is based on personal experiences. A series of representative case reports is followed by a summary of the clinical picture, males and females being treated separately. . It will be seen that the lack of uniformity of response to castration after puberty and, in some cases, a glaring contrast or anomaly, are striking; at the same time, other features will be found to occur as a sequel with some consistency.

Post-puberty Castration in Males

Case S. I, aged 21. This patient was said to have had a normal puberty at the age of 13, and during adolescence had indulged in some masturbation and experienced nocturnal seminal emissions. In April 1945, at the age of 21, he was wounded in battle in the left groin, and both testicles had to be removed surgically. He commenced intermittent treatment with methyltestosterone by mouth in December 1945; but before this, between April and December, he experienced nocturnal dreams with slight fluid emission, and he masturbated once weekly with resulting erection and emission. Frequent hot flushed, followed by cold perspiration, started in September 1945, and were partly controlled by 30 mg. methyltestosterone by mouth daily, and completely by double this dosage. His voice broke at the age of 14, and not changed following the accident. He stated that he did not require to shave until he was 18 years old, and the accident had no effect on his daily shaving. He had a rather scanty moustache, and did not appear to have a hairy face that would require daily shaving. His height was 6 ft. ^{1}h in. and growth had ceased at the age of 18. His span was equal to his height. His weight was 13 st. and he had put on 1 st. since the accident. He attributed the increased weight to lack of exercise through a stiff knee, and his surgical wound.

On examination he was seen to be a tall, slim type, with narrow shoulders and only slightly wider pelvis. The fingers, however, were not long and slender, but rather average. His penis was quite large. No testicular tissue could be felt in the scrotum. The pubic hair was normal in amount, and of male type ascending along the linea alba. Axillary hair was moderate. He also had some hair on the chest.

CRYPTORCHIDISM

Cryptorchidism, or undescended testes, is the term applied to failure of descent of one or both testes into the scrotum.

Incidence

Since the incidence of maldescent is much greater in children than in adults, it is clear that spontaneous descent before or during puberty must take place in a large number of cases. At birth the incidence is about 10 per cent; at puberty about 2 per cent; and in manhood about 0.2 per cent.

Mechanism of Descent

The precise mechanism responsible for descent of the testes into the scrotum is uncertain. In normal development the processus vaginalis, an evagination of the posterior parietal peritoneum, grows downwards along with the gubernaculum, and the testis follows so that by the seventh month of intrauterine life, it lies at the lower end of the inguinal canal. Although it has often been asserted that the gubernaculum is responsible for the descent of the testis, from anatomical considerations this cannot possibly be true.

It has long been known that chorionic gonadotrophin will induce testicular descent in man and since it is only in the human that fully descended testes are normally found at birth, while the human female is the only organism in which chorionic gonadotrophin circulates throughout pregnancy, it seems possible that this hormone plays an important role in bringing about normal testicular descent. Presumably chorionic gonadotrophin stimulates the interstitial cells of the foetal testis to secrete testosterone and this promotes elongation of the structures of the spermatic cord and development of the scrotum. It is doubtful, however, if this can be the whole explanation.

Causes of Maldescent

Although a number of anatomical abnormalities responsible for maldescent have been described, in the majority of cases such abnormalities are not in fact present, since descent eventually occurs spontaneously or can be induced by gonadotrophin therapy. In these cases, therefore, the cause of maldescent is obscure. Thus, it could scarcely be due to lack of hormonal stimulation-and could certainly not be due to such a cause when the maldescent is unilateral. The possibility of failure of the foetal testis to respond to maternal gonadotrophin seems more reasonable, as also that spontaneous descent at or near to puberty may be the consequence of a responsiveness to gonadotrophin acquired later on. Examples of familial maldescent are not uncommon.

Effects of Maldescent on the Testis

There is little doubt that a testis which is not in the scrotum by the time of puberty may suffer irreparable damage to its germinal epithelium with consequent defect or failure of spermatogenesis. The effect on the interstitial cells is much less serious, though testes retained within the abdomen may fail to secrete testosterone under gonadotrophic stimulation. Although it may be true that some ectopic testes have an inherent defect of spermatogenesis, regardless of environmental temperature, the ectopic position is itself harmful, the deleterious effects of cryptorchidism on the seminiferous tubules being due to the temperature of the retained testis remaining similar to that of the body generally. For reasons which are quite unknown, spermato-genesis will occur only at a temperature a few degrees lower, such as obtains in a scrotal testis.

Up until the time of puberty, the seminiferous tubules retain a normal prepuberal appearance, even though the testis be undescended; but at puberty differentiation of the undescended testes fails to occur and degenerative changes appear, consisting in fibrosis and eventual disappearance of the tubules.

Opinions are divided on whether testes suffer damage even before puberty through maldescent. Some authorities believe that irreversible damage may occur in a retained testis before the age of 6; other writers have taken the view that a testis descending or being brought into the scrotum at the time of puberty will be none the worse for its extra-scrotal sojourn until then. Since the fate of the undescended testis clearly influences one's decision on treatment, it is obviously unfortunate, to say the least, that this uncertainty should prevail. It is also necessary to realize that cryptorchidism may represent one aspect of more extensive testicular dysplasia by reason of which normal spermatogenesis fails to occur even in testes effectively brought into the scrotum. This situation is most clearly exemplified by the not uncommon infertility of men with unilateral maldescent, the scrotal testis being relatively small and soft. Of course, many men with unilateral cryptorchidism are fully fertile.

Types of Maldescent

There are two main types of maldescent: (1) retractile testes and (2) retained testes. Retractile testes do not come under the heading of cryptorchidism. They consist of testes which are easily withdrawn from the scrotum into the inguinal region by the cremaster muscle (low retractile), or alternatively of testes which are at times situated in the inguinal region but can be manipulated into the scrotum (high retractile). Many patients referred because of alleged cryptorchidism in fact fall into one or other of these groups. Examination of the patient only in the supine position, and especially if done with cold hands, greatly increases the incidence of this misdiagnosis. For this reason a boy suspected of cryptorchidism should always be examined standing and the examiner's hands should be warmed.

An undescended testis may be (a) intra-abdominal and not palpable, (b) inguinal, not palpable when in the inguinal canal but palpable if it has passed through the canal and reached the superficial inguinal 'pouch', which lies between the external abdominal ring and the upper scrotum, and (c) ectopic, when the testis is found in a position never occupied during normal descent. A testis lying inside the inguinal canal may be palpable in a very thin subject but, with this rare exception, it is correct to state that a testis in this position is not normally palpable. A testis that is clearly palpable, or visible, in the line of the inguinal canal, or to one or other side of it, is usually ectopic. An ectopic testis cannot be made to disappear into the inguinal canla or to pass into the scrotum.

Ectopic testes lie outside the inguinal canal; they may be situated superficial to the aponeurosis of the external oblique muscle (superficial inguinal ectopic testes), in the perineum or in the femoral region. The superficial inguinal ectopic testis is by far the commonest variety of ectopia. Its differentiation from an inguinal non-ectopic testis is important, since surgery is the only treatment for the ectopic testis. In this differentiation, Spence and Scowen have pointed out that the ectopic testis lies more superficially than one which is in the superficial inguinal pouch or within the inguinal canal; that it becomes more obvious when moved upwards and outwards and that it cannot be moved into the canal or into the scrotum.

Course and Complications

Spontaneous descent

This occurs frequently before or during puberty but only rarely after puberty. The retractile testis always descends more completely at puberty but may remain retractile even in adult life. Cases are in fact known in which the inguinal canal remains sufficiently patent for the testis to pass right through it back into the abdomen but nevertheless to descent again spontaneously.

When maldescent is the result of an anatomical abnormality or ectopia, spontaneous descent can never occur. It is very rare that abdominal testes descend spontaneously Testes within the superficial inguinal pouch or actually within the inguinal canal may or may not descend spontaneously. Normal function can generally be expected in a testis descending before or during puberty, but, as previously mentioned, this is not invariably the case.

Hypogonadism

Failure of spermatogenesis is to be expected in the case of persistent bilateral cryptorchidism and such patients are ordinarily sterile. As previously mentioned, this also occurs in some cases of unilateral cryptorchidism. With bilateral abdominal testes eunuchoidism may or may not occur.

Hydrocele and torsion

The undescended testis is more prone to injury than the scrotal testis and hydrocele formation is not an uncommon complication. Torsion of an undescended testis is a rarity.

Psychological disturbances

Boys (and their parents) vary enormously in their emotional reaction to cryptorchidism. Some are compeletely oblivious to the condition. Others may be disturbed because they differ from other boys or because they become the objects of sarcasm or ridicule when seen undressed by other boys. Persistence of cryptorchidism after puberty is more likely to result in emotional disturbance, particularly when bilateral. The realization of sterility adds yet further to the psychological burden of cryptorchidism.

Malignancy

There is no doubt that malignancy occurs more frequently in a cryptorchid than in a scrotal testis, though the precise incidence is a matter of uncertainty. Hinman quotes the frequency as being twenty times more than in scrotal testes—but malignancy is still a rarity.

Treatment

The first decision is clearly as to whether treatment of any kind is necessary. Retractile testes require no attention. There is no uniformity of opinion on the best treatment for retained testes. Ectopic testes can only be brought into the scrotum by surgery, but there is no certainty that such surgical treatment will produce a normal testis, fully functional from the endocrine and spermatogenic points of view. The real difficulty arises when the diagnosis of ectopia cannot be made with certainty, or when the testes are not palpable.

While it is true that the majority of undescended testes reach the scrotum spontaneously by adolescence, the dilemma facing the physician is that no one can say for certain which retained testes will descend spontaneously and which will not, while, on the other hand, the probability of infertility resulting from failure of descent by the time of puberty seriously challenges an attitude of mere watchful expectancy.

It has frequently been asserted that treatment with chorionic gonadotrophin will cause descent only of those testes which would have entered the scrotum spontaneously during puberty; from this it is argued that such treatment is unnecessary. If it were possible to diagnose with certainty cases where spontaneous descent by the time of puberty would occur, and if it were certain that retention of the testis until the time of puberty caused no harm, then the case for never using chorionic gonadotrophin in cryptorchidism would be firm. The situation, however, is otherwise. It is impossible to predict which retained testes will descend spontaneously at puberty, and it is uncertain that retention of the testis until puberty leaves it unharmed. Therefore, a therapeutic trial of chorionic gonadotrophin in young boys with cryptorchidism not definitely diagnosed as ectopic seems thoroughly justified. One cannot be dogmatic on the age at which such therapy should be started. The likelihood of response is greater the nearer to puberty, but, on the other hand, the risk of testicular damage increases with age. Perhaps, therefore, the right time to start is somewhere between 8 aand 11 years of age. The course should consist of 500 or 1,000 I.U. of chorionic gonadotrophin intramuscularly twice a week for about ten weeks. In the vast majority of cases, descent of some degree, if it is going to occur at all, will be seen within this time. If no descent is discernible, a further similar course of treatment, perhaps with an increase of dose to 1,500 I.U., may be given a few months later. Under no circumstances should *continued* therapy-going on for months on end-be permitted. It seems probable that all the cases of precocious puberty resulting from chorionic gonadotrophin therapy have been brought about by excessively prolonged treatment. The writer has never seen any untoward genital development resulting from a single ten-weeks' course of treatment as outlined above. The use of testosterone as an alternative to chorionic gonadotrophin is not recommended.

The real problem arises in the cases of undescended testes which have failed to respond to an adequate course of gonadotrophin. Most surgeons are agreed that the results of orchidopexy are largely very disappointing and this has led some to take the view that:

The boy with unilateral cryptorchidism has a much better chance of having two testes normal in size, consistency, and position if nature is allowed to take its course and surgical treatment is avoided until his 16th year. The question of the best treatment for bilateral non-descent is vital. The boy with this condition also has a much better chance of eventually having normally functioning testes if late spontaneous descent is allowed to take place and surgical interference is deferred until the 16th or 17th year. There is no question in my mind but that the results of most operations that have been performed to correct bilateral non-descent of the testes actually have unintentionally produced what they have been performed to prevent-that is, sterility.

If one could be certain that this view is correct, the course to follow would be clear. Unfortunately, in the writer's opinion, it is not possible to accept this view unequivocally and no doubt each surgeon undertaking the surgical treatment of cryptorchidism will base his opinion whether to operate or to leave alone on his own experience.

When one is confronted with a post-puberal patient with bilateral cryptorchidism, the is no doubt that surgery ought to be undertaken, even though the chance of bringing the testes into the scrotum may be small (especially if they are not palpable) and the chance of fertility even smaller (though not zero). With unilateral cryptorchidisn the necessity for treatment is far less clear. Obviously, gonadotrophin therapy is useless the patient having undergone puberty, his testes will already have been subjected ti effective gonadotrophic stimulation. Since the likelihood is high that a retained testis brought into the scrotum after puberty will fail to produce spermatozoa, there is little point in carrying out such surgery. The risk of malignancy may counsel removal of the undescended testis, but it seems to the writer that this should only be undertaken if the patient, having been fully appraised of the situation, specifically requests it.

GYNAECOMASTIA

Although its Greek derivation indicates a woman's breast, the term gynaecomastia is applied to the development, either bilateral or unilateral, of true breast tissue in the male. The accumulation of pectoral fat in men, unassociated with actual breast tissue, should not be called gynaecomastia and has sometimes been referred to as "pseudo-gynaecomastia".

Incidence

Gynaecomastia, usually mild and transient, is said to occur in about 80 per cent of boys at the time of puberty. Apart from this, gynaecomastia is relatively rare. In a general way, it may be said that when the onset is before the age of 25 it is essentially a manifestation of puberty and of no special consequence (though it may be of sufficient extent to cause embarrassment requiring plastic surgery for its alleviation); but when the onset is after the age of about 25, the likelihood of there being a sinister underlying disease is strong and requires full investigation. Exceptions to this rule certainly occurs: Wilkins described gynaecomastia due to a feminizing adrenocortical carcinoma in a boy of 5, and Holi in a boy of 15.

Aetiology

Oestrogenic stimulation in the male may lead to gynaecomastia and undoubtedly in some cases the condition arises from this cause. But it is far from certain that oestrogenic stimulation alone-or even at all-is responsible for all cases. It is well known, for example, that androgens, as employed in the treatment of eunuchoidism, may sometimes cause gynaecomastia, and so may both deoxycortone and cortical extracts. It is possible that pituitary hormones, particularly prolactin associated with growth hormone, may be of aetiological importance in other cases (for example, the gynaecomastia sometimes seen in acromegaly). Very often the cause of the condition is quite obscure and it may be that in some of these cases an undue sensitivity of the rudimentary breast tissue to normal hormone levels is responsible. Clearly this must be the case in unilateral gynaecomastia; and since the unilateral condition is relatively common, it rather suggests that target-organ sensitivity is one of the more important aetiological factors.

Pathology

The histological picture is identical with that of the gland in normal women, but adenoma formation and fibrosis are occasionally seen. Sometimes the picture is similar to that of fibrocystic mastopathy in women.

Clinical Aspects

Puberty

As already mentioned, puberty gynaecomastia is so common that it may almost be considered as physiological. The condition may be trivial, representing merely a sub-areolar swelling which disappears within a few months to a year, or it may consist of substantial breast development which causes great embarrassment to the boy and may interfere with physical activities. The major variety shows no tendency to regress spontaneously. The condition is often unilateral. In the minor varieties, first one breast and then the other may be affected. Puberty gynaecomastia may sometimes persist.

Exogenous oestrogens

Gynaecomastia is an unfortunate side effect of the treatment of carcinoma of the prostate with oestrogens. It also arises as a result of exposure to oestrogens during their manufacture and has occasionally been reported as arising through the contamination of other drugs, such as certain batches of amphetamine with oestrogenic impurities. The use of digitalis in heart failure has sometimes resulted in gynaecomastia; digitalis contains cardiac aglucones which possess the cyclopentenophenanthrene structure of the steroids.

Male intersexuality

Breast enlargement is found in certain forms of male pseudohermaphroditism.

Hypogonadism

Adrenal Cortical Tumours and hyperfunction. Adipose gynandrism.

Testicular tumours

Gynaecomastia has been reported as an accompaniment of most of the varieties of testicular tumours, such as interstitial cell tumour, chorioepithelioma, teratoma, seminoma, and the one case of so-called Sertoli cell tumour reported in a human male. It is generally supposed that an increased secretion of oestrogens by the tumour cells of the testis, or by hyperplastic Leydig cells stimulated by the chorionic gonadotrophin produced by the tumour cells, is responsible for the breast enlargement in these circumstances.

Hepatic disease

Gynaecomastia may be found in association with severe hepatic disease and is believed to be due to failure of the liver to conjugate oestrogens with the result that relatively high concentrations of free hormone remain in the circulation. The proof that this is true is still lacking. Testicular atrophy is another accompaniment of this condition.

Re-feeding gynaecomastis

Kark, Morey, and Paynter reported the occurrence of gynaecomastia in severely under nourished cirr'hotics during treatment with high-protein, high-calorie diets and suggested that it might be due to the introduction of arginine in adequate quantities in the diet. Re-feeding gynaecomastia was frequently observed among repatriated prisoners of war during the period in which their nutritional status was undergoing rapid improvement. In these cases, the gynaecomastia was transient and consisted of firm sub-areolar *plaques*. In some individuals it was accompanied by *mild orchitis* and return of libido.

Miscellaneous conditions

Tender breast enlargement in men with hypothyroidism has been reported by Berson and Schreiber. In a review by Wheeler *et al.*, attention is drawn to a previously unrecognized or unstressed relationship found between gynaecomastia and the following conditions: chronic generalized dermatitis, rheumatoid arthritis, angioderm-atomyositis, lymphoblastoma, diabetes mellitus, chronic pyelonephritis, chronic glomerulonephritis, and essential hypertension. Gynaecomastia, they point out, may occasionally be the initial manifestation or presenting complaint of serious underlying disease.

Diagnosis

The differentiation from pseudo-gynaecomastia is usually easily made on palpation. A typical granular consistency is found in true gynaecomastia, whereas, where fat alone is present, the tissues have a uniform feel. At puberty, examination of the testes will reveal whether normal enlargement, is occurring or testicular hypoplasia, associated with hypogonadism, is the aetiological factor. Search for other possible endocrine causes or some of the more obscure miscellaneous causes mentioned above should be made.

Treatment

There is no effective endocrine treatment for gynaecomastia which does not regress by itself. Appropriate plastic surgery is required where the condition is extensive and is causing psychological or physical embarrassment.

Male Climacteric

Whereas, for reasons which have been fully discussed in the preceding chapter, every woman who lives long enough must undergo a climacteric, the same is by no means true for men. Gametogenesis in the male, as has already been pointed out, is a function of sexually mature individuals and does not involve the continual loss of primordial germ cells. As a consequence, spermatogenesis may continue until very old age and there are authentic examples of the retention of fertility into the ninth decade. Based on these considerations, and because a syndrome in males corresponding to that of the climacteric in females is relatively rare and its features somewhat vague, it has been widely held that there is no such entity as the male climacteric. This view, however, is untenable, as was argued in the second edition of this book; and a restatement of the case leading to this conclusion has been given by Spence. The fact, which is accepted by all endocrinologists of experience, is that a small proportion of males flo undergo a phase of waning testicular function and may in consequence experience symptoms of the same general character, and no doubt arising in the same way, as are experienced by a far greater proportion of women. It is also generally agreed that when it occurs, the male climacteric appears later in life than the female, usually between the ages of 55 and 65.

The first to produce scientific evidence indicating the existence of a male climacteric were Heller and Myers who found that whereas in 15 men with psychoneuroses or psychogenic impotence, the urinary gonadotrophin excretion was normal, in 23 men whom they considered to be suffering from the male climacteric, the titre was unequivocally higher. Moreover, having performed testicular biopsy on 8 of these 23 man, they found in 5 a reduction in the size and activity of the seminiferous tubules, together with a reduction in the size and number of the Leydig cells; and, in the remaining 3, hyaline degeneration of the tubules. Later observers have reported normal testicular histology, in spite of the presence of increased urinary gonadotrophins, or normal tubules but decreased numbers of Leydig cells with abnormal histological characteristics.

According to Howard *et al.*, the climacteric can be divided into a compensated and a decompensated stage. In the former there is a tendency to decreased production of gonadal hormones, which is met by. overproduction of pituitary gonadotrophin so that gonadal function remains essentially intact. In the decompensated stage this tendency to gonadal failure is not counterbalanced, in spite of increased gonadotrophin production, and so gonadal failure becomes clinically demonstrable. In the female, the compensated stage lasts only a very short time, but in the male, on the other hand, it is the decompensated stage which is seldom encountered. They therefore considered their patients with the male climacteric whose testicular biopsies appeared normal, in spite of increased urinary gonadotrophin titres, still to be in the compensated stage.

Clinical Features

Decrease in both potency and libido is suggestive of declining testicular function, whereas impotence in the presence of normal libido is more likely to be of psychogenic origin. Nevertheless, in some cases of the male climacteric, the libido may remain essentially unchanged even though potency has decreased considerably. In yet other cases, neither potency nor libido undergo a change, yet nervous and vasomotor changes, so characteristic of the female climacteric, may be found. Among the psychic symptoms are anxiety, irritability, impairment of memory, loss of power of concentration, indecision, and insomnia. The principal vasomotor disturbances are not flushes and sweats, palpitation and tachycardia, shortness of breath on exertion and precordial pain. Other symptoms may include easy fatigability, tinnitus, vertigo paraesthesiae, and urinary difficulties-though these latter are more likely to be the direct result of prostatic enlargement. Involutional melancholia and suicidal tendencies are occasionally encountered. Symptoms may persist for several months to several years.

Diagnosis

Differentiation of true male climacteric syndrome from an anxiety state can be made on the basis of the gonadotrophin excretion, though it is seldom necessary to resort to this rather tedious investigation for clinical guidance. A prompt symptomatic response to injections of testosterone propionate would certainly be suggestive of the male climacteric, though the psychological effect of the injections must also be considered. In authentic examples of the climacteric syndrome, the substitution of an inert oil for testosterone propionate leads to relapse, wheras this is not the case when improvement in an authentic cases, testicular biopsy may reveal normal histology, it is clear that resort to this procedure is of no value in the diagnosis of the male climacteric.

Treatment

Testosterone in one form or another is the obvious treatment for the male climacteric. It is difficult to be dogmatic about the dose to be used, since marked individual variation to responsiveness may be encountered. There is something to be said for commencing treatment with relatively large doses, for example, testosterone propionate, 50 mg. thrice weekly, or testosterone phenylpropionate,100 mg., weekly. If with such doses no clinical improvement occurs, the diagnosis of male climacteric can promptly be discarded. On the other hand, a small fraction of these doses may be quite sufficient to maintain the patient in a comfortable state; and occasionally large doses may precipitate prostatic emergencies or cardiac insufficiency, so considerable circumspection should be used. Once a clear response to injection treatment has been obtained, it may prove advantageous to continue treatment by means of implants of testosterone, the appropriate quantity being two to six 100 mg. pellets, renewed as often as the patient feels necessary. These patients usually have no difficulty whatever in deciding when the beneficial effects of the implant have disappeared. Reimplantation becomes necessary every six to eight months, as a rule.

The more serious psychological disturbances are seldom materially affected by testosterone therapy, and for these psychological treatment may be necessary.

10

FEMALE SEX HORMONES

On the basis of present knowledge regarding the biosynthesis and occurrence of gonadal hormones, the terms "male" and "female" hormones are somewhat misleading. Hormones of the gestogen and estrogen as well as those of the androgen group are now known to be produced normally in the gonads of both sexes. In some species estrogens are formed in the male gonad at rates far exceeding those considered normal for the female of the same species. On the other hand, androgen production in the female gonad is in all probability a normal event in estrogen biosynthesis, as is the production of gestogens as precursors for androgens in both ovary and testis.

The term "gonadal" in this connection is also not entirely correct. Gestogens, estrogens, androgens, and relaxin are admittedly produced by the female gonad. But other organs are also partly responsible for the formation of some or all of these hormones. Thus the adrenal cortex normally produces hormones of the steroid groups mentioned; during pregnancy the placenta, at least in some species, takes over the production of these hormones as well as that of relaxin.

Finally, the term "hormones" should also be considered critically as here applied. According to the classical definition hormones are biologically active compounds produced in endocrine glands or structures, from which they are released directly to the blood. This definition would exclude metabolites formed outside the glands of origin, many of which still possess biological activity. In many species a number of such substances related to the sex hormones are present in urine and feces. It is difficult, however, to distinguish sharply between "true" hormones and metabolites. According to the above definition metabolites may well arise in the blood and exercise important regulatory functions in tissue metabolism without being "true" hormones.

In this context the term "female gonadal hormones" will be taken to cover gestogenic, estrogenic, and androgenic hormones, as well as relaxin. Androgens will be dealt with only in connection with estrogen biosynthesis and gestogen metabolism, and only as far as it is necessary for the understanding of certain physiological phenomena related to this group of substances in the female.

In a short treatise like the present one, the coverage of all aspects of the subject is obviously impossible. Preference will be given to recent biochemical contributions in the comparative field. Physiological manifestations of hormone action in different species will be dealt with only briefly.

Chemistry and Biosynthesis of Female Gonadal Hormones

Chemically the female gonadal hormones fall into two widely different categories. The gestogens, androgens, and estrogens belong to the class of compounds designated as steroids, but relaxin is a substance of protein-like nature.

Steroid Hormones

The structural formulas of representatives of the gestogens, androgens, and estrogens are shown in Fig. 10.1. The gestogens (1) contain 21 carbon atoms, the androgens (2) 19, and the estrogens (3) have only 18 carbon atoms. The estrogens are distinguished from the

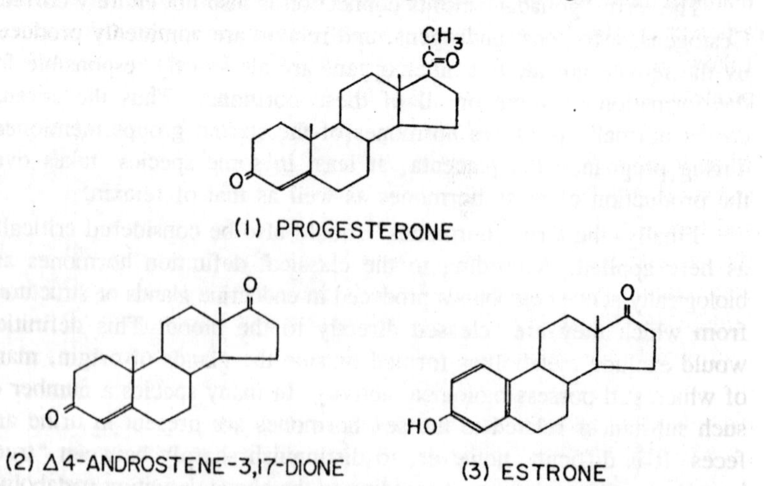

CH_3
C=O
O
(1) PROGESTERONE

O
O
(2) Δ4-ANDROSTENE-3,17-DIONE

O
HO
(3) ESTRONE

Fig. 10.1. Structural formulas of representatives of (1) gestogens, (2) androgens, and (3) estrogens.

two other groups of hormones by their aromatic character. Ring A and/or ring B may be unsaturated. The benzene character of ring A lends an acid function to the hydroxyl group at carbon 3. Very small alterations in chemical structure, such as a reduction of a keto group to a hydroxyl group, or the reverse, often results in marked changes in biological activity. Likewise, the orientation of hydroxyl groups in relation to the steroid nucleus often determines the degree of activity.

In the biosynthesis of steroid hormones in general, cholesterol seems to play an important role. Cholesterol itself is synthesized from acetate through a long series of intermediate reactions. In addition to the liver, which is the main site of formation of endogenous cholesterol, all glands capable of synthesizing steroid hormones are also capable of cholesterol formation.

Progesterone can be produced *in vitro* from cholesterol via Δ^5-pregnene 3β-ol-20-one from organs like the adrenals, testis, ovary, and placenta. Since the two other naturally occurring gestogens may be considered as derivatives of progesterone, this observation explains the biosynthesis of the gestogens.

The wide distribution of the enzymes necessary for the formation of progesterone from cholesterol indicated, however, that this hormone, in addition to its role as a gestogenic hormone, might also play a role as an intermediate in the biosynthesis of other steroid hormones. This has been shown experimentally. Incubation of tissue slices from mammalian testis resulted in the formation of Δ^4-androstene-3,17-dione, a potent androgen which is formed in testis together with testosterone. Likewise, incubations of avian testicular tissue homogenates with progesterone gave rise to Δ^4-androstene-3,17-dione and testosterone.

In 1955 Meyer showed that the bovine adrenal is able to transform Δ^4-androstene-3,17-dione to its 19-hydroxy derivative. 19-Hydroxy-Δ^4-androstene-3,17-dione yields estrone when incubated with human placenta and bovine follicular fluid. Later it was shown that testosterone gives rise to estradiol-17β when incubated with human ovarian slices and with human placental microsomes. It thus seems to be quite firmly established that the biosynthesis of estrone and estradiol-17β goes via androgens. Additional support for this view comes from the recent detection of Δ^4-androstene-3,17-dione in both the human ovary and placenta, and of testosterone in the bovine ovary.

Another problem is the biosynthesis of estriol which in women is the major urinary estrogen in both the nonpregnant and pregnant state. This compound had been assumed to be a metabolite of estrone or estradiol-17β, but recent experiments indicate that during pregnancy estriol is formed by other routes. In *in vitro* studies Ryan (1959) showed that human placenta as well as placental microsomes are able to convert Δ^5-androstene-3β-16α-17β-triol, 16α-hydroxy-Δ^4-androstene-3,17-dione, and 16α-hydroxy-testosterone to estriol. However, estradiol-17β is not converted to estriol by this tissue. The observation lends support to the concept of estriol as a "true" hormone in women.

The horse presents a complicated picture. In mare follicular fluid a new estrogen, 6-α-hydroxyestradiol-17β has recently been found in addition to estrone and estradiol-17β. Nothing is known about the chemical nature of the urinary estrogens of the nonpregnant mare, but during pregnancy, in addition to estrone and estradiols, large amounts of the naphtholic estrogens, equilin, equilenin, and their dihydro derivatives, are found in the urine. These compounds may be formed from acetate. However, injection of C^{14}-labeled estrone to a pregnant mare did not lead to radioactive ring B unsaturated compounds, and testosterone gives rise to estrone, but not to ring B unsaturated estrogens, indicating that these compounds originate through a separate pathway, not via estrone and/or estradiol-17β. As pointed out by Engel (1957) these experiments, although showing that the ring B unsaturated compounds are not peripheral metabolites of estrone, do not exclude the possibility that they are formed via estrone in endocrine tissues. However, this seems unlikely as far as the ovary is concerned, since neither of these compounds has been found in normal follicular fluid. It may well be that they are hormones of pregnancy, produced as such in the placenta, and are analogous to the production of estriol in human placenta.

Relaxin

Relaxin, a water-soluble, protein-like substance originally detected in the blood of various animals during pregnancy, possesses the ability to cause relaxation of the pelvis, the reaction being especially pronounced in ovariectomized, primed guinea pigs. The richest source of this hormone is ovaries from pregnant sows. It has not yet been isolated in pure form, and therefore its chemistry is not completely known. Purification of extracts from pregnant sows ovaries on carboxymethyl cellulose columns has recently been carried out by Paul and Wiquist (1960). Using the uterine relaxing activity as parameter, they found three peaks of which the second and third was active. When rechromatographed after heat treatment the second component appeared in the position of the third component. By this procedure the uterine relaxing activity was increased 8-10 times over that of the "crude preparation." Frieden et al. (1960) have shown that highly active relaxin fractions are associated with polypeptides of a molecular weight of 7500-9000 which contain most of the commonly occurring amino acids except histidine, methionine, and tryptophan.

GESTOGENS

Natural Sources

Progesterone which was isolated in pure form from pig corpora lutea in 1934, was for a long time considered to be the only naturally occurring gestogen. However, discrepancies between results of biological and chemical progesterone determinations in biological material indicated that other gestogens might exist. Recently two previously unknown gestogenic hormones have been isolated from human tissues by Zander et al. (1958). The substances, 20α-, and 20β-hydroxy-Δ^4-pregnene-3-one, are both active in the usual biological tests.

Although the significance of progesterone in reproductive physiology was early established, its isolation has only recently been accomplished from species other than the pig. Of special interest for comparative endocrinology is its presence in the ovaries of the sea urchin *Strongylocentrotus franciscanus* and the mollusk *Pecten hericius*, the starfish *Pisaster ochraceus*, the lungfish *Protopterus annectens* Owen, the dogfish *Squalus suckleyi*, and the hen. Neither of these species has

COMPOUND	MOUSE (HOOKER – FORBES TEST)	RABBIT (CLAUBERG TEST)	MAN
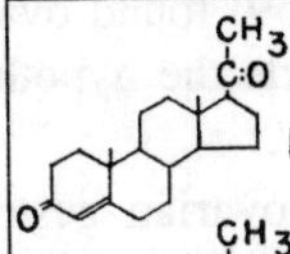 Progesterone	1	1	1
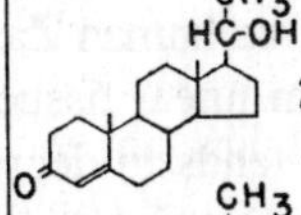 20α-Hydroxy-Δ4-pregnene-3-one	1/5	1/2–1/3	< 1*
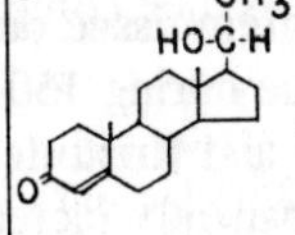 20β-Hydroxy-Δ4-pregnene-3-one	2	1/5–1/10	< 1*

Fig. 10.2. Structural formulas and biological activity of the naturally occurring gestogens (asterisk indicates that compound was administered as cyclopentylpropionate).

an established luteal function. The virtual absence of progesterone from the bovine, caprine, and porcine placenta indicates that in these species the ovaries represent the main source of the hormone throughout pregnancy. This is in accordance with results of ovariectomy during pregnancy.

The occurrence of these hormones is subject to species variation. Although both isomers have been isolated from human tissues, rat and sheep tissues contain the 20α-isomer only. In other species so far examined, only the 20β-isomer is present in addition to progesterone. Their presence in gestogen producing tissues and in blood, as well as their high biological activities, seems to justify the inclusion of these substances in the category of "true" hormones.

Levels in Tissues and Body Fluids

Ovary

In the nonpregnant female the ovary is the main source of gestogens. Of the ovarian components examined, the functional corpus luteum and the follicular fluid show the highest levels. For human follicular wall and/or fluid, Zander et al. (1958) reported values between 1.3 and 150 μg/gm. The average value for mare follicular fluid was 12.4 μg. Edgar (1952) reported values between 0.5 and 2 μg/ml fluid for sheep and 3 μg/ml for cattle. For luteal tissue the following average progesterone values given in μg/gm wet tissue have been reported: human, 14.7; horse, 37.7; cattle, 15.2, 20.2. These determinations were not related to any stage of the sexual cycle. Such observations have been reported for the human by Zander et al. (1958) who found an increased level between 7 and 10 days after ovulation. In hysterectomized guinea pigs Rowlands and Short (1959) found ovarian progesterone levels to be above normal. This supports the hypothesis that the uterus influences ovarian functions.

In the pregnant female the significance of the ovarian gestogen production varies considerably within the species. For the human Zander et al. (1958) reported an average value of 1.1 μg/gm luteal tissue. In cattle the level of progesterone in the corpus luteum tends to decrease during pregnancy. Average values of 12.2 and 10.8 μg/gm tissue during 30-120 days of gestation, and 5.0 and 8.2 μg/gm tissue during 150-280 days of gestation have been reported by Gorski (1958) and Kristoffersen (1960) respectively. In the guinea pig, however, permanently increased levels were observed after 21-23 days of gestation. Evidence for significant ovarian gestogen production was found in late pregnancy in the goat by Raeside and Turner (1955) who found 2.3 μg progesterone/

ml in ovarian vein blood, and in sheep by Edgar and Ronaldson (1958) who reported 1.8 μg/ml. In the blood which drains the ovary of the laying hen Lythle and Lorenz (1958) found 4-5 μg progesterone/100 ml.

Adrenal gland

Evidence for a significant production of progesterone by the adrenal gland in cattle, pigs, and sheep of both sexes was presented by Balfour et al. (1957) who reported values ranging between 7.5 and 46.5 μg/100 ml plasma. A peculiar finding is the transient secretion of comparatively large amounts of 20α-hydroxy-Δ^4-pregnene-3-one by the adrenal in the young calf. The significance of this finding is unknown.

Placenta

Great species variations are encountered with regard to gestogen levels in the placenta. Zander and von Munstermann (1956) found in the human placenta an average of 4.15 μg progesterone/gm tissue during the second and third month of gestation, with a subsequent decrease to 1.65 to 2.09 μg during the remaining months. In the mare placenta 0.073 and 0.25 μg/gm tissue at 120 and 270 days of gestation respectively have been reported, and the levels in sheep placenta are of the order of 0.004-0.009 μg/gm. In the placenta of the cow, sow, goat, and bitch progesterone could not be detected by chemical methods.

Blood

Reports on gestogen levels in the blood of the nonpregnant female are controversial. By bioassy, Forbes (1950) found cyclic variations in human blood gestogen levels with peaks of 1.7-5.2 μg progesterone equivalents/ml plasma on days 19 and 22 of the cycle. Similar results were obtained in the monkey. Using a chemical method Zander (1955) was able to detect progesterone in only 3 out of 16 women. The level was below 0.05 μg/ml. In the sheep Neher and Zarrow (1954) by bioassay found values equivalent to 0.3 to 2 μg progesterone/ml blood at estrus, rising to 6 μg in the luteal phase. In contrast, Short and Moore (1959) chemically determined a level of the order of 0.4 μg/100 ml plasma. The wide discrepancy between the results from biological and chemical assays will be noted.

The reports on the blood gestogen levels in the pregnant female are also controversial. Marked species variations seem to exist. According to Forbes (1951) who used bioassays the peripheral blood of humans and monkeys contained amounts of progesterone not exceeding 2-3 μg/ml. With chemical methods some values per 100 ml have been obtained during late human pregnancy: 14.2 μg, 12.2 μg, and 10-30 μg. Short (1960b) has also identified 20α-hydroxy-Δ^4-pregnene-3-one in human peripheral blood during pregnancy. For the pregnant mare Short (1957) reported negative findings. However, the same author later detected progesterone in the peripheral blood of both nonpregnant and pregnant mares, but only in the presence of a functional corpus luteum. During the second half of gestation when the ovaries become fibrotic, progesterone is no longer found in peripheral blood. The level during the early period of gestation is 0.5-1.4 μg/100 ml plasma. In cattle and sheep the progesterone levels in peripheral blood are very low, with averages of 0.8 and 0.5 μg/100 ml plasma respectively. The Hooker Forbes test indicates high levels of gestogens in peripheral blood during pregnancy in the ewe, rabbit and mouse. A marked drop in the levels was observed after parturition. The large discrepancies between biological and chemical assay results indicate that other gestogens in addition to those known at present may exist during pregnancy.

In the species so far examined it appears that the gestogen levels in fetal blood are considerably higher than in maternal blood. On the basis of his findings in women, Zander (1959) concludes that toward the end of pregnancy about 75 mg progesterone passes from the placenta to the fetus in 24 hours. In human fetal cord blood Aitken et al. (1958) found a level of 45 μg progesterone/100 ml plasma. The corresponding value in the horse is 3.8-6.3 μg. The biological significance of these findings is unknown at present.

Metabolism

In 1929 pregnanediol was isolated from human pregnancy urine. Its relation to the metabolism of progesterone has since been shown repeatedly. Investigations on the metabolism of progesterone have been carried out mainly in the human. However, according to data obtained from other species it seems safe to conclude that considerable species differences exist. The information available is partly based on isolation of the urinary steroids presumably related to progesterone, and partly on direct experimental results.

Urinary steroids related to progesterone

None of the known gestogenic hormones has been isolated from urine. It is generally assumed that pregnanediols constitute the major part of gestogen metabolites in the human, 5β-pregnane-3α-20α-diol being the most important of the different isomers. The same applies to the rabbit. But other isomers have also been isolated. In addition, in some species other compounds have been shown to be metabolites of progesterone. The occurrence of the different metabolites is subject to species variation, as are the quantities in which the compounds are excreted. Thus in late pregnancy women excrete about 50 mg of pregnanediols in 24 hours, but the substances identified in the urine of cows, goats, and sheep, are present only in trace amounts. In the mare the picture is complicated; pregnanediols are present in high amounts during pregnancy, but the relative proportions of the different isomers change with the stage of gestation.

The very low amount of pregnanediols encountered in the urine of some species raises the question whether gestogens in these animals are mainly metabolized to other compounds and/or excreted by ways other than the renal route. This question seems to have been partly answered recently.

In vivo experiments

Using C^{14}-labeled 21-progesterone, Riegel et al. (1950) showed that in rats and mice up to 25% of the injected radioactivity appeared in the expired air. Thus it appears that the side chain at C-17 is split off in the body. High activity was also found in feces. A similar pattern may obtain in ruminants. During pregnancy increased androgen activity in the feces has been reported for cattle. Increased androgen activity in the feces of rams after progesterone injection has likewise been reported. Pure androgens have also been isolated from the feces of cows after progesterone injections. In human subjects it has recently been shown that considerable radioactivity appears in the feces after injections of C^{14}-labeled progesterone. Sandberg and Slaunwhite (1958) have presented evidence for enterohepatic circulation of the hormone. Using C^{14}-labeled progesterone Taylor and Scratcherd (1961) showed that in the cat less than 1% of the injected dose appeared in the urine, whereas up to 67% was excreted in the bile, during the first 6 hours after the injection. Clearly, other compounds in addition to the urinary pregnanes must also be related to progesterone metabolism. This view is supported by the fact that only a low percentage of injected doses of progesterone can usually be accounted for by the metabolites recovered from the urine.

In vitro experiments

The liver has for a long time been considered as a major site of gestogen metabolism, but until recently few investigations had been reported in which metabolites were identified. Using rabbit liver, Taylor (1956) showed that 5β-pregnane-3α-20α-diol was the major metabolite of progesterone. Rat liver homogenate has been shown to convert progesterone into 7 different pregnane compounds. After incubation of progesterone with human liver slices Atherden (1959) isolated 6 pregnane compounds. This indicates that the main metabolic pathway is:

Progesterone → pregnanedione → pregnanolone → pregnanediol

Other tissues are also capable of metabolizing progesterone. Wiest (1959) showed that rat ovarian tissue rapidly transforms the hormone to 20α-hydroxy-Δ^4-pregnene-3-one, and Sweat et al. (1958) found that human uterine fibroblasts cultivated *in vitro* were able to metabolize progesterone to a variety of steroid products.

Biological Functions

Under physiological conditions gestogens and estrogens act synergistically in both the nonpregnant and pregnant female. The difficulties involved in ascribing certain effects to just one of the groups of hormones are therefore obvious, especially since both groups are also produced normally in extragonadal tissues which are not usually removed in most experiments. This fact seems to be of significance when interpretations of ovary ablation experiments are attempted. A full account of the physiological functions of gestogens is beyond the scope of this chapter. Only some of the important aspects will be dealt with.

Effects on the reproductive system

Ovarian function is significantly influenced by gestogens. Large doses of progesterone as well as persistent secretion of this hormone inhibit the release of luteinizing hormone and thereby inhibit ovulation. It may also inhibit the formation of follicles. These effects have been demonstrated in the sow, the cow, and the sheep. However, as demonstrated in the cow, the time of administration of progesterone as well as the dose given is of decisive importance. In this animal small doses of progesterone given at the beginning of estrus hasten ovulation. On the other hand, large doses can prevent estrus. On the basis of these observations and the fact that progesterone is normally present, in follicular fluid, it is reasonable to assume that the hormone plays a role in the events leading up to ovulation. In the laying hen progesterone administration increases ovulation frequency. Induction of ovulation in the hen can also be achieved by injection of less than 5 μg of progesterone into certain regions of the hypothalamus, whereas injections into the pituitary is without effect.

In growing chicks weekly injections of progesterone (2-16 mg) have been shown to cause decrease in size of the testes and comb in males, and delayed sexual maturity in females.

The uterus is affected by gestogens in several ways. In the endometrium progestational changes usually occur a short time after ovulation; these involve enlargement of stromal cells, growth of uterine glands, and increased secretory activity. It is generally accepted that the action of estrogens is necessary before the gestogens can exert their effects during this period. However, gestogen effects vary with the species. In the human female the glycogen content of the endometrium is at its maximum when the luteal function is at a maximum. In the cow it is at its lowest at this period. The changes in alkaline phosphatase levels in the endometrium are also different in the two species. The sow shows another pattern.

The avian oviduct is another target organ for the gestogens. Oviduct growth in young female birds is greatly stimulated by estrogens. An additional effect is seen when either androgens or gestogens are given simultaneously. Maximal albumen production has been reported only when gestogen is given together with estrogen. A peculiar situation exists with regard to gestogen action and the carbonic anhydrase level in the uterus or oviduct. In mammals the concentration of this enzyme is greatly increased by progesterone. In the chick no increase can be produced by progesterone or other steroid hormones tested.

In mammals the cervix shows diminished glycogen content and secretion during the luteal phase. In cattle it responds less to oxytocin during the luteal than during the follicular phase. Progesterone causes marked increase in the consistency of the cervical mucus in the cow.

In the rat, mouse, and guinea pig after ovulation there is a transition from the cornified vaginal mucosa to a mucified condition. In spayed animals estrogens and gestogens in combination, but not singly, can cause mucification. In sheep the postovulatory leucocyte invasion and change in the consistency of the vaginal mucus depend probably upon a combined estrogen-gestogen action.

Normal pregnancy is dependent upon adequate gestogen supply. However, great species variations exist with regard to the quantities required. This is indicated by the great differences in gestogen levels in the placenta and blood in different animals, and by the differences in the requirements of progesterone after ovariectomy in different species in which ovarian gestogen production is normally necessary throughout pregnancy. The following daily doses have been reported necessary in the later stages of gestation: in the goat, 15 mg, in the cow, 75 mg. In women progesterone production has been calculated to be about 250 mg per day during late pregnancy.

In this connection it is also interesting to note the gestogen effect on myometrial contractility. The stimulatory action of oxytocin is inhibited by progesterone in the rabbit and the mouse. In the cow the myometrial response to oxytocin does not vary during the estrous cycle. The cat is peculiar insofar as gestogens and estrogens act synergistically to sensitize the myometrium to oxytocin.

Interesting problems exist regarding the physiological significance of gestogens in lower vertebrates. The ovarian structure is similar in amphibians, reptiles, and birds. But the functional adaptation of the corpus luteum in relation to the reproductive patterns clearly points to variation in the significance of its secretions. Thus in the oviparous animals there is hardly an established luteal function, although gestogen production may be demonstrated. On the other hand, in viviparous amphibians like the toad *Nectophrynoides occidentalis* which is pregnant for 9 months, the corpora lutea persist and are believed to control gestation. Earlier studies had indicated that the corpus luteum was necessary for normal gestation in viviparous snakes. Later, however, observations on the viviparous garter snake and on the ovoviviparous lizard have shown that ablation of the corpora lutea does not lead to abortion. That the corpora lutea nevertheless may have some endocrine function is indicated by the fact that in oviparous reptiles they regress shortly after egg laying, whereas in ovoviviparous and viviparous forms they persist for approximately $^3/_4$ of the gestation period. Progesterone-like activity has been detected by bioassay in the plasma of ovoviviparous snakes, with increased levels during pregnancy. Progesterone has also been shown to stimulate oviduct growth of viviparous lizards.

Extragenital effects

For the normal development of the mammary gland, estrogens, gestogens, as well as pituitary, thyroid, and adrenal hormones, are currently considered to be necessary. In most species progesterone seems to be responsible for lobule-alveolar growth, but normally it acts in conjunction with estrogens. However, large doses of progesterone alone have been shown to cause lobule-alveolar growth in the rat, mouse, and monkey.

Body temperature in women increases at ovulation and after progesterone injections. The same phenomena has been recorded in cows.

An important extragenital function of progesterone is its corticoid activity. It is able to keep adrenalectomized animals in good condition.

This has been shown in the ferret, cat, and rat.

Progesterone also affects sexual receptivity. Injections into the lateral ventricles of the brain in hamsters causes estrus. Psychic signs of estrus have been shown to depend upon combined gestogen-estrogen action in the guinea pig, rat, mouse, and cow. Numerous other extragenital effects of progesterone have been reported.

ESTROGENS

Natural Sources

In 1936 MacCorquodale et al. isolated estradiol-17β from large batches of pig ovaries. Estrone was later isolated from the same source. Estradiol-17β, being the most biologically potent of the naturally occurring estrogens, has since been considered as the "true" estrogenic hormone of the ovary in spite of the fact that until 1958 it had not been isolated from the ovary of any other species. Recent investigations tend, however, to confirm this assumption since it has now been found to constitute the major ovarian estrogen in the cow, women, and mare. Estrone is probably present in small amounts in all species. In the horse a third estrogen, 6-α-hydroxyestradiol-17β is also present in follicular fluid. Smith (1960) has also reported the presence of estriol in the human ovary. In a restricted sense these compounds might be

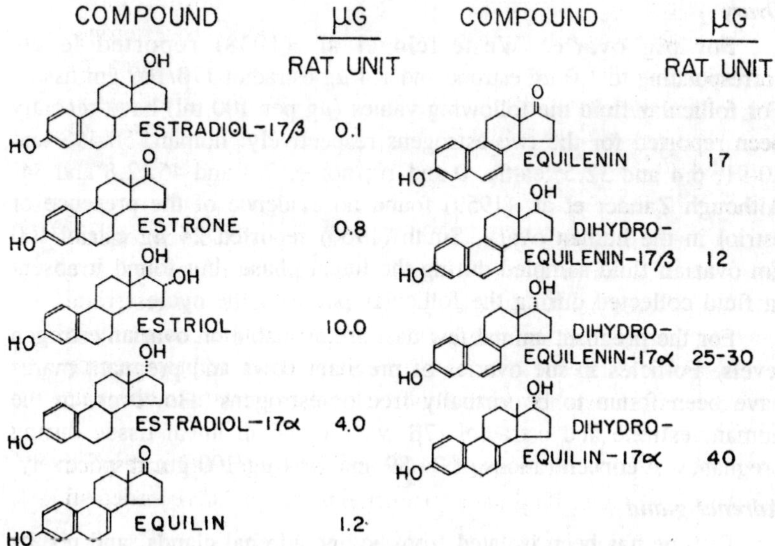

Fig. 10.4. Structural formulas and biological activity of some naturally occurring estrogens.

considered as the genuine female gonadal estrogens. However, when the placental estrogen production is also taken into consideration, the picture becomes more complicated.

In the human placenta estradiol-17β, estrone, estriol, and 16-epiestriol have been detected. The equine placental estrogens have not yet been characterized. But from the studies of the biosynthesis of estrogens in the pregnant mare it seems reasonable to assume a placental origin of the ring B unsaturated compounds as well as of the "ordinary" estrogens.

In the bovine placenta estrone, estradiol-17α, and estradiol-17β have been identified. Estradiol-17α seems to be the major placental estrogen in the goat and sheep, but in the pig only estrone is present in measurable quantities.

Estrogens isolated from ovary and placenta, as well as from testis and adrenal. Recent work identified estradiol-17β in the ovaries of the sea urchin and the mollusk, the starfish, the dogfish, the cod *Gadus callarias*, and also estradiol-17β, estrone, and estriol in the ovary of the laying hen; these findings point to a possible functional significance of these hormones not only in vertebrates but also in some invertebrates.

Levels in Tissues and Body Fluids

Ovary

For pig ovaries Westerfeld et al. (1938) reported levels corresponding to 1.0 μg estrone and 1.4 μg estradiol-17β/100 gm tissue. For follicular fluid the following values (μg per 100 ml) have recently been reported for the two estrogens respectively: human, 50-198 and 10-91, 6.4 and 32.5; cattle, 0 and 10; horse, 3.4 and 46; 2.8 and 34. Although Zander et al. (1959) found no evidence of the presence of estriol in the human ovary, Smith (1960) reported 24 μg estriol/100 gm ovarian fluid sampled during the luteal phase, but found it absent in fluid collected during the follicular phase of the cycle.

For the pregnant animal few data are available on ovarian estrogen levels. Follicles in the ovaries of pregnant cows and pregnant mares have been found to be virtually free of estrogens. However, in the human, estrone and estradiol-17β were found in luteal tissue during pregnancy at concentrations of 24-89 and 7-64 μg/100 gm respectively.

Adrenal gland

Estrone has been isolated from bovine adrenal glands, and results obtained from ovariectomized animals indicate that this gland contributes to estrogen production.

Placenta

For the full term human placenta Diczfalusy and Lindquist (1956) reported mean concentrations of 5.1 μg estrone, 17.0 μg estradiol-17β, and 31.5 μg estriol/100 gm wet tissue, with considerable individual variations. In bovine full term placenta Velle (1958a) found average values of 4.4 μg estrone and 12.4 μg estradiol-17α. In the same species Veenhuisen et al. (1960) reported 3.5 μg estrone, 5.6 μg estradiol-17β, and 9.7 μg estradiol-17α. In bovine placenta levels of 2 μg estradiol-17α, and in porcine placenta 8 μg estrone/100 gm tissue have been found.

Blood

In nonpregnant animals the estrogen levels in peripheral blood are very low. Markee and Berg (1944) using bioassays, found two peaks during the human cycle: one between days 10 and 14 and another between days 18 and 22, corresponding to 0.58 μg estrone/100 ml blood. Svendsen (1960), using a chemical method, found levels of 0.01-0.075 μg/100 ml blood in normally menstruating women.

During pregnancy elevated levels are present in human peripheral blood. Estrone, estradiol-17β, and estriol have been identified in human blood. Slaunwhite and Sandberg (1959) reported values of 3-9, 0-8, and 4-6 μg/100 ml plasma for the three estrogens respectively.

Bile

Biliary excretion of estrogens is indicated by the presence of estrone in the bile of pregnant cows, and by the experimental evidence for enterohepatic circulation of estrogens in human subjects.

Milk

Bovine colostrum has been reported to have estrogen levels comparable to those found in human and bovine pregnancy blood. Turner (1958) concludes from an extensive investigation that the mammary gland cells are relatively impermeable to the estrogens secreted during advanced pregnancy.

Feces

Experimental results obtained in man by the use of C^{14}-labeled estrogens show that 8-14% of injected doses are excreted in the feces. In pregnant cows the fecal estrogen excretion exceeds the urinary excretion. Thus species differences exist in the major routes of excretion of these compounds. Estrone, estradiol-17β, and estriol have been identified in avian droppings.

Urine

Most of the results of urinary estrogen determinations in the nonpregnant female have been obtained from humans. Among other recent investigators Brown (1955) reported average 24 hour values of 20, 9, and 27 μg for the ovulation peak, and 14, 7, and 22 for the luteal maximum for estrone, estradiol-17β, and estriol respectively in normal women. For other species little information is available about levels during the cycle. In the nonpregnant sow Velle (1958a) found peak values of estrone at the time of estrus, but no increase corresponding to maximal luteal activity. Values ranged between 2 and 28 μg/liter of urine.

Urine from pregnant animals has been the most important source of estrogens in most species investigated. In fact almost all naturally occurring estrogens were first isolated from pregnancy urine.

Systematic quantitative studies on urinary estrogen excretion in human pregnancy have so far only been made for the three "classical" estrogens. Brown (1956) showed that the levels of estrone, estradiol-17β, and estriol increased continuously until parturition, after which a sudden decrease was observed. The amounts excreted per 24 hours in late pregnancy are in the range of 1-2, 0.3-0.8, and 20-40 mg respectively. More recent investigations indicate that other estrogens such as 2-methoxyestrone and 2-hydroxyestrone may contribute significantly to the total amount of estrogens excreted.

In the horse the picture is also complicated. In the mare the urinary estrogen excretion increases markedly from the third month to the seventh or eighth month of gestation, and decreases before parturition. For months 7 to 8 Beall and Edson (1936) reported values of the order of 100 mg estrone per liter of urine. Recently the composition of the estrogen mixture present in the urine in the fourth to ninth month of pregnancy has been investigated. Gaudry and Glen (1959) report the following relative figures as per cent of total estrogens present: estrone, 51; equilin, 22; dihydroequilin-17α, 16; estradiol-17α, 6.4; equilenin, 2.6; and dihydroequilenin-17α, 1.3. The authors state that the relative amounts vary considerably with the stage of gestation.

In the pregnant cow the urinary estrogen level starts to increase markedly about 90 days after conception, and the increase continues until parturition, as shown by bioassay and by chemical determinations. In this species as well as in the goat estrone and estradiol-17α are the substances excreted. In sheep even in late pregnancy only trace amounts are present in the urine, as shown by bioassay, and by chemical determinations.

The pig shows a unique urinary excretion pattern; estrogens are present in high amounts during two distinct periods during pregnancy, first during the fourth week and then from about the eightieth day until parturition. As recently shown, estrone is the major compound during both periods, being present at levels of 50-500 μg/liter during the early period and 2-8 mg toward the end of pregnancy. In the pregnant dog urinary estrogens are present in concentrations too low to be detected by chemical methods. Estrone, estradiol-17β, and estradiol-17α were recently detected in rat urine. In other common laboratory animals, mouse, rabbit, and guinea pig, the chemical nature of the naturally occurring estrogens is still unknown. Estrone was recently isolated from the urine of the laying hen in crystalline form.

Estrogens in newborn

Estriol has been identified in human fetal organs. The highest levels were found in the liver. Only small amounts of estrone and estradiol-17β were present. No information is available for other species.

Urine from newborn males contains approximately 7 mg estriol/liter on the second day of life, and decreases rapidly thereafter. Estrone is present in trace amounts; estradiol-17β could not be detected. The excretion pattern is thus different from that of the adult.

In new born calves of both sexes estrone and estradiol-17α have been found in the urine at levels of 0.2 and 0.8 mg/liter of urine respectively; the levels decrease rapidly during the first days of life. It is interesting that estriol and estradiol-17α are the major substances found in humans and cattle, respectively. Presumably these substances represent the major end products in estrogen metabolism in the two species.

The first estrogen isolated from meconium was estriol, found in the human; it contained approximately 100 mg/100 gm wet material. Estradiol-17α has been isolated from bovine, ovine and caprine meconium. The estrogens are present partly in free and partly in conjugated form. Total concentrations measured were of the order of 60, 10, and 650 mg/kg respectively. In equine meconium Velle and Pigon (1960) identified both estrone and estradiol-17α, in yields of approximately 10 and 50 mg/kg respectively. In addition six to eight other as yet incompletely identified phenolic compounds were found in this material. The biological significance of these findings is obscure. But the different proportions between the substances present in the maternal and the fetal excreta points to an active participation in the metabolism of estrogens by the fetus.

Estrogens in male

Very high levels of estrogen activity were recorded in stallions' urine in the early thirties, and estrone was isolated in high yield. Zondek found only trace amounts present in the urine of castrates, indicating the testis as the site of origin. Both estrone and estradiol-17β were isolated from this gland, and the biosynthesis of estrogens in the stallion testis has been demonstrated experimentally. Estradiol-17β and estradiol-17α also have been isolated from the urine of the stallion. Pigon et al. (1961) found levels of 20 mg of estrone and 2 mg of each of the two diols per liter of urine.

From 16,000 liters of human male urine Dingemanse et al. (1938) isolated 7 mg of estrone. Estrone and estradiol-17β have been isolated from human testis by Goldzieher and Roberts (1952).

In the testis of the boar estrone and estradiol-17β were detected by Velle (1958a) and the same hormones were isolated from the urine in levels of the order of 2 and 1 mg/liter respectively. The urinary estrogen levels in the bull, ram, and male goat are extremely low. Estradiol-17β was recently isolated from the testes of a shark, *Scylliorhinus stellaris*, in quantities of 20 μg/kg tissue. The extreme species variation in estrogen production in the male are remarkable, and the biological significance of the high estrogen production in some species remains to be elucidated.

Metabolism

In vivo experiments

In the human species a variety of estrogen metabolites have been isolated from the urine after administration of estrone or estradiol-17β. Based mainly on the work of Marrian and his co-workers, and the investigations on the metabolism of C^{14}-labeled estrogens of Gallagher and his associates, the metabolic scheme shown below may be indicated for the human being.

16-ketoeatradiol-17β   2-hydroxyestrone → 2-methoxyestrone

↑

estradiol-17β → estrone → 16β-hydroxyestrone → 16-epiestriol

↓

2-hydroxyestradiol-17β   16α-hydroxyestrone → estriol → 2-hydroxyestriol

↓

2-methoxyestradiol-17β   16-ketoestrone   17-epiestriol   2-methoxyestriol

For cattle, goat, sheep, and rabbit the following pattern seems to be valid.

Estradiol-17β → estrone ↔ eatradiol-17α

The same pattern is probably valid for the horse. By analogy the following additional reactions may be suggested for the pregnant mare, assuming separate biosynthetic routes for the naphtholic estrogens.

Dihydroequilin-17β → equilin → dihydroequilin-17α

Dihydroequilenin-17β → equilenin → dihydroequilenin-17α

In vivo transformation of C^{14}-labeled dihydroequilenin-17β to equilenin has recently been demonstrated in the pregnant mare.

The simplest picture seems to obtain in the pig in which the only reactions seem to be

Eatradiol-17β ↔ estrone

with the equilibrium toward the right.

Recent investigations indicate a complicated metabolic pattern for estrogens in birds. Thus MacRae and Common (1960), after injection of C^{14}-labeled estradiol-17β to the laying hen, identified estrone, estriol, and 16-epiestriol in the excreta. In a similar study using estrone, Ainsworth et al. (1962) identified estriol, 16-epiestriol, 17-epiestriol, estradiol-17β, 16-ketoestradiol-17β, and 16-ketoestrone. The similarity in metabolic patterns between man and birds is remarkable.

In metabolism experiments with estrone or estradiol-17β in human subjects the recoveries of metabolites in the urine ranged from 13 to 38% and from 43 to 63%. Even though some metabolites are excreted in the feces, a fraction of the injected doses remain unaccounted for, indicating that other metabolites may exist.

In vitro experiments

In vivo experiments give little information about the tissues which are concerned with the metabolism of the hormones. The liver has long been considered the main site of estrogen metabolism, but only recently have the metabolites formed in the liver been chemically identified. Using rat liver, Ryan and Engel (1953a) demonstrated interconversions between estrone and estradiol-17β. The same authors found that these reactions also occur in a variety of human tissues. In these and similar studies, slices or homogenates have been used. Since erythrocytes also cause the interconversions, such studies may not be entirely conclusive. However, the same effects have now been demonstrated in cell cultures from different tissues grown in the absence of blood.

The interconversions of estrone and estradiol-17β take place in the presence of erythrocytes from a large number of species; but species differences exist. Bovine erythrocytes possess the unique ability to transform estrone to estradiol-17α, but erythrocytes from other ruminants such as sheep and goats do not, in spite of the fact that estradiol-17α is the major estrogen metabolite in all of these species.

In addition to the metabolic changes mentioned, a large number of other estrogen transformations have very recently been shown to take place in the liver of different species. These include reversible reduction of keto groups at carbon 17 and 16 to epimeric hydroxy compounds in liver from man, rat, and rabbit, 6-hydroxylation in rat liver, and 2-hydroxylation in rat and rabbit liver. Using C^{14}-labeled estrone, Jellinck (1959) found evidence for the formation of nonsteroid, water soluble compounds after incubation with rat liver. Of special interest to comparative endocrinology is the recent demonstration of 16-hydroxylation of estradiol-17β to estriol in human fetal liver, in rat liver, and in avian liver.

The liver also actively participates in the conjugation of estrogens and plays an important role in mediating estrogen-protein binding. The latter phenomenon may have an important bearing on problems connected with the mechanism of action of estrogenic hormones.

Recently attention has also been called to the intestinal wall as a major site of estrogen glucosiduronate formation in the rat and in man.

Biological Functions

The major property ascribed to the estrogenic hormones is the ability to cause estrus and characteristic changes in the female genital tract. In addition numerous other effects of estrogens have been reported. The literature in this field is overwhelming, and no attempt will be made to cover all aspects.

Effects on the reproductive system

In experiments including a variety of species estrogenic hormones have been shown to possess the ability to influence sex differentiation and development. This is most clearly demonstrated in lower vertebrates. In the fish *Lebistes reticularis*, for example, feeding of Progynon tablets containing "estrogenic substances" to young, sexually undifferentiated males caused the suppression of secondary sex characteristics and of spermatogenesis. Estrogens fed in large amounts to adult males of the same species caused no observable change. On the other hand, inclusion of estrone (1250 IU/gm) and stilbestrol (5000 IU/gm) in the diet to genotypic males of another fish *Oryzias latipes* from the time of hatching to about 8 months of age caused complete feminization and sex reversal: fully grown females of this male genotype produced offspring after mating with normal males. In the brown trout (*Salmo trutta* L.) no sex reversal took place when estradiol-17β was added to the water in concentrations of 50-300 μg/liter, but there was strong inhibition of the development of germinal tissue.

In amphibians estrogens exert marked effects on sex differentiation, but the responses vary considerably with species, time of administration, dose, and the nature of hormone administered. Thus, estradiol-17β has a feminizing effect in low doses and a masculinizing effect in high doses on *Rana*, whereas high doses will cause feminization in *Discoglossus*. Both estradiol-17β and testosterone cause feminization in *Pleurodeles waltlii*.

Chang and Witschi (1955a) showed that *Xenopus* larvae of male genetic sex developed into phenotypic females when kept in water containing 25-1000 μg estradiol-17β per liter. Feminized genotypic males bred to normal males produced only male offspring, showing also that in *Xenopus* the males are the homozygous sex.

Another peculiar estrogen effect in amphibians is the experimentally produced adrenogenital syndrome in the frog. When estrogen is administered to larvae, the adrenals become hyperplastic. Moderate doses cause feminization of males, but high doses cause masculinization of females. The adrenal hyperplasia is not produced in the absence of the pituitary.

In reptiles the administration of estrogens to males causes reduction of testis size and also of epididymis. In the female stimulation of sexual receptivity and of oviduct growth have been reported. But, as stated by Miller (1959) "the specific role of ovarian hormones whatever their nature or source, remains undetermined in the reptiles."

In birds also dramatic effects of estrogens on sex differentiation have repeatedly been reported. In females the ovaries and the internal genitalia are generally not affected to any extent. But application of estrogens to young, genetically male embryos leads to the development of an ovotestis on the left side; the right testis may or may not be affected, depending upon the dose given. Generally the degree of sex reversal of genetic males is roughly proportional to the estrogen doses applied. The whole question of modifications in sex and secondary sexual characteristics is treated in detail by Domm (1939) and Witschi (1961).

On the ovary of the adult female estrogens exert both a direct and an indirect effect. Small doses may cause direct stimulation of follicular maturation, but large doses may interfere severely with the normal ovarian cycle. Normally the estrogens are believed to cause a release of the luteinizing hormone from the pituitary, resulting in maturation of the follicle, and ovulation.

The effects of estrogen on the uterus are numerous. The normal proliferation of the endometrium preceding ovulation is considered to be a genuine estrogen effect, and has been experimentally demonstrated in humans, monkeys, and rodents. The manifestations of estrogen action are most pronounced in the surface epithelia and in the endometrial glands, but great species variations exist with respect to the histochemical picture of the different structures, indicating that other factors may modify the action of the estrogens. The composition of the uterus as a whole, is profoundly affected as shown by Mueller et al. (1958) in the rat. Characteristic are increases in water imbibition, glycogen deposition, and incorporation of amino acids into nucleic acids and proteins.

Using rat uterine growth as a parameter for estrogen potency, Velardo (1958a) tested a number of naturally occurring estrogens, and found that the activity decreases in the following order: estradiol-17β, estrone, 16α-hydroxyestrone, 16β-hydroxyestrone, estriol, 16-epiestriol.

In birds the oviduct responds markedly to estrogens, even in low doses. Prolific growth of this organ can be brought about by estrogens alone, but additional effects are seen when either androgen or gestogen is given simultaneously.

Myometrial contractility is similarly influenced by estrogens; the effect being generally a stimulatory one. In the rat, uterine contractions are rhythmic during estrus, and uterine contractions are more frequent in cattle during estrus than during other stages of the cycle. Estrogens may act directly, but may also sensitize the myometrium to the action of oxytocin.

In many species secretion of cervical mucus is stimulated by estrogens. Its consistency is lowered by estrogens.

In rodents cornification of the vaginal epithelium is a result of estrogen action. In other species cornification is less pronounced, both during estrus and after estrogen administration.

It is generally assumed that the combined action of estrogens and gestogens is necessary for the establishment of pregnancy, and for the normal functions of the pregnant uterus. From a comparative point of view it is interesting to note the apparently large species variations which exist in regard to qualitative and quantitative aspects of hormone production during pregnancy. One might assume that the high amounts of estrogens produced during pregnancy in women and in mares by far exceed the quantities necessary for the adequate supply of target organs.

Extragenital effects

Among the extragenital effects of estrogenic hormones the influence on mammary development and function is well known. Duct formation and growth are in most species considered to be effected by estrogens alone. This is especially marked in the mouse. In the guinea pig lobule-alveolar growth is also stimulated by estrogens. Other species range between these extremes. According to Reece (1958) the dog, cat, and rabbit show a pattern similar to that of the mouse, and the rhesus monkey, goat, and cow react more like the guinea pig. Also, mammary function is profoundly affected by estrogens. Moderate doses can provoke milk secretion in nonlactating animals. High doses or high production suppress milk formation, as demonstrated in the pregnant cow, in which milk production decreases markedly from about the fifth month, concomitant with a rapid increase in estrogen production. Estrogens also influence the composition of milk. A marked increase in free butyric acid combined with off-flavor of the milk has been observed after administration of estradiol-17β to a lactating cow.

Numerous effects of estrogens on metabolism are known in mammals and birds. In ruminants implantation or injection of long-acting estrogen preparations are used commercially to improve feed gain and weight increase. Most dramatic perhaps are the effects in birds, in which marked changes in blood and tissue composition take place during periods of ovarian activity and following estrogen administration. Among the most marked changes are lipemia and the elevation of blood calcium and phosphorus concentrations. Lipemia has been recorded in hens, doves and pigeons, ducks, and turkeys. The lipemic response as well as the increased levels in blood minerals are thought to be related to the formation of the eggs. Along with the increase in blood minerals produced experimentally, hyperossification may also occur. This has been demonstrated in pigeons chickens, ducks and sparrows. These responses in birds are contrary to those seen in mammals, in which estrogens tend to suppress the levels of both lipids and calcium in the blood.

In birds which have sex-dimorphic feathers estrogens are responsible for the normal female plumage, influencing both the distribution of pigment and the shape of the feathers.

Sexual receptivity in many species is induced by estrogens. But the manifestations and duration of this action are subject to species variation. In rodents continuous estrus can be produced by frequent administration of an estrogen. In the cow continued estrogen injections fail to cause estrual behaviour for more than a short period, indicating that other factors are also important for development of heat.

ANDROGENS

As already mentioned androstenedione has recently been isolated both from ovarian and placental tissues, and the available evidence suggests that this androgen is a very important precursor in the biosynthesis of estrogens. The role played by progesterone as intermediate in the production of corticosteroids, and its presence in significant amounts in adrenal venous blood, is analogous to the possibility that during periods of high ovarian estrogen production, androstenedione and/or testosterone may be released into the ovarian veins in amounts sufficient to exert physiological effects.

Experimental evidence points to ovarian androgens as physiological substances. Ovarian grafts in male mouse castrates are capable of maintaining full functional activity of the accessory glands. In the hen there is a correlation between comb size and ovarian activity; in the nonlaying hen the comb undergoes atrophy, whereas in the laying hen the comb size increases. This phenomenon can be reproduced with androgens, but not with estrogens or gestogens.

RELAXIN

Natural Sources

Bioassays show that relaxin is widely distributed among vertebrates, having been detected in humans, horse, cattle, pig, dog, cat, rabbit, guinea pig, rat, mouse, whale, chicken, and shark. Since it is usually considered to be a hormone of pregnancy, it is interesting that it has now also been detected in the ovary of the nonpregnant sow, and rat, in the serum of estrous dogs, and in the testis of the rooster. Its main sources in the pregnant animal appear to be the ovary and the placenta. The relative role of these organs for the production of relaxin seems in some way to be related to the degree of necessity of the ovaries for the maintenance of pregnancy. Thus in the mouse, rat, and sow, which require the ovaries throughout most of the gestation period, relaxin is probably mainly produced in the ovaries. In animals like the guinea pig, on the other hand, in which ovariectomy may be performed at midterm without interruption of pregnancy, the placenta is probably the main source, since relaxation of the pelvis occurs also in the absence of the ovaries.

Levels in Tissues and Body Fluids

The bioassay of relaxin is usually carried out by palpation of the pelvic symphysis of estrogen-primed guinea pigs after relaxin administration, 1 guinea pig unit (GPU) being taken as the least amount which causes a palpable relaxation in 9 of a group of 12 castrated animals weighing between 350 and 800 gm.

Very few data are available for nonpregnant animals. For pregnant sow ovaries values between 675 and 9600 GPU/gm tissue have been reported. Values per gm tissue have been recorded for other species: rabbit, 25-30; rat, 98-720; mouse, 114-200; fin whale, 24; blue whale, 536; shark, 4-10. Concentrations in the placenta vary also considerably. In most species the levels are low; values in the range 0.5 to 4 GPU/gm tissue having been reported for the sow, cat, fin whale, and blue whale. The rabbit placenta on the other hand shows values in the range of 50-75 to 137 GPU/gm. In this species high levels are also found in blood serum (of the order of 10 GPU per ml). In the sow, cat, mouse and guinea pig the reported values lie between 0.1 and 2 GPU/ml. For women Zarrow et al. (1955) reported levels of 0.2 GPU/ml serum at 7-10 weeks of pregnancy, increasing to 2 GPU at 38-42 weeks. The hormone disappeared within 24 hours after delivery.

Biological Functions

Pelvic changes occur during pregnancy in a wide variety of mammalian species, but the degree of relaxation during pregnancy and labor is subject to great variation. Very pronounced changes take place in the guinea pig and mouse. In the guinea pig relaxation of the pelvis can be brought about experimentally by estrogens, gestogens, a combination of the two, or by relaxin. The length of treatment required depends upon the combination of hormones given. The estrogen as well as the relaxin effect is direct, but that of progesterone in all probability is indirect. The latter hormone is only active in the presence of a functional uterus. It is therefore assumed that it induces relaxin formation. In the mouse also pelvic relaxation can be induced by either estrogens or relaxin, but progesterone shows an inhibitory effect.

In immature rats uterine water uptake is largely stimulated by relaxin. A transient increase in glycogen, nitrogen, and dry weight in the uterus of estrogen-primed, spayed rats has also been observed.

In vitro spontaneous uterine motility is inhibited by relaxin in the guinea pig, rat, and mouse. In human pregnancy relaxin has been reported to inhibit premature uterine contractions. The relaxin effect on the uterine cervix is important. Experimental evidence for cervical softening has been presented for the cow, the woman, and the rat. In the sow the reaction was accompanied by changes in water and mucopolysaccharide content, and in the rat by changes in water and glycogen content.

Extensive investigations by Steinetz and his co-workers indicate that relaxin plays a role in balance with the steroid hormones in the maintenance of normal pregnancy. It is also suggested that it may play a role in the initiation of parturition in the rat and mouse, since changes in the balance of relaxin and progesterone alter the response to oxytocin injections.

The extragenital effects upon the mammary gland must be mentioned. In the guinea pig, mouse, and rat relaxin influences lobule-alveolar growth. In both intact and castrated mice relaxin plus estrogen effectively stimulates lobule-alveolar growth. Ten times more (by weight) progesterone than relaxin is required to cause the same effect. The authors suggest that progesterone may affect mammary development by inducing relaxin formation.

Concluding Remarks

The first appearance of the female gonadal hormones in phylogeny and the stage at which they began to exert physiological functions are questions of major importance in comparative endocrinology.

Progesterone and estradiol-17β, which had previously been shown to occur naturally in a long series of vertebrates ranging from fishes to primates, are also present in the ovaries of some invertebrates like the sea urchin and the mollusk. Relaxin, the occurrence of which has been demonstrated in a variety of mammals, was recently detected also in shark ovaries. These observations indicate a biological role for the female gonadal hormones over a considerably wider range of animal species than previously anticipated. The chemical nature of gonadal hormones in amphibians and reptiles is unknown. However, since female gonadal hormones of mammalian origin are able to cause the expected effects when given to lower vertebrates, it seems reasonable to expect the presence of hormones of similar or identical nature in these forms.

Since, as far as is known, progesterone is a precursor for androgens and estrogens, it is tempting to assume that the enzyme systems for the biosynthesis of this compound were developed earlier in evolution than those needed for the formation of the other two groups of hormones. Although it is known that luteal bodies can be found in representatives of all classes of vertebrates and in some protochordates, an established luteal function is unusual except in mammals. However, progesterone is also produced by follicular elements, and the functional corpus lutcum represents a specialized structure in which the ability to transform progesterone to C_{19} and C_{18} steroids may to some extent have been lost.

Little is known about the physiological functions of the female gonadal hormones of lower animals. It has been suggested that the presence of estrogens in fish ova may have a bearing on the maintenance of a large hyperemic uterus after ovulation. In amphibians gonadal hormones are believed to maintain the secondary sexual characteristics. In reptiles ovariectomy causes regression of the oviducts. This can be prevented by the administration of estrogens or androgens, indicating a possible natural role for these hormones.

Sexual behaviour is greatly influenced by female gonadal hormones, but generally it seems that their importance in this respect decreases, as other factors assume more importance during ontogeny.

Finally, it seems that the same steps in the biosynthesis of female gonadal hormones occur in all species studied. Although this indicates a great similarity in the distribution of the enzymes responsible for the biosynthetic processes, there are marked quantitative variations. Furthermore, there are marked qualitative species differences in the distribution of the enzymes involved in the catabolism of these hormones. The physiological and phylogenetic implications of these observations require elucidation.

11

Ovarian Diseases

Oophoritis

Primary *amenorrhoea* or a premature menopause are often described in women with autoimmune disease, particularly 'idiopathic' *Addison's disease*, *myxoedema* or *hypoparathyroidism*. Histologically, the ovaries show lymphocytic infiltration, as do the other target organs in autoimmune endocrinopathies. These women sometimes have steroidal cell antibodies which react with Leydig cells, ovarian granulosa and theca interna cells. The presence of such antibodies predicts ovarian failure, especially in patients who have Addison's or other autoimmune diseases, yot who still have normal menstrual function. The pathogenic significance of ovarian antibodies in autoimmune oophoritis remains to be determined.

Infertility

Case

A 29-year-old builder had been married for 6 years but had no children. His wife had been extensively investigated; she ovulated regularly with a normal menstrual cycle, and had patent fallopian tubes and normal endocrine function. He had normal levels of luteinizing and follicle stimulating hormones and testosterone. He had no past history of orchitis or testicular trauma. On examination, he was a well-virilized, healthy-looking man with normal sized testes. A semen sample showed a low sperm count with sluggishly motile sperm and sperm-associated immunoglobulin (IgA and IgG). The sperm-cervical mucus contact test was abnormal and the use of normal donor sperm and normal cervical mucus confirmed that only the husband had antisperm antibodies. Antibodies to fresh donor sperm were detectable in the serum to a titre of over 1/1000, and in the seminal plasma to a titre of 1/32. The patient was treated with high dose steroids on days 1-10 of his wife's menstrual cycle. His wife became pregnant in the cycle following the fourth course of treatment and subsequently gave birth to a healthy baby girl.

Immunology of infertility

Human spermatozoa and seminal plasma contain strongly imonunogenic material: some of these antigens are unique to sperm or seminal plasma (semen-specific antigens), but others are shared with other fluids, secretions and organs. Five to 14% of infertile couples show evidence of spermantibodies. *These antibodies may be produced by the man, the woman, or both.*

Experimental male animals can be made sterile by active or passive immunization against testicular or seminal antigens. In man, damage to the seminal tract by surgery, *accidental trauma*, occlusion or infection may trigger autoimmunity to testicular and seminal antigens. For example, antisperm antibodies appear in the serum in 50% of vasectomized men within 6-12 months of surgery. Antisperm antibodies seldom appear in seminal plasma following vasectomy as local antibody production occurs proximal to the operation site. High titres of antisperm antibodies may appear in the semen after reversal by vasovasostomy and *modify the success of the reversal.*

Autoantibodies to sperm antigens may cause infertility in otherwise normal men by: (1) immobilization and agglutination of spermatozoa; and (2) inhibition of mucus and/or egg penetration by sperm, possibly by blocking specific receptors on the sperm surface.

Investigations of possible *autoimmune infertility* include a postcoital test. Poor mobility of sperm in this test suggests the existence of antisperm antibodies. Serum from both partners, cervical mucus and seminal plasma are tested for sperm antibodies using normal donor sperm and cervical mucus. When semen is mixed with cervical mucus, spermatozoa normally move rapidly and unidirectionally; IgA antibodies to spermatozoa prevent this type of movement.

Before considering treatment, it is important to make sure that there is no additional cause of infertility. Prostatitis has been found in about one-third of men with antisperm antibodies, and prolonged antibiotic treatment may be accompanied by a significant fall in antibody titres and pregnancy in a proportion of the wives. Manipulative techniques such as in vitro fertilization (IVF) or gamete intrafallopian transfer (GIFT) have been used with limited success but are not widely available. High dose intermittent steroid therapy has many side-effects and its use is therefore debatable but it can be successful.

The harmful effects of antisperm antibodies in women is unclear. Since the female genital tract is well endowed with immunocompetent cells, local isoimmunity is probably important in infertility. Where antibodies are found only in the female partner, treatment has been disappointing though controlled studies are lacking. *Immunosuppressive therapy with steroids is contraindicated* as exposure of the zygote and early embryo to high dose steroids may result in congenital abnormalities.

Hypogonadism in Females

Primary Hypogonadism

As in the male this can be either congenital or acquired. Congenital primary *hypogonadism* is sometimes called ovarian infantilism. Two main groups can be distinguished:

(1) developmental inadequacy of the ovaries; (2) ovarian agenesis.

Albright, Smith, and Fraser were the first to describe patients with sexual infantilism accompanied by an increased excretion of gonadotrophins and not associated with significantly decreased stature. They suggested that the condition could be explained on the basis of a 'premenarchal menopause praecox', the process of follicle atresia which normally commences at birth having proceeded to such a degree that no further responsive follicles were left by the time pituitary gonadotrophin secretion commenced.

Ovarian agenesis is a fascinating form of congenital primary hypogonadism which is usually associated with a group of other congenital anomalies giving rise to a fairly clear-cut syndrome. Decreased statural growth associated with a rather typical stocky habitus is almost invariably found. Webbing of the neck is another typical feature, but is not uncommonly absent. Cubitus valgus, giving rise to an increased carrying angle of the arms, is seen to some extent in nearly all cases. Other anomalies include cardiovascular abnormalities—in particular, coarctation of the aortadigital deformities, squints, and so on.

The urinary gonadotrophin output is elevated once the age of normal puberty has been passed. Laparotomy reveals the uterus and tubes to be of infantile dimensions, and the ovaries to be represented merely by a fibrous cord, the continuation of the ovarian ligament. Histological examination of this ovarian remnant reveals a normal stroma with rete ovarii and medullary canals, but no trace of follicles in any stage of development. Clumps of hilus cells (large pale staining polygonal cells disposed along the course of nerves and strongly resembling the *Leydig cells* of the testis) are sometimes prominent.

The growth failure in these patients may conform with one or other of two distinctive patterns. In the first, there is a conspicuous smallness from early infancy, and in the second, growth continues apparently normally until the age of about 9 or 10 years, whereafter it remains at a level appropriate to this age. It is probable that it is in this group of cases alone that the absence of the normal prepuberal and puberal growth spurt is contributory to the growth failure. The fact that some of these patients begin to grow again when treated with oestrogens conforms to this view. Since primary hypogonadism is not always associated with growth failure, it seems clear that the growth failure in this syndrome is an associated defect, either genetically determined, or induced as a result of a mutually injurious influence operating in the 5- to 17-mm. phase of embryogenesis. It is at this stage that the cortex of the primitive genital ridge undergoes organization for the process of penetration by the primordial germcells, the sex of which is already determined genetically, which migrate from the endoderm of the yolk-sac. Upon this penetration further sexual differentiation of the gonads would seem to depend, so that with failure of this process the cortical elements specific to the sex of the gonads would not emerge, whereas medullary rudiments would remain relatively unaffected.

This hypothesis of the pathogenesis of ovarian agenesis would seem to receive considerable support from two different sources. The first relates to experimental work in which it has been shown that removal or destruction of the gonads of embryos in mice and rabbits in the sexually indeterminate stage leads to the development of foetuses all of which are apparently female, the males having undergone intersexualization. The .second is the demonstration, by means of the skin biopsy technique for differentiating the chromosomal sex, that some, at least, of the patients with Turner's syndrome are of the male chromosomal sex. It is of interest that Polani *et al.* were led to suspect this possibility from the consideration that male examples of Turner's syndrome are very rare; that coarction of the aorta is common in the 'female' cases of *Turner's syndrome* but is otherwise decidedly commoner in males than in females; and from these facts they suspected that some of the apparent female cases of Turner's syndrome might actually be completely intersexualized genetic males. The three 'female' patients with Turner's syndrome associated with coarctation of the aorta on whom they determined the chromosomal sex by skin biopsies all proved to be genetic males. Wilkins *et al.* reported eight patients with the ovarian agenesis syndrome; six of these had male-type epidermal nuclei and two had the female type.

Acquired primary hypogonadism is rare and due either to surgical trauma or removal, or to local pelvic diseases which is itself rare in prepuberal girls. It is possible that long-standing and debilitating diseases can also lead to intrinsic ovarian failure and so to acquired primary hypogonadism.

Clinical Features of Primary Hypogonadism in Females

Genital Organs

The presenting complaint on account of which girls with primary hypogonadism are brought to the physician is usually either failure of the onset of menstruation or failure of sexual development. The external genitalia remain of infantile character and proportions and the vaginal epithelium fails to undergo the changes normally seen in postpuberal girls as a result of endogenous oestrogenic stimulation. The vaginal smear, therefore, consists almost exclusively of small, rounded basal cells with relatively large vesicular nuclei, and it may contain numerous leucocytes. The breasts, including the nipples and areolae, are either completely undeveloped, or at most show no more than trivial enlargement. The uterus remains of the infantile size, and the passage of a sound demonstrates that the cervical canal is considerably longer than the uterine cavity proper (infantile proportions; in the adult uterus these proportions are reversed).

Hair

There is considerable variation in the amount of pubic and axillary hair. This may be quite absent or represented only by scanty hairs on the labia majora, whereas in other patients the growth is more abundant. Yet a further pattern is for the pudendal hair to be very scanty, while the axillary hair of normal amount. This latter arrangement depends upon the fact that growth of the axillary hair is more largely due to adrenal androgens which may be produced in normal quantity in these patients, whereas the pudendal hair is more oestrogen dependent.

Skeletal changes

These resemble those of eunuchoidism in the male, and typically eunuchoidal proportions (span exceeding height and lower measurement exceeding upper) are commonly found even when there is decreased overall stature as in Turner's syndrome. The fingers and toes may be long and slender, though this is less commonly so in Turner's syndrome. The shape of the pelvis approximates to that of the male. Dentition may be delayed. The radiological bone age is less than the chronological age and there is delay in epiphyseal union. Osseous retardation is less marked in Turner's syndrome than in those patients with primary hypogonadism without decreased stature.

Sexual behaviour

These patients are normally completely lacking in libido and have no attraction towards the opposite sex. They usually remain somewhat infantile in outlook and present an immature mentality. They may also feel a sense of inferiority as a result of the knowledge of their physical shortcomings.

Special features in Turner's syndrome

Several of these have already been mentioned. Shortness of stature and associated congenital anomalies distinguish this condition from primary hypogonadism without decreased stature. Generalized osteoporosis of the skeleton has been reported in some patients with this condition. It may give rise to *scoliosis* and *lordosis*. *Chondrodystrophia* of the dorsal vertebrae has also been described.

The commonest congenital deformities are webbing or apparent shortening of the neck, and cubitus valgus, giving rise to an increased carrying angle of the arm. A characteristic shield-shaped chest has also been described, the thorax being prominent anteriorly and broader than normal with an increased antero-posterior diameter. Other defects which have been reported include per cavus, increased pronation of the feet, syndactylism, spina bifida, congenital deafness, and ocular disturbances, such as bilateral ptosis, slight exophthalmos, internal and external squint, cataract, tubular vision, and lack of retinal pigment. Some patients have shown mental deficiency. The occurrence of coarctation of the aorta has already been the subject of comment. Hypertension with a systolic blood pressure between 130 and 150 and a diastolic pressure between 90 and 112 has been noted in many patients with *Turner's syndrome*, independently of the existence of coarctation of the aorta.

Hormone excretion

Typically, patients with primary *hypoovarianism* have an increased excretion of pituitary gonadotrophins. The extent, however, of this increase is very variable, and in not a few patients the excretion has been found to be within the normal range. It seems clear, moreover, that day-to-day fluctuations of considerable magnitude may be encountered. The excretion of 17-ketosteroids is usually somewhat diminished but figures within the normal range have been encountered. Only a few measurements of oestrogen excretion have been made in these patients; the expected low values have been found, but oestrogen are not entirely absent from the urine. The microgram or two presumably arise from the adrenal cortex and not from the gonads.

Diagnosis

The differentiation of primary *hypogonadism* without decreased stature from secondary hypogonadism due to idiopathic deficiency of pituitary gonadotrophin can be made only on the basis of the urinary gonadotrophin excretion, and this is not always reliable. The two forms of primary hypogonadism are distinguished by the shortness of stature and congenital anomalies found in Turner's syndrome but not in the other variety. The differentiation of Turner's syndrome form pituitary infantilism is summarized in the following comparative table (below), adapted from del Castillo, de la Balze, and Argonz.

Secondary Hypogonadism

Pituitary dwarfism is one cause of *primary amenorrhoea*, the failure of sexual maturation corresponding with that seen in the same condition in the male. Perhaps the commonest cause of primary amenorrhoea is idiopathic deficiency of pituitary gonadotrophin secretion, without evidence of failure of production of the other pituitary trophic hormones. Girls with this condition are usually of normal or slightly increased stature and may be either thin or adipose. In spite of the rudimentary state of the genital organs and absence of mammary development, pubic and axillary hair may be present in relatively normal amounts. In a few of these patients the clitoris is enlarged, perhaps as a result of relative adrenocortical hyperfunction. Secondary hypogonadism may also be the result of cretinism, juvenile myxoedema, and milder forms of hypothyroidism, toxic goitre, and diabetes mellitus. When primary amenorrhoea is the result of adrenocortical hyperfunction, virilizing changes are also found. Secondary hypogonadism may result from severe and long-standing disease in other systems, such as anaemia, chronic nephritis, sepsis, and malnutrition.

Diagnosis

The differentiation of secondary from primary hypogonadism can only be made on the basis or urinary gonadotrophin excretion. In

Rudimentary Ovaries	*Hypophyseal Dwarfism*
Women of short Stature.	Dwarfs.
Infantile mammary glands and genital organs.	The same.
Development of pubic and axillary hair.	Lack of pubic and axillary hair.
Well-nourished and strong.	Weak and easilyu tired.
Bone age some years retarded.	Very marked delay in bone age.
Late closure of teh epiphyses.	Lack of closure of the epiphyses.
Very frequently vertebral chondrodystrophia.	The same.
Follicle-stimulating hormones increased in the urine.	Lack of follicle-stimulating hormones
17-ketosteroids some what diminished.	17-ketosteroids considerably dimished.
Normal insulin curve.	Persistent hypoglycaemia after intravenous insulin.
Congenital abnormalities.	Not observed.
Diffuse osteoporosis and early senility.	Not observed.
Normal sella turcica.	Pathological modifications may be observed.
Visual fields : some functional alterations.	Abnormalities in teh presence of a neoplastic lesion.

secondary hypogonadism the excretory level is too low to be measured by the available clinical tests.

Treatment of Hypogonadism in Females

The treatment of primary *hypo-ovarianism* can clearly only be substitutive, since it is impossible to replace the functionless or missing ovarian endocrine tissue. Theoretically the ideal treatment of secondary hypo-ovarianism would be the administration of the appropriate pituitary gonadotrophins. In practice, however, this cannot be achieved, and in general it may be said that treatment with the gonadotrophin preparations available for clinical use is mainly disappointing. Consequently the treatment of both primary and secondary hypogonadism in the female resolves itself into the administration of oestrogens so as to bring about growth of the oestrogen-sensitive tissues-principally the genitalia and breasts - and to produce cycles of uterine bleeding. The necessity, or even desirability, of achieving either of these ends is sometimes questioned, but there can be little doubt that most patients afflicted with hypogonadism are very grateful for the changes which can be brought about by oestrogen therapy, and experience great satisfaction in having regular bleeding even though they understand these are essentially artificial and do not indicate their normality from the reproductive standpoint.

Under physiological conditions oestrogens would appear always to be secreted in cyclically fluctuating fashion. It is therefore reasonable to assume that oestrogen substitution therapy ought also to be cyclic. Once a responsive endometrium has been built up, it is imperative that oestrogen administration be discontinuous, since otherwise endometrial hyperplasia with resultant prolonged and excessive irregular bleeding will be the consequence. Some authors consider that continuous oestrogen administration may be employed in the initial stages of treatment so as to produce quicker results. It is doubtful, however, whether there is any advantage or even justification for this.

The simplest form of oestrogen therapy is oral; stilboestrol 2 mg., or ethinyloestradiol 01 mg., daily, is a reasonable average dose and may be given for courses of twenty days. Ten days should elapse between each successive course, unless an oestrogen-withdrawal bleeding occurs in the meantime. When this happens, the succeeding course may begin on the fifth day of the 'cycle' so established, counting the day of commencement of uterine bleeding as day number one of the new cycle. Once begun, these courses of treatment must be continued indefinitely or at least until the age at which a climacteric might have been expected has been attained. This certainly is true for patients with primary hypogonadism, and is probably true for most of those with secondary hypogonadism also. For some of the latter, however—that is, those patients wherein the defect appears to be an idiopathic deficiency of pituitary gonadotrophin secretion—the hope may be entertained that eventually normal menstrual function might occur, just as in some male eunuchoids of the same type normal testicular function appears to be able to continue after initial treatment. In these cases, therefore, it is reasonable to stop treatment after several months in order to see whether spontaneous *menstruation* might occur thereafter. On theoretical grounds there is something to be said for combining progesterone of ethisterone with the second half of the oestrogen course in the hope that these hormones will together influence the anterior pituitary in such a way as to evoke gonadotrophin secretion. A method which has sometimes proved successful is as follows:

After several cycles of treatment with oestrogen alone, a daily dose of 40 mg. of ethisterone is given for the last ten days of the twenty-day course of oestrogen (stilboestrol 2 mg., or ethinyloestradiol 0.1 mg., daily). On the fifth day of the next cycle another twenty-day oestrogen course is started, this time at half the previous dose, and on the fifteenth day of that cycle it is combined with 60 mg. daily of ethisterone for ten days. On the fifth day of the next cycle a final oestrogen course is started, the daily dose again being halved (stilboestrol 0.5 mg., or ethinyloestradiol 0.025 mg.), and on the fifteenth day of that cycle a daily dose of 80 mg. of ethisterone is started for ten days. All treatment is then stopped and the patient is observed to see whether spontaneous menstruation will occur.

No useful purpose is served by giving progesterone or ethisterone to patients with primary hypogonadism.

Some girls are unable to tolerate oral oestrogens, complaining of nausea and vomiting; ,for these, and for these alone, oestrogens must be given by intramuscular injection. Oestradol benzoate or oestradiol dipropionate, 5 mg. twice weekly for three weeks out of every four, is a convenient regime; a further possible alternative is oestradiol monobenzoate in microcrystalline suspension, in a dose of 10 mg. every four weeks. It is claimed that oestrodiol valerianate provides prolonged oestrogenic stimulation when administered intramuscularly, in oily solution, 10-20 mg. every three or four weeks being an appropriate dose. Being in solution, this long-acting ester preparation does not suffer from the disadvantages of microcrystalline suspensions.

In addition to.stimulating growth of the genitalia and breasts, oestrogen therapy for hypogonadal females may be expected to have important psychological effects. The patient loses her childlike mannerisms and psyche, becoming more adult in mentality and outlook. Whether this is a direct effect of oestrogens on the psyche, or whether it be a psychological response to the morphological maturation, is unknown. Oestrogens appear to stimulate growth in some patients with primary hypogonadism and decreased stature. This is expecially true of those cases where growth was relatively normal until what should have been the age of puberty, but then ceased. It is less likely to be true for those cases where growth was always deficient. As part of the growth-promoting picture, however, oestrogen therapy leads to epiphyseal closure, whereupon, of course, growth ceases. The maximum growth increment is therefore strictly limited and unlikely to exceed 3 or 4 inches. Even this however, is much welcomed by these patients.

CLIMACTERIC

The *climacteric* or '*critical period*' or '*change of life*' covers the phase of waning ovarian function which is a consequence of the continual loss or degeneration of potential ova and the absence of any provision for their replacement. When no more primordial follicles remain in the ovaries reproductive ability obviously comes to an end, but, on the other hand, the endocrine activity of the ovary undergoes a more gradual phase of diminution. This phase begins before the menstrual periods cease and continues for some time afterwards. The *menopause*, or cessation of menstruation, is merely a single incident of this climateric period.

Physiology of the Climateric

In contrast to spermatogenesis, which is essentially a function of sexually mature males, oogenesis—the actual production of oocytes—is probably purely a foetal activity. Though the more or less classical view as put forward by Swezy and Evans was that the production of new oocytes from the germinal epithelium continued throughout life, the idea was opposed by Simpkins and the accurate observations of Zuckerman and his colleagues have failed to substantiate it. At birth the two ovaries contain some 400,000 (more or less) primordial oocytes, but of these, only a small proportion, perhaps 400, are destined to take part in ovulation. The remainder 'disappear through the process of atresia. Loss of bocytes by atresia begins at least as early as birth, and is indeed most active before puberty, for the ovary of the new-born has many thousands more primary follicles than that of the adolescent girl. During the proliferative phase of each menstrual cycle several follicles commence to grow but only one reaches the stage of ovulation. The remainder, outstripped by the 'chosen' follicle, regress and become atretic. In this way, during each cycle some 30 or 40 follicles are lost by atresia for each one by ovulation. Since the ovarian hormones are secreted by the follicles or their derivatives, it is clear that when few or no follicles remain, ovarian hormone production must fall to a low level or cease altogether.

The age at which the climacteric commences varies in different women in much the same way as does that at which the menarche occurs; it is in fact more or less normally distributed. Various factors-racial, hereditary (other than racial) general health, sociological and so on—no doubt do determine it, and diseases of various kinds directly affecting ovarian physiology can accelerate it.

It has often been supposed that there is a relationship between the age of the menarche and that the earlier the occurrence of the former, the later is that of the latter. This is probably not true: in an investigation of the menopause of 1,000 women it was found that the average age at the menopause for women whose menarche occurred at 13 years was 47.3 years; while that for women whose menarche occurred at 18 years was 47.5 years. Occasionally, the menopause occurs at a very early age—even before 20 years; in these cases there is nearly always an underlying endocrine abnormality. There have also been reports of the continuation of menstruation until very advanced years-up to the age of 104 in one instance. In general, however, prolongation of the menopause after the age of 55 calls for gynaecological examination to exclude the possibility of genital malignancy. There is good reason to suppose that many of the patients with delayed menopause reported in ancient literature had oestrogen-producing tumours of the ovary (such as granulosacell tumours). It is, of course, equally important, or even more so, to make a gynaecological examination if, after the menopause has definitely occurred, genital bleeding should reappear. Although this is frequently of benign cause (particularly if injudicious oestrogen treatment for menopausal symptoms has been given) all too commonly malignant disease of the cervix or body of the uterus if found to be responsible.

When the number of ovarian follicles has become significantly reduced, ovulation ceases to occur in every cycle although more or less regular cyclic activity may continue for some time. Later, ovulation ceases altogether and by this time some irregularity of the cycles has usually become apparent. Anovular cycles may continue for some time-interspersed with an occasional ovular cycle perhaps—the bleeding becoming more infrequent and scantier, eventually to cease altogether. In other women, the *menopause* may take the form of an abrupt cessation of previously regular periods. Yet a further variant is that in which the alterations in ovarian hormone production lead to the development of irregular, prolonged and often heavy ***bleeding—climacteric menorrhagia***. It is sometimes considered that in these circumstances a phase of increased oestrogen production precedes the termination of ovarian endocrine activity. It is doubtful, however, if this is true and more probable that the menorrhagia is the result of continuous, as opposed to discontinuous, oestrogen production, albeit on a decreasing scale.

The gradual lessening of ovarian endocrine activity during the climacteric leads to many secondary changes. Foremost among these are regression of the genital organs-uterus, vagina, vulva, and breasts. But in many women the changes are almost imperceptibly slow and long after the menopause, little evidence of genital regression may be seen. In particular, the vaginal smear in post-menopausal women often reveals little evidence of significant oestrogenic deficiency. However, atrophy of the vaginal epithelium may be the cause of post-menopausal bleeding, or pruritus. The tendency to gain weight, which is commonly seen at this stage of life, leading to so-called 'middle-age spread' has been thought to result from failing ovarian function, but this is not likely. As seen in the previous chapter, bilateral ovariectomy by no means invariably leads to the development of adiposity. A more probable explanation is that, as part of the general ageing process, less energy is expended; however, because the appetite remains unchanged, a previous balance between energy intake (in the food) and that expended in the day's activities becomes converted into an excess of energy intake, thereby leading to deposition of fat. It is said that this fat has a peculiar distribution, accumulating mainly around the abdomen, hips, and thighs. The possibility must not be overlooked that the distribution is the same as would have occurred, given a comparable weight gain, in earlier years. A suggested explanation for the alleged special distribution (assuming that it is in fact special to this time of life) is that the normal ageing process is accompanied by a relatively greater loss of adipose tissue cells in other parts of the body and that, when new fat is laid down, most appears where the adipose cells remain in greater abundance.

Further secondary effects of ovarian failure are changes in the activity of the endocrine glands. These are most important for a complete understanding of the physiology of the climacteric. Because of the progressive failure of ovarian response to pituitary gonadotrophic stimulation, the modifying effect of the ovarian hormones on the activity of the pituitary gland becomes less and less. With the reduced inhibitory effect of oestrogen, the production of pituitary gonadotrophin (mostly of follicle-stimulating type) increases, so that in post-menopausal women a great excess can usually be detected in the urine. Sometimes this gives rise to false pregnancy-diagnosis tests, thus confusing an already delicate issue if the woman fears (or hopes) that the delayed period caused by the climacteric is due to pregnancy. Along with this increase in pituitary gonadotrophin production, there is probably also an increase in the output of thyrotrophic and adrenocorticotrophic hormones. These find responsive target organs, and so hyperactivity of the thyroid and adrenal cortex may ensue. It is probably the combination of falling oestrogen level and increased output of thyroid and adrenal hormones which is responsible for most of the untoward effects which may be experienced by women at this stage. An alternative view has supposed that the pituitary overactivity itself is responsible for the climacteric symptoms; but since these are often suppressed by very small doses of oestrogen which have no detectable effect on the pituitary hyperactivity, it seems far more probable that the falling oestrogen level is the more important factor.

The increased activity of the adrenal cortex tends to restore the endocrine balance by taking over some of the functions of the ovary. Thus, there is little doubt that most of the oestrogen which circulates in post-menopausal women, when the ovaries have become quite functionless, is secreted by the adrenal cortex. This oestrogen, and the other adrenal cortical hormones produced in increased amounts, serve to depress the excessive pituitary activity and so reduce the thyroid overactivity. With still further advance in age, it is likely that some degree of refractoriness occurs in both the adrenal and the thyroid gland, so that in these older women the clinical picture may suggest deficiency, rather than excess, of function of both of these glands.

Clinical Features

Certain special features relating to the altered endocrine function of the climacteric will now be considered.

Virilism

Because of the increased adrenal output of androgens, the male characteristics, represented to a slight degree in all females, may become accentuated. Thus, the down on the upper lip and chin tends to become thicker and sometimes the hair is sufficiently coarse and luxuriant to form a moustache and beard, causing great mental anguish. Hair may also increase at the sides of the face, and the pubic hair may extend upwards along the linea nigra towards the umbilicus. It is rare, however, that hirsutism assumes the proportions and distribution of pathological virilism. Nevertheless, in one young woman with a pre-existing tendency to virilism, bilateral ovariectomy, for a gynaecological condition, produced severe hirsutism. The pubic hair tends to become uncurled at the climateric and the development of facial hirsutism may be accompanied by loss of scalp hair. The voice may become deeper and more powerful, so that singers sometimes find themselves able to reach notes lower than any previously possible, the upper notes, on the other hand, becoming more difficult to obtain.

A change in mental outlook with an approximation to 'male' characteristics may take place, greater resolution, command, initiative, originality, and administrative capacity being shown with the attainment perhaps of considerable commercial or public success in later life. These changes are more likely to develop in the later part of the climacteric, after the menopause itself has occurred. As further evidence of increased adrenocortical activity, it may be noted that patients with Addison's disease may show considerable amelioration, or even cure, at the climacteric, although this is preceded by an earlier phase in which the disease appears to be aggravated. The occurrence of pigmentation in some women at the menopause may well be directly due to the increased pituitary production of melanocyte stimulating hormone, along with adrenocorticotrophic hormone.

Hypertension

This is a common accompaniment of the climacteric and may be of a labile type, disappearing spontaneously after a year or two. In such circumstances it is presumably the direct consequence of vasomotor lability. When, however, it progresses to a more severe and permanent type, with secondary changes in the vessels, it seems doubtful if the climacteric itself can be held responsible and more probable that the hypertension is a consequence of general ageing processes. This conclusion follows from the fact, already pointed out in the preceding chapter, that ovariectomy or radium castration does not necessarily lead to hypertension.

Impaired carbohydrate tolerance

Although only a small percentage of women develop clinical *diabetes mellitus* at the climacteric, it can be shown by investigation of the carbohydrate tolerance that an impairment to some degree occurs in a considerable proportion. It is likely that this is due to pituitary overactivity, since suppression by sufficient doses of oestrogen, so as to cause disappearance of urinary gonadotrophin, may lead to a normal carbohydrate tolerance curve, while cessation of oestrogen treatment is followed by a return to the original condition. Presumably, both pituitary growth hormone and the increased production of adrenal glucocorticoids must be held responsible for the impaired carbohydrate tolerance shown by many climacteric women.

Thyroid changes

The excessive production of thyrotrophic hormone by the pituitary gland may initiate *exophthalmic goitre* at the climacteric, or lead to an exacerbation of pre-existing mild or latent hyperthyroidism. Sometimes a symptomless and long-standing goitre or small adenoma is driven into activity, with the development of symptoms of toxic goitre (*secondary thyrotoxicosis*). Climacteric hyperthyroidism may be progressive or there may be a gradual return to normal in mild cases. Sometimes this involution is excessive and goes on to myxoedema, but the latter, however, may appear without an obvious preceding hyperthyroid phase.

Breasts

At the climacteric, the mammary glands tend to undergo atrophy but this may not be evident owing to the deposition of fat which indeed, can produce an actual enlargement. As a result of loss of cyclic oestrogen production during the climacteric, there may be changes in the ducts. These may undergo lengthening and tortuosity, associated occasionally with epithelial proliferation; and various symptoms, such as tenseness, paraesthesiae, hyperaesthesia (sometimes erotic.) and pain may develop. Occasionally, the breast secrete a thin fluid, probably as the result of the superimposed activity of the pituitary lactogenic hormone (*prolactin*), which in turn is a consequence of the general hyperactivity of the pituitary; as also may be the prolonged lactation which is sometimes observed in women who become pregnant towards the end, of the reproductive period.

Acromegaly

Mild, or fugitive *acromegaly* sometimes occurs at the climacteric and can be explained by an increased output of pituitary growth hormone.

Hypopituitarism

Occasionally, following the climacteric, and particularly when there is a history of multiple pregnancies, an indefinite syndrome, with some features of *Simmonds' disease*, may be encountered. Presumably, in these patients, the pituitary gland enters upon a phase of exhaustion, following the phase of hyperactivity, in much the same way as hyperthyroidism may be replaced spontaneously by hypothyroidism.

Climacteric Syndrome

The precise frequency with which climacteric symptoms of more than minor degree are experienced is uncertain. Hamblen quotes one authority who estimated 75 per cent. of all women as suffering from distressing symptoms at the climacteric, and other authorities who held that 70 to 90 per cent of climacteric women experience no symptoms materially interfering with general health, or with domestic or social activities.

The symptoms of the climacteric syndrome consist mainly of psychological disturbances and vasomotor instability. There is no doubt that the former are the more important, for it is just those women who have, or through force of circumstances develop psychological inadequacy, who are most liable to suffer from climacteric symptoms and who do so most severely. Two classes of women provide the majority of climacteric sufferers. At one end of the social scale is the woman of wealth and social standing, for whom the menopause is a remainder which cannot be ignored of advancing years with their attendant loss of good looks, sex appeal, and consequent dominance in her social circle. At the other end if the woman who has led a frustrated existence, deprived of good looks and the activities open to the more fortunate, who has been unwanted and unloved; for her the expectation and realization of the approaching end of reproductive function mean the abandonment of all hope of fulfilling her natural childbearing destiny. It is not surprising, therefore, that she should show evidence of despair or, instead, a protest reaction. The least susceptible women are those who have had a healthy, happy life, who have made a successful marriage and have raised a healthy family. For them the menopause is merely another milestone in life passed, with every reason to expect the next phase to be no less worth living than those which had gone before.

The psychic symptoms are largely conditioned by the circumstances which have evoked them. In all instances, however, irritability and depression and predominant. The former makes the woman short-tempered and intolerant; any slight deviation from the expected course of affairs provokes an exaggerated protest. There may be associated anxiety with vague forebodings of ill health or economic disaster in the future. Depression may be associated with tearfulness or with sadness, indifference, and apathy. The woman may lose all ambition, ceasing to care for her husband, her home, and her appearance. As part of the climacteric syndrome, changes in libido may occur. There may be an increase as a compensatory reaction to waning reproductive ability, or as a 'last fling' effect; or it may disappear abruptly as in the case, for example, of an unhappily married woman for whom the menopause can serve as an excuse for avoiding distasteful sexual activity. In the well-balanced woman the climacteric need cause no change in libido and many women continue to have normal and satisfying sexual relations long after menstruation has ceased.

The commonest manifestations of vasomotor instability in the climacteric are hot flushes (*vasodilatation*) often starting in the face and travelling all over the body. They occur spontaneously, or may be induced by emotion. They may be infrequent, or may be repeated many times during the day and, typically, even more frequently at night. The flushes are often followed by a wave of chilliness (vasoconstriction) and profuse perspiration. The blood pressure may fall during the flushes and rise with the chilliness. Paraesthesiae, cold extremities, tremors, palpirations, colonic spasm, cardiospasm, angioneurotic oedema, and pseudoangina are other features of vasomotor instability.

Migraine is sometimes very troublesome at the climacteric. It appears to be influenced by endocrine factors, as witness its association with puberty, menstruation, and the climacteric, together with a tendency to disappear during pregnancy, lactation, and postclimacteric life. Though often aggravated in the earlier phase of the climacteric many women lose their migrainous symptoms after this period. Its exacerbation at the climacteric might be explained by vasomotor spasm of the cerebral vessels, since cervico-thoracic sympathectomy relieves the condition. An alternative explanation is that it is due to enlargement of the pituitary gland, since migraine is a feature of pituitary syndromes, with or without neoplasm. It has been thought that migraine occurs more commonly in those women whose suprasellar diaphragm is calcified, so preventing pituitary expansion. The pituitary theory is supported by the beneficial effects of doses of oestrogen which are sufficient to suppress pituitary activity to the extent of causing the disappearance of urinary gonadotrophin. Since benefit may also result from injections of gonadotrophin, which would itself depress pituitary hyperfunction, it is clear that the migraine cannot be due to the excessive production of that hormone. About half the cases of idiopathic migraine obtain relief from an artificial menopause but in the remainder the condition is made worse.

General or local pruritus may be very troublesome and various forms of dermatitis and impetigo occur. Pruritus, leucoplakia, kraurosis, and even superimposed carcinoma may affect the vulvar skin and cause much suffering. These abnormalities are aggravated by glycosuria, but also occur in its absence. It is uncertain to what degree the endocrine changes of the climacteric themselves should be held responsible for those various dermatoses, and to what extent they should be attributed to general ageing processes.

Treatment

The first point to be stressed is that the climacteric itself requires no treatment, being a physiological condition. It is only when symptoms supervene, of a degree sufficient to interfere with general health, or with domestic or social activities, that treatment is required. The second point to be stressed is that the successful treatment of the climacteric syndrome demands common sense and delicacy. The routine administration of oestrogen, often for long periods and in excessively high dosage, is strongly to be condemned; just as is the unhekpful, nihilistic approach which taking its stande on teh physiological nature of the climacteric, refuses any form of treatment. It is perfectly true that mild cases can be treated effectively merely by reassurance and perhaps moderate sedation, such as is provided by phenobarbitone, ½ gr. twice a day; but it is equally certain that the more severe cases will require oestrogen therapy and, is occasional instances, psychotherapy as well.

The basic principles of intelligent oestrogen therapy for the climacteric are as follows: first oestrogen is never produced continuously under physiological conditions, so that oestrogen treatment should always be in interrupted courses; second, the object of the treatment is to convert an abrupt fall in oestrogen level into a more slowly declining one. It follows, therefore, that if comparatively large doses are given, so that the total oestrogen level is restored to the previously normal value, symptoms will be certain to reappear every time treatment is stopped. Ignorance of this situation is responsible for most of the so-called 'difficult cases' who have been on oestrogen therapy for long periods at a time, but who relapse miserably each time the treatment is stopped. The view is held by some that large doses of oestrogen are required in order to suppress the pituitary hyperactivity-doses which, in fact, are sufficient to bring about the disappearance of *urinary gonadotrophin*. It is, however, a matter of clinical experience that relief of climacteric symptoms may be obtained with oestrogen doses which are a small fraction of those necessary to produce tangible pituitary depression; and since the drawbacks to the use of large oestrogen doses are serious and numerous, their exhibition in the climacteric seems to be indefensible. Chief among these drawbacks are: nausea; gastrointestinal disturbance; backache and pelvic congestion; uterine bleeding, which may be prolonged, excessive, and extremely difficult to control, except by surgery; mastopathy; and, possibly, carcinogenesis.

In general it is convenient to start treatment with stilboestrol, 0.2 mg. daily. The treatment should be continued for twenty-eight days and then should be stopped for a fortnight, when it can be resumed at the same dose level. With this dose, most, if not all, of the symptoms will be relieved. Indeed, it is best if an occasional hot flush still occurs while the patient is under oestrogen treatment. Generally some return of symptoms will be experienced during the fortnight off treatment. However, as these monthly courses of treatment proceed, the stage will be reached when symptoms no longer return on omitting treatment. This is the time to reduce the dose of stilboestrol to 0.1 mg daily. Two or three further interrupted courses of treatment at this level will often be sufficient to see the end of all significant symptoms. Occasionally, 0.2 mg. will prove inadequate for controlling the symptoms; if this be so, the dose may be increased to 0.5 mg. daily but it is seldom necessary to exceed this figure.

Ethinyloestradiol may be used instead of stilboestrol; the corresponding dose is about one-twentieth that of stilboestrol so that 0.2 mg. of the latter would be represented by 0.01 mg. of ethinyloestradiol.

It has been claimed that natural *oestrogens* (oestradiol, oestrone sulphate, oestriol) are superior to synthetic oestrogens in controlling climacteric symptoms since they 'promote a sense of well-being' which the synthetic oestrogens do not. There is no truth whatever in this contention.

In recent years, a vogue has developed for the use of mixed hormone preparations, usually containing ethinyloestradiol and methyl-testosterone. The rationale for such treatment is a little dubious but the general idea is that the two hormones are synergistic in many of their desirable characteristics, while being mutually antagonistic as far as some of the undesirable side effects are concerned. By the use of such a mixture, one is often able to control climacteric symptoms with far smaller oestrogen doses than would be possible without the admixture of the androgen. Generally speaking, there is little need to use these mixed preparations in the average case. It does appear, however, that for the more difficult patient who is responding poorly to oestrogen alone, the mixture may be of real value. The general principles mentioned above apply equally to the use of mixed hormone preparations, the starting dose for which should be about two tablets per day. Some patients prove unduly sensitive to the androgen moiety of these mixed preparations, developing troublesome hirsuties.

Recently, a synthetic oestrogen of the allenolic acid series called methallenoestril (*Vallestril*), for which certain special properties have been claimed, has been advocated for the control of climacteric symptoms. It is claimed for this oestrogen that, whereas in a dose of 3-9 mg. per day climacteric symptoms may be fully controlled and an atrophic vaginal mucosa restored to normal, remarkably little effect is produced on the endometrium. The consequent advantage is that the risk of inducing uterine bleeding is minimized. Although this is of no importance in the majority of cases, if treated with stilboestrol in the manner described above, it is on the other hand true that occasionally women are found to have withdrawal bleedings following treatment even with as little as 0.2 mg. of stilboestrol daily. For these patients, *Vallestril* may be the drug of choice. Another recently introduced oestrogenic substance, chlorotrianisene (tri-p-anisylchlo-roethylene, marketed under the name TACE) has interesting properties by reason of which, it is claimed, it has particular advantages in the treatment of the climacteric symdrome. It is stored in the body fat, from which there is slow prolonged release. *TACE* itself is a pro-estrogen, without direct oestrogenic activity, but is converted in the liver into a true oestrogen, the nature of which is unknown. The recommended course of treatment is two capsules (each of 12 mg. of chlorotrianisene dissolved in oil) by mouth daily for thirty or sixty days. This is stated to ensure complete relief in 50 percent of patients. A second course can be given; not more than 1 per cent. of patients are said to require a third course.

Some physicians feel the need for elaborate laboratory investigations, such as repeated vaginal smear studies or even urinary gonadotrophin determinations, as indices to progress in the treatment of the climacteric syndrome: Common sense alone shows that these studies are totally unnecessary, since the treatment is purely a symptomatic one, the climacteric itself being a physiological state, as already stressed. The clinical state of the patient is the only guide needed in the treatment of this syndrome. The uselessness of vaginal smear studies is underlined first by the fact that many post-menopausal women, even though suffering from climacteric symptoms, may still show a relatively well-oestrogenized smear and second vaginal that the production of a fully oestrogenized smear by oestrogen therapy is suggestive not of correct treatment but of over dosage.

Where marked psychological aberrations are present, it is obviously necessary to adopt psychiatric measures in addition to such hormonal therapy as may be required. Since the syndrome is a self-limited one, the prognosis is usually fairly good and it is not often that elaborate psychotherapy is needed.

Cimacteric Menorrhagia

Except in mild and transient cases, it is a wise rule to subject every patient with *cimacteric menorrhagia* to a diagnostic dilatation and curettage. This procedure appears to be curative in a fair proportion of patients; estimates vary between 30 and 60 per cent. If the curettage fails to reveal any endometrial pathology, but severe haemorrhage nevertheless persists, a complete menopause, with cessation of bleeding, can be produced by external radiation, insertion of radium into the cervix, or by hysterectomy. Both external radiation and radium are relatively simple measures but may precipitate cimacteric symptoms and adiposity in some patients. A carcinoma of the body of the uterus may also be overlooked. Hysterectomy does not involve the ovaries and does not therefore disturb the endocrine system. It is, however, a major surgical procedure not entirely free from risk, or from anxiety on the part of the patient, but it is probably the method of choice in severe menorrhagia not responding to hormone therapy.

After a diagnostic curettage it is reasonable to try medical treatment before deciding on surgery or radiation. Androgens, progesterone, and oestrogens are used for their local action on the uterine endometrium, as well as for their inhibition of pituitary gonadotrophic activity. Nevertheless, treatment remains empirical. Ethisterone, 15 mg., together with methyltestosterone, 5 mg., as two tablets or in a combined tablet (*Androgeston*) may be given twice daily sublingually, commencing a week before the period is expected and continuing until the third day of bleeding. This treatment will often restore the heavy period to more normal proportions. When the bleeding is prolonged and irregular, the choice lies mainly between treatment with progesterone to produce 'medical curettages' or with oestrogens.

A daily intramuscular injection of progesterone, 25 mg., and testosterone, 50 mg., for four days will usually terminate a prolonged bout of bleeding. A few days later a self-limited progesterone withdrawal bleeding will occur. Repetition of the progestrone injections (without the testosterone) at approximately monthly intervals will usually produce regular and normal 'periods'. Oestrogen therapy depends upon the fact that a sufficient dose of oestrogen will usually terminate a bout of bleeding. It may be given conveniently as stilboestrol, 2 mg., or ethinyloestradiol, 0.1 mg., daily, continued for a total of twenty days. If there has been no noticeable effect on the extent of the bleeding within forty-eight hours, the dose should be doubled and continued at this higher level for the remainder of the twenty days. Within a week or ten days after stopping the oestrogen treatment, a withdrawal bleeding will occur and on the fifth day another course of oestrogen therapy should be begun. In this way the regular bleeding is converted into regular cycles and the dose can gradually be reduced in succeeding months. Eventually a stage will be reached when the dose will be too small to provoke oestrogen withdrawal bleeding and the menopause will have been established. If during the course of oestrogen administration the bleeding stops but begins again before the course is completed, further oestrogen should be withheld for five days and then a new course commenced as before.

Fears have been entertained that administration of oestrogens to women at the menopause may lead to carcinogenesis. There is no acceptable evidence to substantiate such a fear, and in any case, provided the oestrogen is given discontinuously and in minimal effective doses as recommended above, the likelihood, even on theoretical grounds, of its exerting a carcinogenic action seems to be remote.

Failure to respond to hormonal treatment is not uncommon in cases of cimacteric menorrhagia; it is obviously unwise to persist in such treatment if it is proving ineffective, and recourse should then be had to surgical measures without undue delay.

Abnormalities of Menstruation

In a textbook of major endocrine disorders, only limited aspects of abnormalities of menstruation call for attention. The gynaecological implications have always to be borne in mind, and the information on these should be sought in textbooks of gynaecology. Moreover, abnormal uterine bleeding or amenorrhoea may be part of obvious clinical disorders of the endocrine glands (e.g. adrenogenital syndrome, hyperthyroidism, myxoedema) and as such are considered elsewhere in the present work. This chapter will deal only with certain conditions in which the disorder of menstruation appears to be the only evidence of endocrine dysfunction.

AMENORRHOEA

Absence of menstruation for long intervals of months or years is termed *amenorrhoea*. It is said to be primary when the patient has never menstruated, or secondary if it supervenes after some years of more or less normal menstruation. Physiological amenorrhoea occurs during pregnancy.

Primary Amenorrhoea

Apart from purely gynaecological causes, such as imperforate hymen or uterine aplasia, *primary amenorrhoea* is of two main types: the first, due to primary hypogonadism, consists in failure of response of the ovaries to adequate gonadotrophic stimulation and has already been discussed; in the second there is secondary hypogonadism, that is, a lack of gonadal function due to absence of gonadotrophic stimulation. This may be the result of a hypothalamic lesion or ill-understood functional disturbance. It may result from a lesion of the anterior pituitary gland, such as *craniopharyngioma* or *chromophobe adenoma*. It is also a characteristic feature of pituitary infantilism, or may be the result of an apparent selective deficiency of gonadotrophin secretion. In hyperpituitarism due to an acidophil tumour resulting in giantism, primary amenorrhoea is caused by destruction of the basophil cells which are considered to be the source of the pituitary gonadotrophins.

Other endocrine conditions in which primary amenorrhoea may occur are cretinism, juvenile myxoedema, and milder forms of hypothyroidism, toxic goitre, diabetes mellitus, and adrenocortical tumours. It may also result from severe organic disease of other systems, such as anaemias, chronic nephritis, chronic sepsis, and malnutrition.

It may be very difficult clinically to differentiate between delayed puberty and primary amenorrhoea. Consequently the continuation of normal menstruation after the exhibition of therapy for primary amenorrhoea cannot necessarily be regarded as *prima facie* evidence of cure of the primary amenorrhoea, since one may have been dealing merely with delayed puberty. Reference has already been made to male pseudohermaphroditism as a possible cause of apparent primary amenorrhoea.

Secondary Amenorrhoea

Without doubt, the commonest cause of cessation of the periods more or less normal menstruation has been established, apart from the occurrence of pregnancy, is psychological. Sometimes such a cause is of major proportions and obvious in its implications, but in many instances the precise psychological factors may be obscure and difficult to elicit.

As in the case of primary amenorrhoea, *secondary amenorrhoea* may be caused by severe general diseases, such as anaemias, advanced tuberculosis, chronic nephritis, malignant disease, chronic infections, and malnutrition. It is uncertain by what mechanism these conditions bring about the cessation of normal pituitary-ovarian activity.

Secondary amenorrhoea also arises in a large variety of endocrine disorders. These include hypothalamic disease affecting anterior pituitary function; *Simmonds' disease* and other forms of hypopituitarism; ecromegaly, as a result of destruction of basophils by the eosinophil adenoma and Cushing's syndrome. Amenorrhoea may occur in both hyper- and hypothyroidism, in diabetes mellitus, and in Addison's disease. It is present in the adrenogenital syndrome and is also a symptom of virilism due to other causes, such as arrhenobl-astoma and adrenal rest tumours of the ovary. It arises when destructive lesions, of whatever kind, of both ovaries destroy, a sufficient amount of oestrogen-producing tissue.

The normal cimacteric is a natural form of secondary amenorrhoea, resulting from the loss by ovulation and atresia of Graafian follicles. If this happens at an unusually early age, and is not accompanied by typical cimacteric symptoms, the secondary amenorrhoea which results may be clinically indistinguishable from that which may occur from other causes; in such cases the finding of a raised gonadotrophin excretion would indicate the climacteric nature of the condition. Other possible causes of amenorrhoea are a functional failure of the pituitary gland to secrete adequate amounts of gonadotrophins, or a loss of the ability of the ovaries to respond to stimulation by gonadotrophins (for reasons other than those which apply at the climacteric), or of the endometrium to respond to the action of ovarian hormones.

Treatment of Amenorrhoea

The treatment of *amenorrhoea* is largely unsatisfactory and a long digression on the methods which have been employed is not warranted. Suffice it to say that the use of gonadotrophin preparations and of steroid hormones, though occasionally apparently successful, is essentially unreliable and good results cannot be predicted. In primary amenorrhoea the principal indication for long-continued cyclic oestrogen therapy is the effect that this has on the development of secondary sex characters and the psychological advantages to the patient of having apparently normal menstrual bleedings. In secondary amenorrhoea the value is again mainly psychological, except in those instances where, on cessation of a fairly prolonged course of such interrupted therapy, normal menstruation is re-established. For many intelligent women for whom sterility is not a complaint, reassurance that secondary amenorrhoea is not due to underlying disease is often sufficient, and for these hormone therapy is apparently quite unnecessary.

ABNORMAL UTERINE BLEEDING

Regular menstruation which is nevertheless excessive in the amount of blood loss, and often at the same time in the duration of the bleeding, is usually referred to as *menorrhagia* or *hypermenorrhoea*. Irregular uterine bleeding is usually called metrorrhagia, and when it is associated with hyperplastic changes in the endometrium, the condition is described as metropathia haemorrhagica. The too frequent occurrence of menstrual periods is called polymenorrhoea and their too infrequent occurrence, oligomenorrhoea. Unusually scanty menstrual periods constitute hypomenorrhoea. In the elucidation of the causes of abnormal menstrual bleeding, proper gynaecological examination, including, in most cases, diagnostic curettage, is an essential preliminary.

Menorrhagia

In the absence of organic cause, such as fibroids, or pelvic inflammatory disease, this condition most commonly has a psychological basis and would seem to consist essentially in an abnormality of neurovascular control in the endometrium. Biopsy of the endometrium usually reveals an entirely normal secretory pattern in the immediately premenstrual phase, and may show no abnormalities if taken during bleeding itself. In such circumstances it is difficult to imagine any possible endocrine basis for the disturbance and it is not surprising that hormone treatment often has little success.

In some cases, however, an endometrial biopsy taken about the fifth day of bleeding shows the pattern of incomplete shedding of the endometrium, in which mixed proliferative and secretory changes are found alongside each other. It is possible that in these cases there is a delay in the involution of the corpus luteum, with a corresponding prolongation of the secretion of progesterone, though in suboptimal amounts; the continued elimination of pregnanediol during part of the bleeding episode (it usually disappears from the urine, except for traces, before the onset of bleeding) supports this supposition. Unfortunately no treatment based on such an hypothesis has succeeded in dealing effectively with the condition.

Polymenorrhoea

This is often combined with menorrhagia, and when not due to organic cause is again most commonly of emotional origin. It may also arise as a result of early ovulation, and in some cases ovulation may occur even before the menstruation has ceased, thus leading to involuntary sterility. Occasionally it may arise as a result of premature degeneration of the corpus luteum with a corresponding reduction in the post-ovular phase of the cycle. The fundamental disturbance may then be one of anterior pituitary function.

Metrorrhagia

Irregular uterine bleeding, which may also be prolonged and heavy, can result from organic pelvic lesions, but commonly occurs in the absence of any such obvious causes. Diagnostic curettage may reveal a normal proliferative endometrium, a hypoplastic one or a hyperplastic one, showing typical cystic glandular hyperplasia (the '*Swiss cheese*' hyperplasia of Novak). Occasionally mixed hyperplastic and secretory patterns have been described.

This abnormality of menstruation may arise at any age between the menarche and menopause being relatively rather common close to both of these epochs. The essential feature appears to be absence of ovulation, with a consequent lack of cyclic production of oestrogen and progesterone by the ovary. There is no certainty that the typical picture of metropathia haemorrhagica, in which cystic ovaries are found, in association with cystic glandular hyperplasia, is due to an excessive oestrogen production, and the precise hormonal derangements are not known. Some women appear to have an essentially unstable menstrual rhythm, so that they may alternate between phases of regular menstrual function. and of irregular metrorrhagia. A wholly endocrine cause of metrorrhagia is the occurrence of an oestrogen-secreting tumour of the ovary, such as the granulosa-cell tumour. These tumours occur most frequently in the climacteric phase and later.

Treatment of Abnormal Uterine Bleeding

The diagnostic curettage itself provides effective treatment in. a proportion of patients with abnormal uterine bleeding. This proportion, however, seems to vary with different observers. Menorrhagia of emotional origin is unlikely to respond to any kind of hormonal treatment and logically should be dealt with on psychiatric lines. Unfortunately this approach too is often fruitless, and in severe cases hysterectomy has seriously to be considered. In other cases of regular menorrhagia not due to organic disease, the results of any kind of treatment are unpredictable. However, in the writer's experience, a regime worth trying consists of methyltestosterone, 5 mg., and ethisterone, 15 mg., twice daily sublingually, beginning a week before the period is expected and continuing until the third day of bleeding. A combined tablet containing these steroids in the above proportions is marketed as *Androgeston*. The methyltesterone can be replaced by *Androstalone*, 25 mg., daily, and for some patients this regime is more effective.

Hormone treatment is most likely to succeed in cases of metrorrhagia without orgainc cause and may take various forms. For the arrest of prolonged and excessive uterine bleeding, the injection of progesterone and testosterone propionate for a few days is often the most effective procedure. If the endometrium is known or suspected to be atrophic the injection should also include oestradiol benzoate or dipropionate. The doses of these steroids are not very critical and something like progesterone 50 mg., testosterone propionate 25 mg., and oestradiol benzoate 2 mg., daily for three or four days is usually sufficient to bring about a complete arrest of bleeding or a very marked amelioration. A few days later a progesterone withdrawal bleeding, which may be likened to a normal period, begins and is self-limited, ending after five or six days. It is of course necessary to inform the patient that this will happen, as otherwise she will fear that the treatment has been ineffective.

It is usually a wise plan to follow this treatment with oestrogen therapy, beginning on about the fifth day of the withdrawal bleeding referred to above. Stilboestrol, 2 mg., or ethinyloestradiol, 0.1 mg., daily should be given orally for a total of twenty days, when an oestrogen withdrawal bleeding will begin some days later.

An alternative method for securing haemostasis, where the bleeding is not severe, is to begin with oestrogen. The dose may be similar to that mentioned above and if there is no reduction in the bleeding within forty-eight hours the dose should be doubled and maintained at that level for the rest of the twenty-day course. Difficulties are sometimes encountered because the dose of oestrogen necessary to secure haemostasis leads to nausea or vomiting as a side reaction. A further possibility is that of the bleeding, having ceased, beginning again before the twenty-day course is completed. If this happens treatment should be stopped for five days and then begun again when effective control will usually be obtained.

Haemostasis having been secured, it is then necessary to re-establish cyclic bleeding, and this is conveniently done by repeating the courses of oestrogen, commencing on the fifth day of each withdrawal bleeding and continuing as before for twenty days. After three such courses there seems to be an advantage in combining the oestrogen with progesterone or ethisterone, since by this means the chances of including ovulation seem to be increased. A method which has been found effective in a fair proportion of patients is as follows: after three controlled cycles with oestrogen alone a fourth cycle is started as before with oestrogen on the fifth day of the cycle (stilboestrol, 2 mg., or ethinyloestradiol, 0.1 mg., daily for twenty days). On the fifteenth day of the cycle, that is, after oestrogene alone has been given for ten days, ethisterone 40 mg. daily, is given for ten days. In this way the ethisterone course, overlaps the second half of the oestrogen course. On the fifth day of the next cycle the dose of oestrogen is reduced to half its previous value, and on the fifteenth day of the cycle ethisterone is begun again, this time at a dose of 60 mg., daily for ten days. In the next and final cycle the dose of oestrogens is again halved (stilboestrol, 0.5 mg., or ethinyloestradiol, 0.25 mg., daily) and on the fifteenth day of the cycle ethisterone is again started, this time at a dose of 80 mg., for ten days. Evolation has been known to occur during the final cycle of treatment and pregnancy to ensue. Satisfactory results can be expected in some 80 or 90 per cent. of patients with irregular, prolonged, and excessive uterine bleeding after treatment in this way, though some of these may later relapse. In about 40 per cent. of the patients ovulation will be re-established following this regime.

An alternative form of treatment for this kind of menstrual disorder is with progesterone, as originally described by Scowen. Once haemostasis has been secured, a few injections of progesterone (the precise dose seems to be relatively unimportant: 20 mg. on alternate days for three injections, or even 50 mg. in a single injection seems to be satisfactory) will be followed by a withdrawal bleeding. These injections are repeated monthly so as to bring about regular progesterone withdrawal bleedings, which, accompanied as they are by endometrial shedding, prevent the building up of a thick vascular proliferative endometrium from which future irregular bleeding could occur. In some patients it is possible, after a few courses of such injections, to substitute ethisterone (e.g., 50 or 60 mg. daily for about five days, repeated monthly) and to obtain equally satisfactory results.

Other forms of treatment, using androgens and gonadotrophins have often been advocated. The use of androgen combined with progesterone has been mentioned already and is very effective for initial haemostasis. Androgen alone, however, is, in the writer's opinion, much less satisfactory, the dose required to produce a satisfactory effect often being such that, when given for even moderately prolonged periods of time, it tends to cause undesirable virilizing effects. The use of chorionic gonadotrophin has little to recommend it, but it is possible that *Synapoidin*, which is a combination of chorionic and pituitary gonadotrophin, may sometimes prove effective where other measures have failed.

STEIN-LEVENTHAL SYNDROME

Although it was as long ago as 1929 when Stein performed his first wedge resection of bilateral '*polycystic ovaries*' for the relief of the accompanying amenorrhoea, it is only within comparatively recent years that general interest has been aroused in what has now come to be known as the '*Stein-Leventhal syndrome*', the literature on which, up to 1954, has been summarized by Bishop. The fact that some abnormality of menstruation is usually (though not invariably) a part of the syndrome is the justification for considering it in this chapter.

Bilateral ovarian enlargement is the only invariably finding in this syndrome; however, there is usually amenorrhoea or oligomenorrhoea, and there is commonly also hirutism of varying degree. Though some patients are obese, many are not and there is no justification for including obesity as part of the syndrome.

Ovaries

As mentioned above, bilateral enlargement is invariable and is usually palpable on pelvic examination, if necessary under an anaesthetic. However, cases have been reported in which the ovaries, though found to be enlarged on laparotomy, were not palpably enlarged on pelvic examination. Indeed, Stein, Cohen, and Elson stated that in only about half of their series of 75 patients could the ovarian enlargement be detected on pelvic examination and they therefore felt that radiological investigation, following the induction of a pneumoperitoneum, which outlines the shape and size of the ovaries, should be an essential procedure in attempting to establish the diagnosis. Other procedures which have been advocated for this purpose are culdoscopy and peritoneoscopy, and combined *pneumoperitoneum* and *hysterosalpingography* ('*gynaecography*').

Although the condition has been referred to as 'polycystic ovaries', the cysts, which are not always present, are in fact no more than slight or moderate enlargements of multiple follicles. Sometimes the ovaries are quite solid, there being no cystic change at all. On microscopical examination, the most striking feature is marked overgrowth of the ovarian stroma. Numerous follicles may be found in all stages of maturation, and in general, there is evidence of increased atresia also. In some cases masses of luteinized cells are found, but these are rather unusual. Hyperplasia of the hilus cells is a feature of some of these ovaries, but again others fail to show it. Although some authors have stressed hyperplasia of the theca interna cells and Fraenkel coined the term '*hyperthecosis ovarii*', such hyperthecosis is by no means invariably and it seems doubtful if the term is at all justifiable.

Menstrual Disturbance

Most commonly this is *secondary amenorrhoea*, though one case associated with primary amenorrhoea has been described. Some patients have oligomenorrhoea, in which the interval between menstruation may be anything from a matter of several weeks to several months. In yet others, occasional or more frequent irregular heavy bleeding occurs, the clinical picture resembling that of metropathia haemorrhagica. Most often there is absence of ovulation, but some patients certainly do ovulate, and Stokhuyzen reported a case with normal regular cycles and a seceretory endometrium at menstruation. In one of the author's patients a fresh corpus luteum was found at laparotomy, and the endometrium obtained at the same time showed a normal secretory pattern.

Other Signs and Symptoms

Hirsutism is a very common, though again not invariable, component of the syndrome. In a few cases it has been of a very severe degree and even accompanied by the other signs of virilism, such as enlargement of the clitoris, voice changes, acne, and a muscular physique. These, however, are extreme examples. In some the hirsutism may be so mild as to give rise to no complaint.

Sterility is one of the commoner complaints, though of course would be expressed only by those patients who are married. Obesity occurs in some patients, but many are of normal weight, and some individuals with the syndrome may be markedly underweight. It seems probable in fact that any significant variations from the normal in weight are more likely to be due to psychological causes than to underlying endocrine changes.

Hormone Studies

Very little has been published about hormone studies in these patients, but certain points have been established. In the first place, the 17-ketosteroid output is almost invariably within the normal limits, and attempts to demonstrate the excretion of excessive amounts of androgen in the urine do not appear to have met with success. Occasionally pregnanediol may be found in possibly significant amounts in the urine, but it is certainly not in excess in other cases. Oestrogen excretion studies have not been reported, but on clinical grounds it may be concluded that, although in some patients the oestrogen level is normal or even high, in others it is relatively low. The output of gonadotrophins is uncertain in these cases; in the author's patients no excessive secretion of urinary gonadotrophin has been revealed by techniques which, though in current use, are known to be of limited sensitivity.

Although there has been a good deal of theorizing about the precise nature of the disorder and its hormonal implications, it is fair to say that all this has been purely speculative and that we really do not know the true nature of the condition or how the signs or symptoms are brought about. It may be that in some cases there is an abnormally high output of progesterone or some related hormonal substance with mildly androgenic properties, but this could scarcely hold true for all. It has been suggested that over-production of luteinizing hormone by the pituitary might be the cause of the ovarian changes, but again this fails to account for those cases in which there is no luteinization in the ovaries.

The remarkable response to treatment does not help in any way to explain the genesis of the condition, and the conclusion seems inescapable that we really are still very ignorant about the precise nature of this syndrome.

Treatment

Whatever may be doubted about the cause and nature of the *Stein-Leventhal syndrome*, the value of bilateral wedge resection in treatment is undoubted, though the manner in which this procedure produces the beneficial effects is just as mysterious as the cause of the syndrome itself. In point of fact, Stein did his first wedge resection solely for diagnostic purposes, and was very agreeably surprised to find that normal menstruation was restored following the operation. Such restoration of normal ovular menstrual cycles can be expected in about 80 per cent. of cases, and conception is common in those patients who are complaining of sterility. The effects on the hirsutism, however, are a good deal less spectacular. Nearly always there is some reduction in the rate of hair growth, but it is rare for the excessive hair to disappear and the results cannot be compared with those which follow removal of a virilizing adrenal tumour.

The precise duration of the benefits of bilateral wedge resection is uncertain, but it would seem that in general there is little likelihood of relapse, and it may well be that most patients after treatment remain normal until their natural menopause.

In a recent paper Stein reports that in the past twenty five years he has done wedge resections on 88 carefully selected patients, in 95 per cent of whom menstrual function was restored. Fifty-four became pregnant with a total of 118 pregnancies, and there were no recurrences of bilateral polycystic ovaries.

12

UROHYPOPHYSEAL GLAND

That peculiar morphologic elements suggestive of certain endocrine activities are present at the caudal aspect of the fish spinal cord is by no means a new finding; however, biologists have largely ignored their presence. As early as 1827 Weber described clearly the existence of a warty appendix ("Knoten") at the end of the spinal cord proper of the carp, with additional remarks that similar structures were present in other kinds of fishes such as the catfish and the burbot. According to this classic author, "Zwar ist die Endigung des Ruckenmarks in einen Knoten schon langst bekannt." Later authors have cited Arsaky (1813) and Serres (1826) as preceding Weber in the exploration of the unusual terminal morphology of the spinal cord. Early observations were confirmed by several subsequent anatomic studies, and, beginning with Rauber (1877), histologic investigation was pursued, most significantly by Verne (1914) and by Favaro (1926). Verne called the terminal swelling "renflement caudal," and Favaro designated it as "ipofisi caudale" (hypophysis caudalis). It is of particular interest that both authors, being impressed with the analogy in terms of gross histologic pattern, but without taking into account characteristic features of the respective nervous organizations, arrived at essentially the same view that the structure in question was glandular in nature and comparable to the posterior pituitary, as well as to the epiphysis, in the diencephalic region.

Another important approach was made in the United States in studies of a specialized type of cell in the posterior end of the fish spinal cord. Speidel (1919, 1922), having confirmed Dahlgren's discovery (1914) of enormous cells of peculiar appearance in the posterior part of the spinal cord of skates, worked out details of their cytology and extended research to other kinds of elasmobranchs as well as to a number of teleosts. From these studies it became evident that a morphologically discrete material was elaborated by these cells; Scharrer and Scharrer (1945) refer to Speidel's work as the earliest demonstration of a neurosecretory phenomenon. Unfortunately, Speidel's investigations were not specifically concerned with the neural relations of the secretory cells, and the fate of the secretory material was left undetermined; however, Speidel did report that, in the flounder, the spinal cord is provided with a dorsally overlapped "terminal enlargement" and that the secretory cells are most numerous and of greatest size in close proximity to this structure. This suggested at least a close spatial relationship between the Dahlgren cells and the characteristic terminal body extensively studied by European workers.

Concept of a Caudal Neurosecretory System

Recently, the older observations, integrated with newer findings, have been reinterpreted. The significant morphologic elements described above became recognized as the principal components of a definite neurosecretory system, for which the term "caudal neurosecretory system" was proposed.

This development was based on a chance observation, in the caudal region of the spinal cord of the Japanese eel (*Anguilla japonica*), of a remarkable nervous organization represented by a number of large and small secretory neurons and their secretion-bearing axons with enlarged bulbous endings. In routine histologic preparations, such as those stained with Mallory's, Mann's or Masson's technics, the neuron cell bodies show pronounced unclear polymorphism, changes in basophilia in the cytoplasmic periphery as well as around the nucleus, and granules or droplets showing affinity for acid fuchsin, eosin, iron-hematoxylin, etc. Affinity of the cytoplasmic granules for acid fuchsin is much more pronounced in histologic sections pretreated with alkaline thioglycollate. The cell bodies are located in a well-vascularized area and are often in direct contact with capillaries. Along their entire extent the axons that run posteriorly to terminate in enlarged bulbous endings show the same tinctorial response as that of the cytoplasmic secretion, indicating apparent continuity of the stainable material between the site of origin and the axon. The end-bulbs filled with colloid showing the same stainability as that of the secretory material are adjacent to capillaries running along the ventral and lateral aspects of the terminal region of the spinal cord. Here the cord tissue slightly bulges out because of

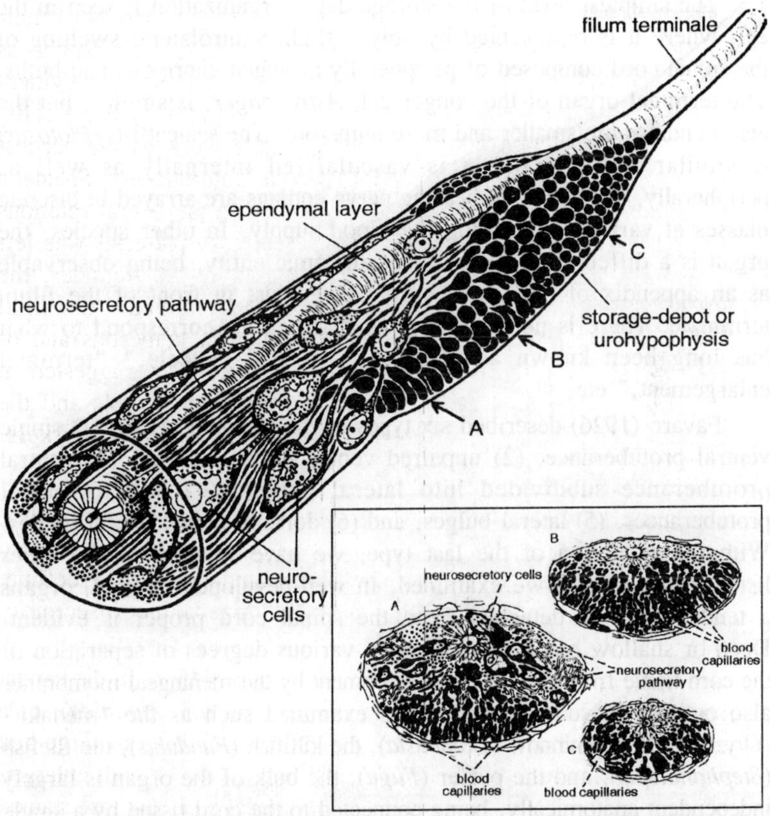

Fig. 12.1. Pattern of arrangement of the caudal neurosecretory system of the eel.

the massive aggregation of nerve endings. Occasionally large or small droplets, resembling the colloid in the end-bulbs, are found within the lumen of contiguous blood vessels. Thus, the characteristic neuron in the caudal spinal cord appears to be involved in production of a kind of secretory material, its axon serving as the neurosecretory pathway, and the axonal end-bulb responsible for storage and release of the secretory material.

With this background of information, histologic research was extended to nearly 100 species of fresh-water and marine teleosts. This survey demonstrated that in all teleost fish so far examined the characteristic terminal nervous system is developed on the same basic plan as that seen in the eel. However, there are variations in morphologic pattern, particularly in the relation of the secretion-bearing nerve endings to the circulatory system.

The simplest form of the storage-depot organization is seen in the eel, where it is represented by only a slight ventrolateral swelling of the spinal cord composed of peripherally arranged enormous end-bulbs. The terminal organ of the conger eel, *Astroconger*, is similar, but the nerve endings are smaller and more numerous. The sea-catfish (*Plotosus*) is similar, but the organ is vascularized internally as well as peripherally, and secretion-bearing nerve endings are arrayed in discrete masses at various sites along the blood supply. In other species, the organ is a differentiated prominent anatomic entity, being observable as an appendix of the spinal cord located just in front of the filum terminale. There is no doubt that such structures correspond to what has long been known as "Knoten," "ipofisi caudale," "terminal enlargement," etc.

Favaro (1926) described six types of the terminal organ: (1) single ventral protuberance, (2) unpaired ventrolateral swelling, (3) ventral protuberance subdivided into lateral halves, (4) paired ventral protuberances, (5) lateral bulges, and (6) dorsally-united lateral bulges. With the exception of the last type, we have seen all of the types listed in the species we examined. In well-developed terminal organs a tendency to be detached from the spinal cord proper is evident. Deep or shallow notches bring about various degrees of separation of the cord tissue from the organ; envelopment by the meningeal membranes also occurs. In some of the fishes examined such as the "medaka" (*Oryzias*), the topminnow (*Gambusia*), the killifish (*Fundulus*), the filefish (*Stephanolepis*), and the puffer (*Fugu*), the bulk of the organ is largely independent anatomically, being connected to the cord tissue by a single stalk or by a pair of stalks composed of all the neurosecretory axons. In his comparative study of the mammalian neurohypophysis, Hanstrom (1954) pointed out a parallelism between the evolution of the organ and the degree of facilitation of contact of the hypothalamic neurosecretory fibers with the circulatory system. This appears to be the case also with the terminal organization of the caudal neurosecretory system. The development of the blood supply and of terminal ramifications of the secretion-bearing nerve fibers parallel the tendency to be separate from the spinal cord and may be related to the increase of efficiency of the storage and release function.

Local accumulation of secretory material at various spots along the axon have been seen. This corresponds to the characteristic beaded appearance frequently noted in the neurosecretory fiber tracts in many kinds of animals. Also, a variety of peculiar structures resembling the

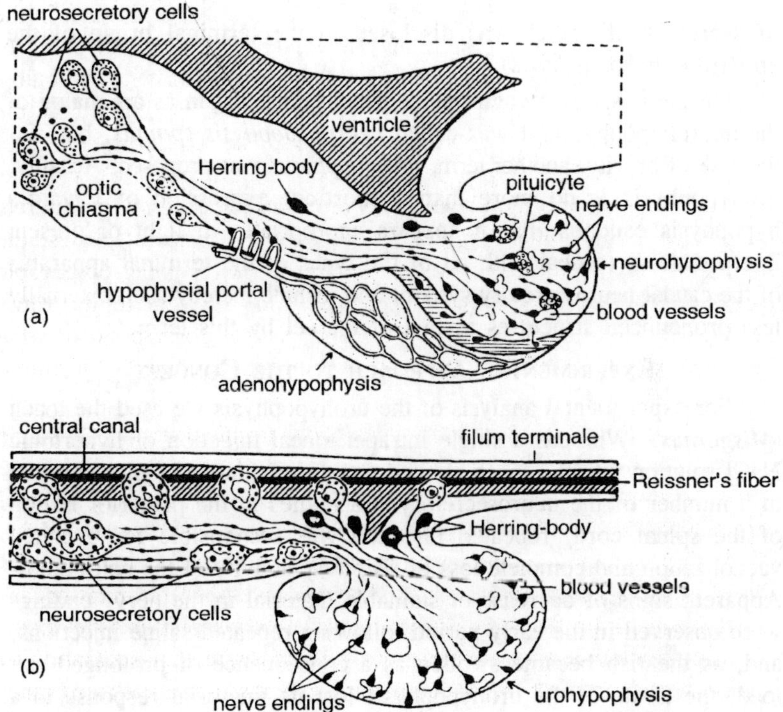

Fig. 12.2. Comparison of the pattern of organization of the caudal neurosecretory system (b) with that of the hypothalamo-hypophysial system.

so-called Herring bodies in the hypothalamo-hypophysial system is found distributed mainly at the dorsal aspect of the terminal organ. The nature of such figures is yet unknown, but some of them suggest either a degenerative change, or a kind of holocrine activity of the posteriorly distributed small neurosecretory cells.

Thus, the terminal organ of the caudal neurosecretory system is reminiscent of the neural lobe of the pituitary both in shape and basic histologic pattern. It conforms to the current concept of neurosecretion. In connection with this similarity, it is of interest to observe the presence of a short outpocketing of the central canal at the dorsal side of the terminal organ in the catfish (*Parasilurus*) and the saury (*Cololabis*), which bears some resemblance to the infundibular recess of the third ventricle near the posterior pituitary, generally, the glial elements included in the tissue of the terminal organ are not specifically differentiated into such characteristic cells as the parenchymatous pituicytes of the neurohypophysis. However, a kind of cell suggestive of secretory character was disclosed in the terminal organ of the gurnard (*Chelidonichthys*).

On the basis of recognition of the terminal organ as an analog of the neurohypophysis, it was called *neurohypophysis spinalis*, but, for the sake of brevity, another term, *urohypophysis*, was proposed. Actually, urohypophysis is no more than a short-cut expression of Favaro's hypophysis caudalis, but it appears appropriate, in light of present interpretation, to deal with all of the types of the terminal apparatus of the caudal neurosecretory system, and include thereby such externally less pronounced structures as that of the eel by this term.

Experimental Approach to the Concept

For experimental analysis of the urohypophysis we used the loach (*Misgurnus*). Whereas a single intraperitoneal injection of hypertonic NaCl solution into the loach elicited a transient state of hypersecretion in a number of the neurosecretory cell bodies in the posterior region of the spinal cord, repeated injection was responsible for eventual vacuolization and complete loss of secretory activity of the cell bodies. Apparent signs of decrease of stainable material in the nerve endings were observed in the early period following repeated saline injections, and, as the fish became swollen as a consequence of prolonged salt load, the tissue of the urohypophysis lost its tinctorial response to a great extent.

When the spinal cord was transected in front of the urohypophysis and the fish was subjected to repeated salt loading, these events occurred: At first a rapid depletion of stainable material took place in the urohypophysis, followed by a state of increased secretory activity in the small cells posterior to the site of operation. In rough parallel to the increase of activity of the cells in question, stainability of the nerve terminals was partially restored, suggesting that the accelerated release of secretory material from the nerve endings became counterbalanced by increased supply of newly formed secretion from the neighbouring small cells. Nevertheless, both the small cells and the urohypophysis eventually became depleted in the later phase of profound impairment of osmoregulation. On the other hand, the anteriorly distributed large cells, whose axonal connection with the urohypophysis was disrupted, began to show increased secretory activity after most of the posterior small cells attained their peak of activity. Occasionally, a marked picture of accumulation of secretory material was observed in the fiber tracts arising from the large cell bodies. This was particularly pronounced proximally to the level of operation,

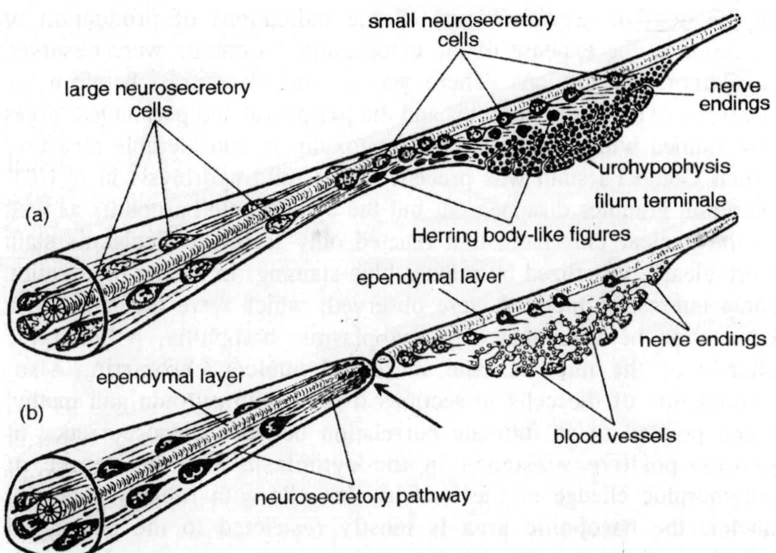

Fig. 12.3. Diagrammatic representation of the result of experiments on the loach. (a) Average appearance of the caudal neurosecretory system in fish. (b) Situation after repeated injection of hypertonic saline in fish with the spinal cord transected.

where secretion-bearing axons developed enlarged terminal swellings. This unusual accumulation of secretion of the large cells was brought to an end as signs of edema became apparent.

These observations indicate that the neurosecretory cells in the caudal spinal cord and the nerve terminals in the urohypophysis are significantly correlated so as to work in unity in response to the same physiologic demand. As in the well-explored hypothalamo-hypophysial system of vertebrates, as well as analogous organizations in crustaceans, insects, etc., peripheral movement of neurosecretory material by way of the axoplasm also occurs in the caudal neurosecretory system.

Some Cytologic Aspects of Secretion-production

The enormous caudal neurosecretory cells of the eel provide a suitable object for the cytology of secretory activity. As mentioned above, secretory material originating in the cells stains with acid fuchsin, etc., but unlike that elaborated by hypothalamic neurosecretory cells, it does not show affinity for chrome alum hematoxylin, taking phloxine pronouncedly in preparations stained after Gomori's technic. The same tinctorial response to Gomori's stain was reported by Scharrer and Scharrer (1954b) for the secretory material of the Dahlgren cell in the dogfish (*Scylliorhinus*). Some indications of production of secretion at the expense of the cytoplasmic basophilia were observed in Gomori preparations. There was a marked contrast between the nonreactive secretion granules and the peripheral and perinuclear areas that stained with chrome alum hematoxylin in considerable measure. When Gomori's stain was preceded by a mild hydrolysis in *N* HCl, secretion granules disappeared, but the cytoplasmic basophilia as well as the nuclear chromatin that reacted only feebly to Feulgen's stain were clearly visualized by intense blue staining. In such a preparation some interesting pictures were observed, which were indicative of a correlation between increased cytoplasmic basophilia, polymorphic change of the nucleus, and increased nuclear chromatin. Also, examination of the cells in sections treated with pyronin and methyl green pointed to an intimate correlation between the occurrence of pyronin-positive substance in the cytoplasm and the degree of polymorphic change of the nucleus. In cells with roughly spherical nuclei, the basophilic area is mostly restricted to the peripheral cytoplasm, but, in cells showing pronounced nuclear polymorphism, significant amounts of basophilic substance are also found adjacent to the nucleus. Generally, the perinuclear basophilia is located at large or small depressions in the nuclear membrane, being practically absent along smooth and convex portions of the nucleus. There are cells showing extremely lobate small nuclei with little or no pyronin-positive substance in either peripheral or perinuclear areas.

It appeared significant that the nucleolus, which stained intensely with pyronin, was occasionally much elongated and was in contact with the nuclear membrane. This was demonstrated also by sections impregnated moderately with protargol and counterstained with pyronin, in which the nucleoli appeared as double-structured figures consisting of an argentaffin interior and a pyronin-positive exterior. Of the elongated nucleolus, the slender apical portion that extended to the nuclear membrane was mainly represented by the pyronin-positive material. The argentaffin internal material was concentrated in the basal enlargement. By use of Hamazaki's so-called KES technic (1951), it was observed that a small aggregate of reaction-positive granules (purple-stained) was located just opposite to the attachment point of the apex of a given elongated nucleolus at the nuclear membrane. Also, seemingly spherical figures rimmed with the stainable substance were found attached to the surface of the nuclear membrane and lying in shallow depressions of the latter. Hamazaki believes that his technic stains various derivatives of the nucleic acids that cannot be observed by means of conventional histochemical methods. Whether or not such an opinion is tenable, the present finding suggests at least an intimate relationship between the nucleolar material and a certain cytoplasmic material. Although it is far from established, such a line of observation implies a possible functional link between nuclear activity and cytoplasmic basophilia in the synthesis of secretory product.

Electron Microscopy of Secretion

A recent electron-microscope study supplemented the light microscopic observation of secretory material of the caudal neurosecretory system. Within the neuron cell bodies in the eel, spherules and ellipsoids of high electron density occur, measuring approximately 1000-2500 Å in diameter. Those located at the axon hillock as well as at the proximal portion of the axon are more or less enlarged and of somewhat irregular contour, as compared with the granules at the interior of the perikaryon. As one proceeds along the axon posteriorly, large-scaled deformation of the secretion is encountered. Unlike the particles observed in the neuron cell body, those in the distal portion of the axon appear enlarged, measuring approximately 2500-4200 Å in diameter. Such enlarged particles may be of uniformly high or low electron densities; they may be of a vesicular nature, being of high density externally and relatively low density internally; some are suggestive of coalescence of different particles. Finally, within the end-bulb in the urohypophysis, secretory granules are markedly uniform, of a much diminished size of approximately 500 Å in diameter. Most of these fine granules show low electron density, but intermingled with them are granules of similar size, which are comparatively electron-dense. It is apparent that packets of such minute granules represent the stainable colloid deposit in the nerve terminals observed with light microscopy. Thus, it appears likely that the secretory material synthesized within the neuron soma becomes altered in its physicochemical properties in the course of migration by way of the axoplasm and is reduced to uniform particles by the time it reaches the nerve ending. A certain degree of "maturation" might be ascribed to the observed modification of the secretory material, although no further information of significance has been obtained in this respect.

Association of Zinc with Secretion

One of the significant properties of the secretory material of the caudal neurosecretory system in the eel is its association with zinc. Supravital staining with alkaline dithizone solution selectively demonstrates the neurosecretory system, showing specifically stained secretory material which corresponds to the picture observed in routine histologic preparations. Polarographic determination of zinc in different portions of the spinal cord showed principal concentration of the element in the caudal portion, which includes the bulk of the neurosecretory system. Measurements of fractions resulting from extraction with water, ethanol, acetone, chloroform, or petroleum ether provided data indicating that, although both soluble and insoluble forms of zinc are contained in large amounts in the caudal spinal cord, it is the former that is especially abundant, amounting to several times the quantity detected in anterior portions of the spinal cord. When the caudal spinal cord was fixed in ethanol for histochemical demonstration of zinc in sections, a loss of the tinctorial response occurred both in the axons and in the nerve endings. However, a positive response remained in sonic of the perikarya, with reactive granules or droplets simulating the secretory material to some extent. Since ordinary histologic preparations of ethanol-fixed tissues do not show any residual secretory material throughout the entire caudal neurosecretory system, the presence of zinc-reactive material unaffected by ethanol does not seem to indicate specific insolubility in ethanol of the secretory material itself at the site of origin. Rather, this observation suggests that the secretory material within the perikaryon, unlike that in the axon and in the end-bulb, exists in combination with ethanol-insoluble zinc compounds. In this respect, the electron-microscopic observation of changes during the course of peripheral migration would appear to be of particular interest.

FUNCTIONAL SIGNIFICANCE

Possibility of Participation in Sodium Exchange

The result of experimental salt loading in the loach stimulated an inquiry into the possibility of production by the caudal neurosecretory system of a hormonal principle concerned directly or indirectly with osmoregulation. The small cyprinodont "medaka" (*Oryzias*) exhibited some signs of disturbance in adaptive sodium-regulation, after removal of the part of the spinal cord lying posterior to the dorsal fin, or after transection of the cord just anterior to the base of the fin. In contrast to virtually no change in potassium-content of the body (excluding the viscera), total sodium-content was increased in comparison to that of the intact animal, after repeated transference of the operated fish from fresh water to physiologic saline and thence to half-concentrated artificial sea water, and in reversed direction. Intraperitoneal injections of homogenate or crude extract of eel caudal spinal cord were followed by a decrease of total sodium-content of the operated fish, when the recipient was preadapted to isotonic physiologic saline or hypertonic 50 per cent artificial sea water. On the contrary, an appreciable increase in sodium-content was observed as the result of such injections when the recipient had been kept in fresh water.

These observations suggest the possibility of a relation between the caudal neurosecretory system and certain target organ(s) responsible for excretion and uptake of sodium. Concerning the endocrine control of adaptive osmoregulation in fishes, it is not possible to synthesize the various views and provisional conclusions presented to date and thus to delineate at least the outline of osmoregulatory coordination. Much more research is needed to obtain any clear-cut picture of the problem. Future work might consider the possibilities raised by the experiments reported above.

Relation to Gas Metabolism

The chance observation that the goldfish (*Carassius*) loses buoyaney on removal of the tail posterior to the base of the anal fin made possible a fruitful line of research. Intraperitoneal injection of a crude extract of eel caudal spinal cord results in a marked increase of buoyancy, which appears as a kind of forced floating of the tailless fish. By measuring the change of buoyancy of the fish body as a whole according to the principle of cartesian diver manometry, events resulting from injection of effective extracts can be easily followed. In practice, the test goldfish is deprived not only of its tail but also of all fins with the exception of the dorsal one, in order to eliminate as far as possible spontaneous movements of the body. Postoperative mortality is less than 5 per cent for at least 7 days. Usually, experiments are carried out on the eighth day, when the fish shows the characteristic state of continuous submergence.

By injection of a crude eel extract, two successive phases of increase of buoyancy are evoked. That the two responses are dependent on different substances was learned from experiments involving injection of various eluates of a paper-chromatogram of a crude extract developed with *n*-butanol-acetic acid-water (40 : 10 : 50). Of the eluates obtained from four conveniently separated regions of the chromatogram as visualized by staining with bromphenol blue, ninhydrin, and dithizone, the eluate prepared from the topmost section elicited the first response, which attained its maximum in approximately 120 minutes at 20°C, whereas the eluate of a lower section was responsible for the second

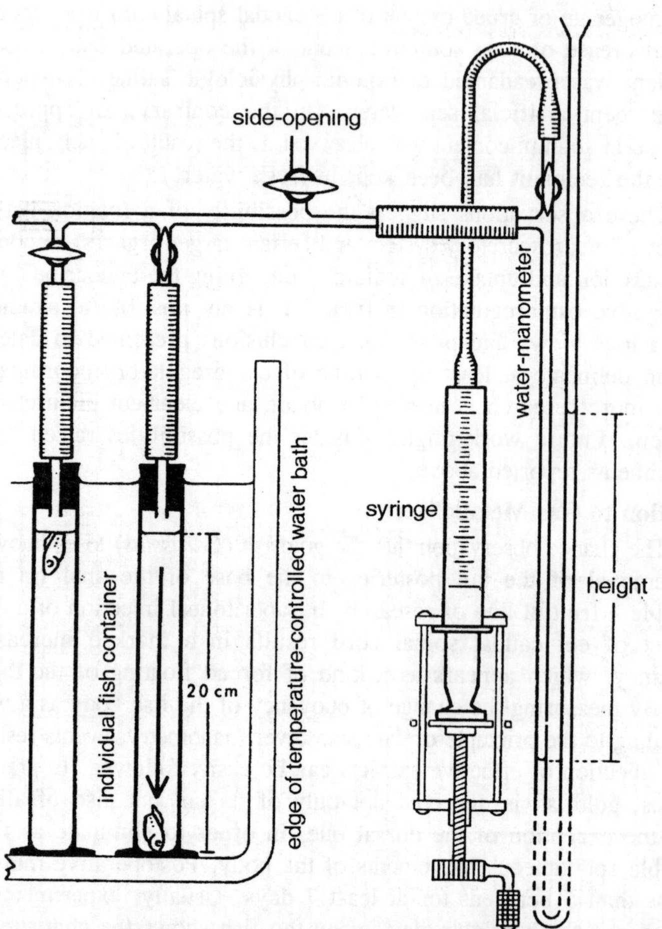

Fig. 12.4. Apparatus for measuring change in buoyancy of tailless and finless goldfish according to the principle of certesian diver manometry.

response initiated about 150 minutes after injection. It is of interest that the test goldfish exhibited some signs of respiratory disturbance (diminution of amplitude and increase in frequency of the opercular movement) following injection of either the original crude extract, or the eluate of the topmost section of the paper chromatogram; no external sign of respiratory impairment was observed for at least 4 or 5 hours following injection of the eluate of the lower effective section. The phenomenon seems to be related to the occurrence of dithizone-positive bands in the topmost section, which appear to be identified with the presence of zinc, in view of the observed abundance of the soluble

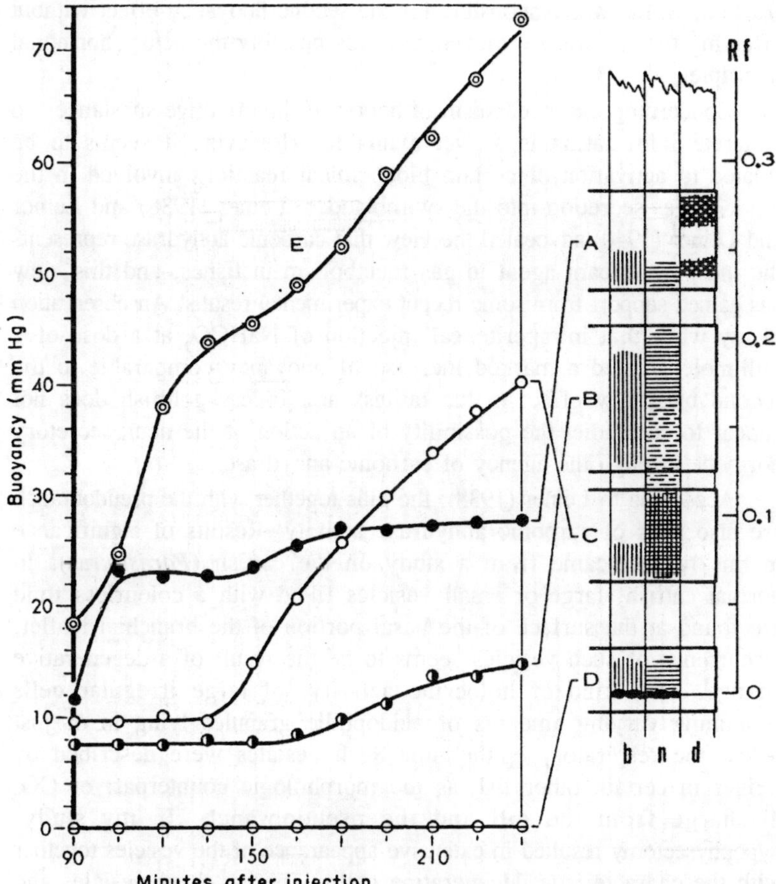

Fig. 12.5. Effects of injection of fractions obtained by paper-chromatographic development of a crude extract of eel caudal spinal cord, in comparison with that of the original extract (curve E). b, bromphenol blue-treatment; n, ninhydrin-treatment; d, dithizone-treatment.

form of this element in the caudal spinal cord of the eel. Paper chromatography of $ZnSO_4$ has shown a dithizone-positive band at Rf 0.25 under the same experimental conditions. Intraperitoneal injection of $ZnSO_4$ at doses in the range of 0.1-0.01 millimoles simulates the result of injection of the eluate of the topmost section both in regard to buoyancy and respiratory effects. Because of this similarity, the observed activity of the topmost section of the paper chromatographic development of the crude extract was considered to be due to its zinc ingredient; it was conjectured that a substance included in the lower fraction, which was responsible for the second buoyancy effect without affecting the respiratory movement, was possibly the actual hormonal principle.

Concerning the mechanism of action of the effective substance, no concrete information is as yet available. However, it seems to be related to activation of certain biochemical reactions involved in the work of gas secretion into the swimbladder. Leiner (1938) and Leiner and Leiner (1940) advocated the view that carbonic anhydrase represents the most significant agent in gas metabolism in fishes, and this view has gained support from some recent experimental results. An observation in my work that intraperitoneal injection of $NaHCO_3$ at a dose of 1 millimole elicited a marked increase of buoyancy comparable to the second buoyancy effect in the tailless and finless goldfish does not appear to contradict the possibility of an action of the neurosecretory hormone through the agency of carbonic anhydrase.

According to Leiner (1938), the gills together with the pseudobranch are also sites of carbonic anhydrase activity. Results of significance in this respect came from a study on the catfish (*Parasilurus*). In normal catfish, large or small vesicles filled with a colourless fluid are found at the surface of the basal portion of the branchial leaflet. Occurrence of such vesicles seems to be the result of a degenerative change, or a kind of holocrine activity, of large irregular cells containing varying amounts of acidophilic granules lying in or just below the respiratory epithelium. Such vesicles were described by Leiner in certain other fish as the morphologic counterpart of CO_2 discharge from the gill and the pseudobranch. In my study, hypophysectomy resulted in extensive appearance of the vesicles together with the characteristic degenerating cells, whereas these vesicles and relatively intact large cells wholly disappeared after removal of the caudal spinal cord or after transection of the cord at levels anterior to the site of the caudal neurosecretory system. The effect of surgical operations on the caudal spinal cord was reproduced to some extent by intramuscular implantation of excess pituitaries: on the other hand, subcutaneous injection of the eluate of the significant lower section of the paper chromatogram of the extract of eel caudal spinal cord imitated the effect of hypophysectomy. Application of the active fraction of the eel extract to the branchial leaves maintained *in vitro* also resulted in pronounced appearance of the characteristic histologic picture. Furthermore, faradic stimulation of the spinal cord, injection of acetylcholine, adrenalin, noradrenalin, pilocarpine or physostigmine, and injection of concentrated NaCl solution were equally responsible for an increased appearance of the special cells and vesicles.

Such observations may be interpreted in accordance with Leiner's contention, or may be used in support of another kind of explanation. The alternative explanation is based on interpretation of the characteristic changeable cell as the so-called "chloride cell," in line with the opinion of Keys (1931) and Keys and Willmer (1932) and substantiated by Copeland (1948, 1950). These authors insisted that the large acidophilic cell in the gill is charged with dual activities, being responsible for excretion of chloride in a hypertonic medium and for absorption in a hypotonic medium. Such a view would provide a pertinent rationale for the changes of total Na-content observed under the influence of the crude extract of the caudal spinal cord. Nevertheless, it must be stated that the functional significance of the "chloride cell" is still being disputed. In fact, the recent work of Burden (1956) has shown no obvious change of the cell in the killifish (*Fundulus*), despite extensive loss of serum chloride in fresh water after hypophysectomy. Burden is inclined to ascribe the observed profound impairment of chloride-exchange to a loss of the protective role of the mucus, stating that the mucous cells in the gills underwent extensive atrophy in parallel with disturbances in osmoregulation. In my study, the mucous cells in the gills and in the skin of the catfish were brought to hyperactivity (accelerated holocrine activity) by various experimental treatments, but none of the latter seems to be correlated specifically with the osmoregulatory phenomenon.

In general, many problems remain unsettled in the analysis of the coordinating mechanism(s) concerned with the phenomena of Na-exchange, gas-metabolism, and gill morphology. However, if one emphasizes the significance of carbonic anhydrase, the activity of which is known to be related also to the so-called chloride shift, a common basic mechanism underlying the three kinds of phenomena might be imagined.

Isolation of the Effective Fraction

Very recently an attempt at isolation of the caudal neurosecretory hormone was made, employing the buoyancy effect as a valid reference for assay. Preliminary information derived from extractions of nearly 6000 isolated spinal cords of the eel, has shown that the expected hormonic substance can be separated from the fraction responsible for respiratory disturbance by precipitation with 85 per cent acetone. It is soluble in water as well as in various concentrations of ethanol, is stable in acid media but liable to lose activity in alkaline media, is dialyzable through collodion and cellophane membranes. In one experiment 2458 isolated caudal spinal cords of the eel were subjected to the following procedure.

Essentially the procedure consisted of (1) removal of lipids, (2) precipitation of the active principle with 85 per cent acetone, (3) fractionation with 45 per cent ethanol, and (4) further fractionation by means of paper-electrophoresis. On assay, the ethanol-soluble fraction designated as Fraction A was found to be exclusively active, others designated as Fractions B, C, D, and E being without observable effect. Paper-electrophoresis of Fraction A in phosphate buffer of pH 7.8 and $\mu = 0.6$ showed development of roughly 5 ninhydrin-positive bands, of which the one moving to the anode was found to include the activity. The eluate of this band (A1) was run again on paper in the same buffer but of lower ionic strength ($\mu = 0.2$), resulting in its separation into two ninhydrin-positive bands, of which the one proceeding to the anode (A2) contained most of the activity. On further paper-electrophoresis of the active eluate in acetate buffer of pH 3.0 and $\mu = 0.01$, two ninhydrin-positive bands developed showing similar mobilities in opposite directions. The one that moved to the anode (A3) was found to be effective, in contrast to the practically ineffective one moving to the cathode (A4). Paper chromatography of A 3 with a mixture of *n*-butanol-acetic acid-water (40:10:50) showed a significant

1st Run (Fraction A): phosphate buffer (pH 7.8, μ=0.5), 7 v./cm., 120 min.

O, ninhydrin, +, A 1, A 1-0, 1 cm

2nd Run (A1, eluted with water of pH 5.0):
phosphate buffer (pH 7.8, μ=0.2), 5 v./cm., 180 min.

O, ninhydrin, +, A 2, A 2-0, 1 cm

3rd Run (A2, eluted with ethanol):
acetate buffer (pH 3.0, μ=0.01), 3 v./cm., 25 min.

O, ninhydrin, dithizone, +, A 3, A 4, 1 cm

Fig. 12.6. Paper-electrophoretic fractionation.

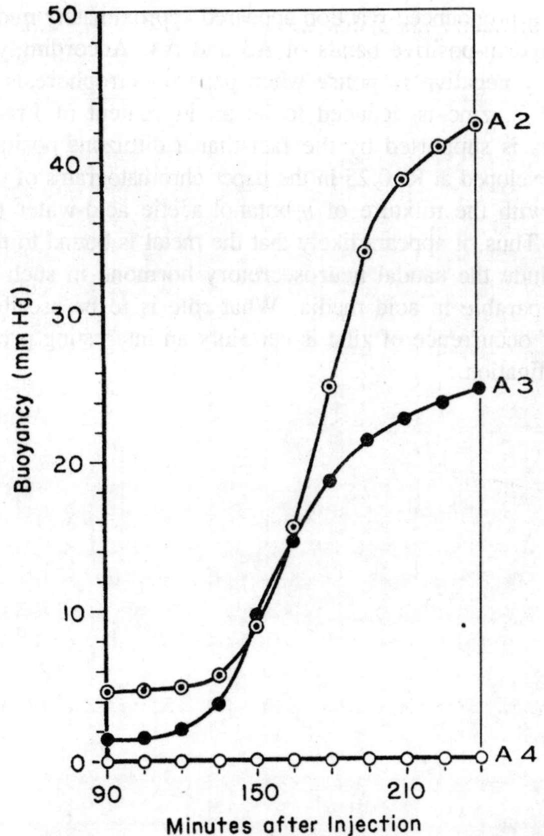

Fig. 12.7. Comparison of results of assay of fractions A2, A3, and A4.

ninhydrin-positive spot around Rf 0.13, being associated with less marked ones detected at Rf 0.07 and Rf 0.18, respectively. Further efforts to obtain an ultimately purified substance in sufficiently large amounts should be made before any attempt is made to ascertain the chemical properties of the hormone. For the time being, it can be stated that the practically isolated effective fraction appears to include polypeptide-like substances.

As regards the significant occurrence of zinc in association with the secretion of the caudal neurosecretory system of the eel, fractions obtained in the process of extraction described above provided some interesting information. In the paper-electrophoresis of Fractions A and A 1 at pH 7.8, no characteristic dithizone-positive reaction was observable. However, in the paper-electrophoresis of Fraction A2 at pH 3.0, a pronounced reaction appeared approximately midway between the ninhydrin-positive bands of A3 and A4. Accordingly, despite the seemingly negative response when paper-electrophoresis is conducted near pH 7, zinc is deduced to be an ingredient of Fractions A and A1. This is supported by the fact that a dithizone-positive band was found developed at Rf 0.25 in the paper chromatograms of these fractions treated with the mixture of *n*-butanol-acetic acid-water (40:10:50) at pH 2.5. Thus, it appears likely that the metal is bound to the complexes that include the caudal neurosecretory hormone in such a manner as to be separable in acid media. What role is to be ascribed to such a mode of occurrence of zinc is certainly an interesting problem worthy of investigation.

13

ROLE OF HORMONES IN MIGRATION

Migration in birds is a complex behaviour pattern. It occurs periodically with remarkable temporal precision and comprises a sustained orientated movement over great distances from a breeding to a non-breeding area. Its temporal precision and its orientation pose the intriguing problem of regulation, and the studies of the past 35 years have demonstrated clearly the regulatory role of internal, or physiologic, factors and of external or environmental, factors. An analysis of bird migration is in reality a study of the ecologic and physiologic factors which initiate, maintain, and inhibit a complex behaviour pattern. The behaviour is thought to be adaptive, since it results in the seasonal use of favourable habitats in the higher latitudes for feeding and breeding.

Experimental studies of the regulation of migratory behaviour were initiated by Rowan (1925, 1929), who discovered that gonadal recrudescence could be induced out of season, in late fall and winter, by subjecting slate-coloured juncos (*Junco hyemalis*) to artificial increases in day length. On the assumption that migratory behaviour in the spring, when birds are flying north to their breeding grounds, was a phase of sexual behaviour. Rowan tested the effect of experimentally induced gonadal recrudescence on migratory behaviour by releasing juncos in winter, many months ahead of the normal time of their spring migration and with their gonads at various stages of development. From the results of these experiments. Rowan concluded that in the slate-coloured junco, and other fringillids, the stimulus to migrate in the spring was regulated by external and internal factors.

The external factor was the increasing day length after December 21. The internal factor was the production of sex hormones, which he correlated with the recrudescing gonads of spring and the regressing gonads of fall. The minimal gonad of winter and the maximal gonad of the breeding period showed little or no hormone-producing interstitial tissue and, hence, would not stimulate migratory behaviour. From the results of later work with crows (*Corvus brachyrhynchos*), Rowan (1932) concluded that the southward migration in fall appeared to be independent of the influence of the gonads.

Rowan's epochal work defined the basic problems and stimulated an experimental attack on the regulation of bird migration and gonadal cycles. The studies performed since Rowan's discovery have centered around the following basic problems: (1) the nature of the external factor, (2) the physiologic changes induced by the external factor, and (3) the mechanisms whereby the external factor induces the physiologic changes which precede spring migration. The problems of how the change in physiologic state induces migratory behaviour and what terminates the behaviour have yet received little attention.

The first extensive experiments designed to test Rowan's hypothesis utilized the Oregon junco (*Junco oreganus*), a species closely related to the slate-coloured junco. These experiments corroborated Rowan's observation that birds could be stimulated to migrate northward months ahead of time by subjecting them to artificial increases in day length in the late fall and winter. However, they also demonstrated that birds already in breeding condition would migrate, a finding contrary to Rowan's conclusion. In addition to gonadal growth preceding migration, there was also a marked increase in body weight caused by large deposits of subcutaneous and intraperitoneal fat, which appeared to be a better criterion of a readiness to migrate than the condition of the gonad. Later studies demonstrated that the pituitary was also involved in this premigratory change in physiologic state. Members of a non-migratory race of the same species, which were exposed to the same environmental conditions in nature and in the laboratory, differed from migratory individuals in not showing marked deposition of fat nor increase in body weight and in having a much faster rate of gonadal development which resulted in an earlier breeding season.

The experiments of Rowan and Wolfson pointed clearly to the increasing day lengths of winter and spring as the environmental stimulus for spring migration, but there was a serious weakness in this aspect of Rowan's theory. Birds that wintered in the north temperate

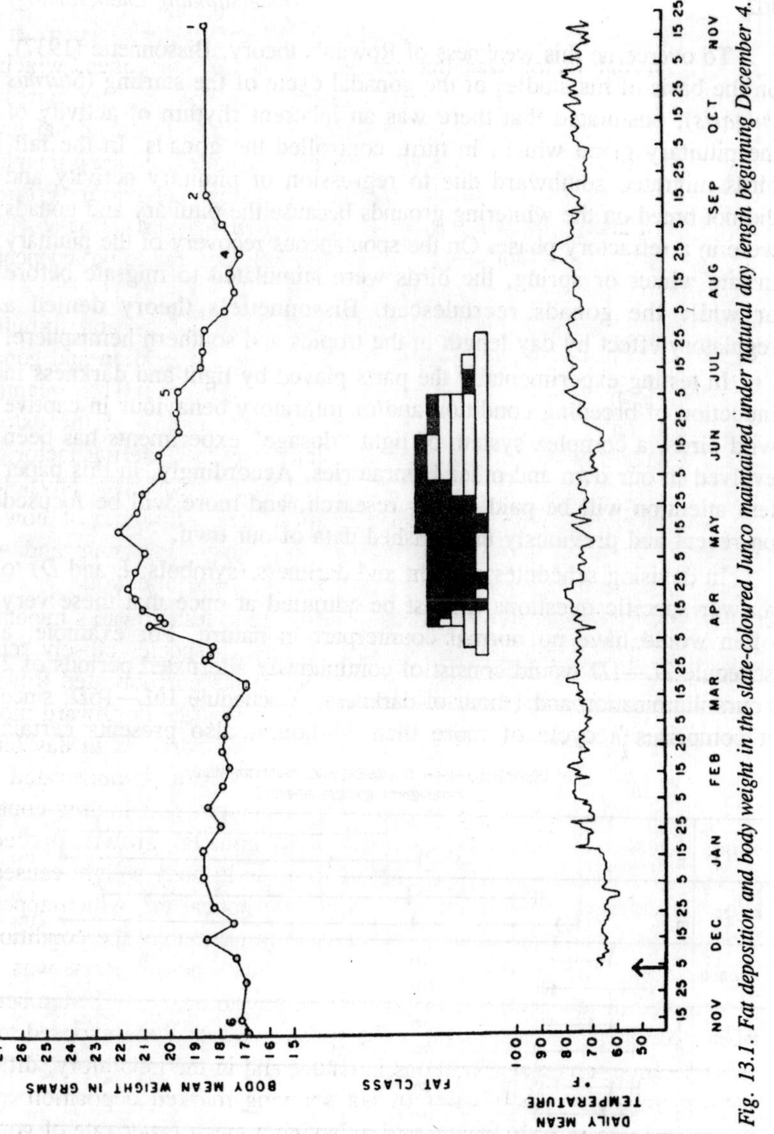

Fig. 13.1. Fat deposition and body weight in the slate-coloured Junco maintained under natural day length beginning December 4.

zone or northern subtropical zone would experience substantial increases in day length after December 21, but not the birds which wintered in the tropics, on the equator, or in the southern hemisphere. Increasing day length, obviously, was not a feature of the environment on the wintering grounds of all migratory birds.

To overcome this weakness of Rowan's theory, Bissonnette (1937), on the basis of his studies of the gonadal cycle of the starling (*Sturnus vulgaris*), postulated that there was an inherent rhythm of activity of the pituitary gland which, in turn, controlled the gonads. In the fall, birds migrated southward due to regression of pituitary activity and did not breed on the wintering grounds because the pituitary and gonads were in a refractory phase. On the spontaneous recovery of the pituitary in late winter or spring, the birds were stimulated to migrate before or while the gonads recrudesced. Bissonnette's theory denied a regulatory effect by day length in the tropics and southern hemisphere.

In testing experimentally the parts played by light and darkness in induction of breeding condition and/or migratory behaviour in captive wild birds, a complex system of light "dosage" experiments has been evolved in our own and other laboratories. Accordingly, in this paper less attention will be paid earlier research, and more will be focused on recent and previously unpublished data of our own.

In devising schedules of light and darkness (symbols: *L* and *D*) to answer specific questions it must be admitted at once that these very often would have no normal counterpart in nature. For example, a schedule 2*L*—1*D* would consist of continuously alternated periods of 2 hours illumination and 1 hour of darkness. A schedule 16*L*—16*D*, since it comprises a cycle of more than 24 hours, also presents certain

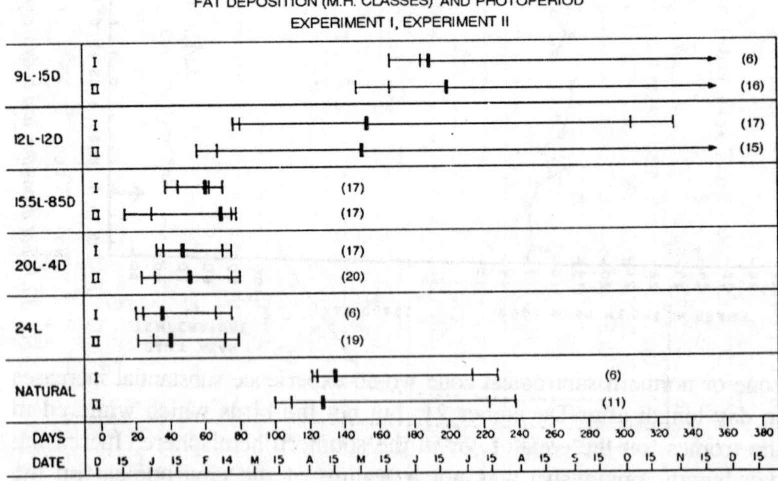

Fig. 13.2. Time of fat deposition in relation to photoperiod, beginning December 4. The upper horizontal bar in each photoperiod represents experiment I, the lower one experiment II.

difficulties of interpretation. However, despite these inherent difficulties some clear answers do seem to be offered, and it is our purpose here to bring these out. As will emerge from the ensuing discussion, a lively debate centers about the relative contributions of alternated light and dark to gonadal maturation. With present criteria it has been impossible, obviously, to separate the contributions of each of these two physical states when one must always be tested in the presence of the other. In interpreting the results of "light dosage" experiments it is useful to bear in mind the possible mechanisms through which the experimentally altered physical conditions exercise their ultimate effects upon breeding condition. As is indicated by work such as that of Benoit and Assenmacher (1955) the nervous receptor in this case is the eye, and the nervous pathway probably leads through the hypothalamus in

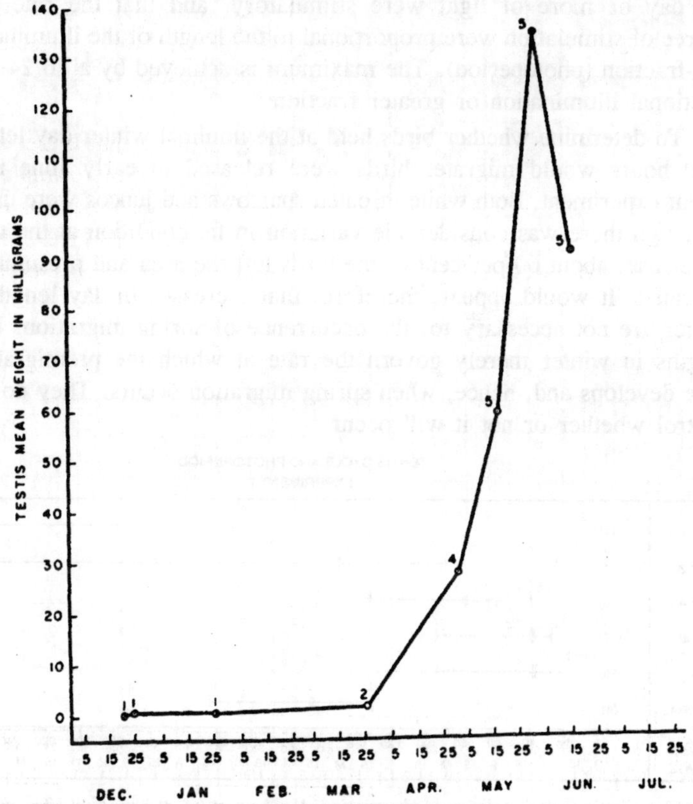

Fig. 13.3. Weight and development of testis in the slate-coloured junco under natural day lengths beginning December 4.

some way to the anterior pituitary. The eventual gonadal changes then follow gonadotropic hormone release.

The species investigated most extensively in our laboratory are the slate-coloured junco and the white-throated sparrow, *Zonotrichia albicollis*. The criteria we have found most useful in judging positive responsiveness in a given experiment are body weight, size of depot-fat deposits, and gonadal size and gametogenetic stage.

Day length and Initiation of the Migratory and Reproductive Responses

In agreement with others we were able to show with our birds that fat deposition and gonadal growth and maturation may be precociously stimulated by unseasonally increased illumination. The significant conclusions drawn from these experiments were that 9 hours per day or more of light were stimulatory, and that the rate and degree of stimulation were proportional to the length of the illuminated day-fraction (photoperiod). The maximum is achieved by a 16/24 day fractional illumination or greater fraction.

To determine whether birds held at the minimal winter day length of 9 hours would migrate, birds were released in early June in a recent experiment. Both white-throated sparrows and juncos were used. Although there was considerable variation in fat condition at the time of release, about 60 per cent of the birds left the area and presumably migrated. It would appear, therefore, that increases in day length in winter are not necessary for the occurrence of spring migration. Day lengths in winter merely govern the rate at which the premigratory state develops and, hence, when spring migration occurs. They do not control whether or not it will occur.

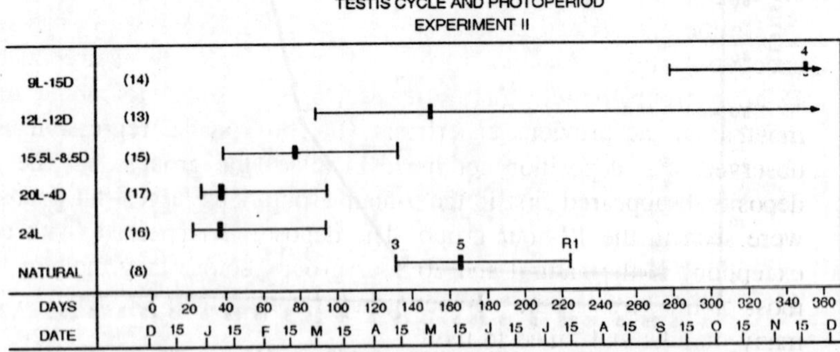

Fig. 13.4. Testis cycle in relation to photoperiod. Numbers at left of horizontal bars indicate the number of birds at the beginning of the experiment.

On the basis of the results obtained in the junco and the interpretation given above, day length could regulate the onset of migration in birds wintering in the tropics and in the southern hemisphere. Even though the days are constant on the equator and are decreasing in the southern hemisphere after December 21, they could be well above the length of the daily effective photoperiod for the birds wintering there, judging from the studies of north temperate species.

Day length and the Maintenance of the Migratory and Reproductive Responses

This first series of experiments demonstrated primarily a relation between photoperiod in a 24-hour cycle and the initiation of the fat and gonadal responses. Does the daily photoperiod have a role in maintenance (or duration) of these responses once begun? If so, then the daily photoperiod would be significant in the regulation of the entire annual cycle, and not just in its initiation. Experiments were designed specifically to answer this question. Birds caught during the spring migration in April and May were subjected to constant day lengths of 9, 12, and 20 hours (photoperiods in a 24-hour cycle), and to natural day lengths. At the beginning of the experiment the gonads were partly developed, and the birds showed also heavy fat deposits indicative of the migratory physiologic state.

In the 9-hour group the gonads regressed almost immediately, in contrast to the previous experiment in which a gonadal growth response was induced with a constant 9-hour photoperiod. Was the reduction in day length as such responsible for the regression in the 9-hour group? Perhaps not, for in the 12-hour group, which also experienced a reduction in day length (from natural conditions) of approximately $2^1/_2$ hours at the start of the experiment, gonadal growth continued.

In the natural and 20-hour groups the gonads regressed (with one exception) after a few months of activity, as occurs in nature. In the 12-hour group, activity of the testes was maintained for about nine months; in the previous experiment (II), no gonadal regression was observed. Fat deposition continued in all of the groups, but the fat deposits disappeared first in the 9-hour group. The largest fat deposits were seen in the 12-hour group. The deposits disappeared (with one exception) in the natural and 20-hour groups before they molted, but most of the birds in the 12-hour group retained their fat deposits for many months and failed to molt.

The general similarity in the response of these groups and the natural group suggests that the gonadal cycle is operating at close to

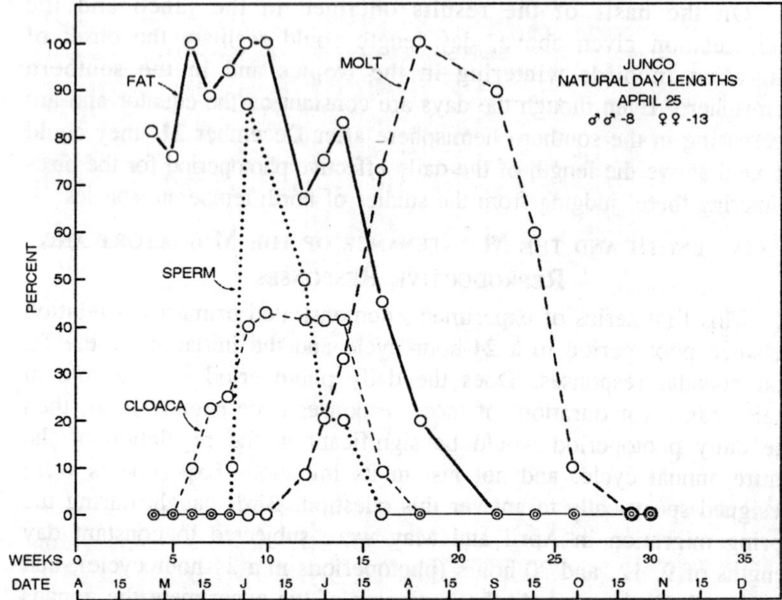

Fig. 13.5. Fat, reproductive, and molt responses in the slate-coloured junco under natural day lengths.

its maximal rate by late April and a marked increase in photoperiod (from about 5 hours on April 25 and gradually decreasing to about $3^1/_2$ hours on June 21) had little if any effect. However, the birds treated beginning April 6 seem to show a slight acceleration of the gonadal and molt cycles. The results of the first series of experiments showed that the daily photoperiod determined the time at which the response occurs. The results of the second series demonstrated that the daily photoperiod regulates also (a) the maintenance of the response, (b) the rate at which the response develops, and (c) the duration of the gonadal and fat cycles (or the time of gonadal regression).

Role of Light and Darkness in the Progressive phase of the Migratory and Reproductive Responses

At about the time the second series of experiments was concluded. Kirkpatrick and Leopold (1952, 1953) and Jenner and Engels (1952) demonstrated the importance of the dark period in the gonadal response to the daily cycle of light and darkness. Further experiments were undertaken by us to learn more specifically the role of the dark period.

If the effective stimulus is the total quantity of light which a bird receives daily, then birds receiving either 20 hours of light per day in

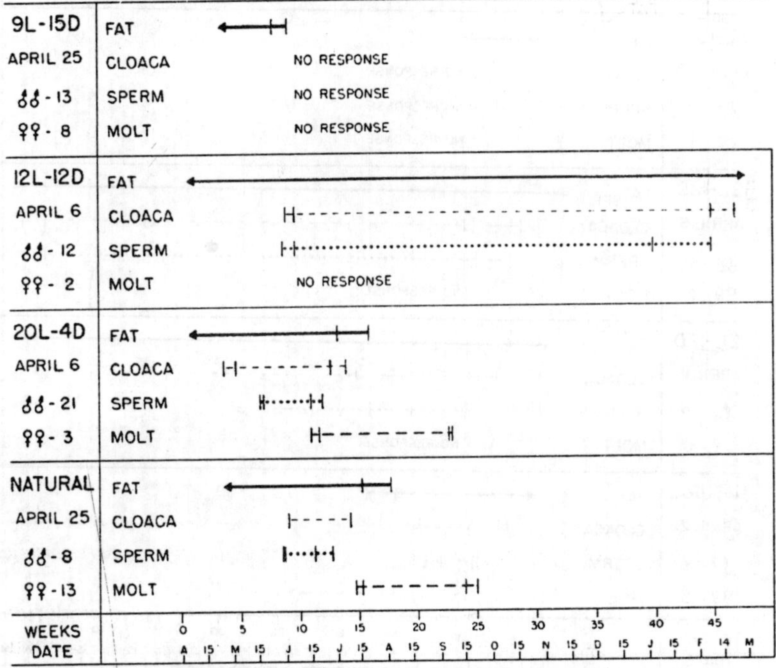

Fig. 13.6. Fat, reproductive, and molt responses in relation to phosoperiod.

one dose or four equal doses of 5 hours should respond equally. In the first experiment the birds were subjected to 5-hour photoperiods followed by 1-hour dark periods (*5L*— *1D*). The schedules of Kirkpatrick and Leopold, *9L*—*7D*—*1L*—*7D*, and Jenner and Engels 8.25*L*—*7D*—1.75*L*—*7D* both involve interruption of a 14-hour night by a brief light period. In our test of this situation we employed a 8*L*—7.25*D*—1.5*L*—7.25*D* schedule resembling the preceding "interrupted night" experiments and an 8*L*—8*D* schedule. Though the daily ration of light was 9.5 hours in one case and alternately 8 and 16 hours in the second, both of these schedules were highly stimulating. Thus, breaking the long night period precludes its inhibitory regulatory influence. An alternative proposal has been put forth by Farner, Mewaldt, and Irving (1953a, b). They suggested that the daily dark period *per se* has no positive function; rather, there is a persistent carry-over period which follows the end of each photoperiod, and the effective part of a photoperiodic schedule is the photoperiod plus the carry-over period. They envisioned the duration of the carry-over period to be a function of the duration and

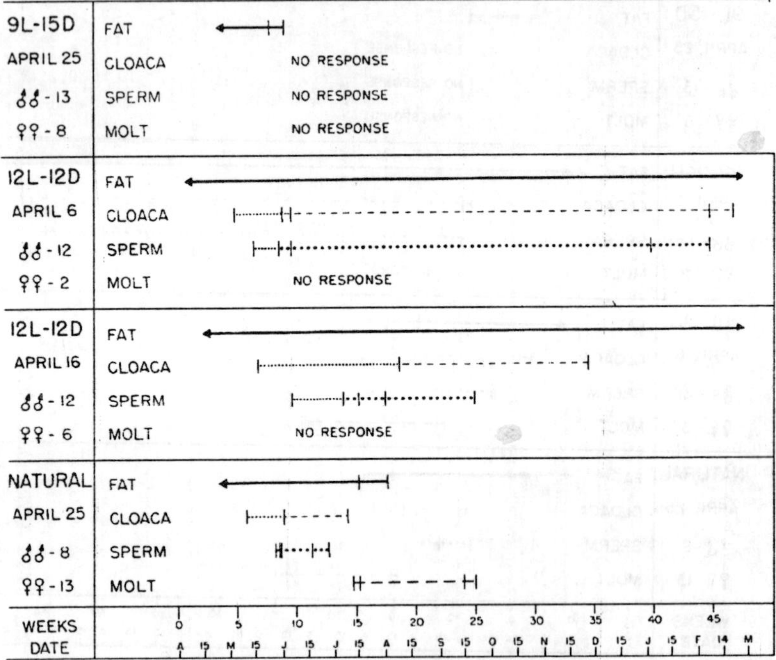

Fig. 13.7. Comparison of fat, reproductive, and molt responses in birds treated with 12L-12D, beginning April 6 and April 16.

nature of the preceding photoperiod. Their interpretation of the rate of testicular response as a function of the summated daily gonadotropic effects was in agreement with the summation hypothesis.

Experiments designed to test this thesis with *Z. albicollis* in our laboratory have been, in general, confirmatory. The schedule 1*L*—2*D*, which, in sum, equals the nonstimulatory 8*L*—16*D* simple light ratio, was strongly stimulating. This appears to rule out the mere ratio of light to dark as a controlling factor. However, it does not favour either of the two alternative interpretations that can be drawn from it: the carry-over hypothesis of Farner et al., or the interruption of inhibitory long nights as suggested by Kirkpatrick and Leopold, and others.

As a further test of the intrinsic properties of long nights we tried a (1*L*—0.25*D*) × 7—1*L*—14.25*D* schedule. This, in 24 hours, provided as much light as the stimulatory 1*L*—2*D* schedule, but it proved nonstimulatory. Neither did this schedule maintain breeding condition in birds already in the reproductive phase. Unfortunately,

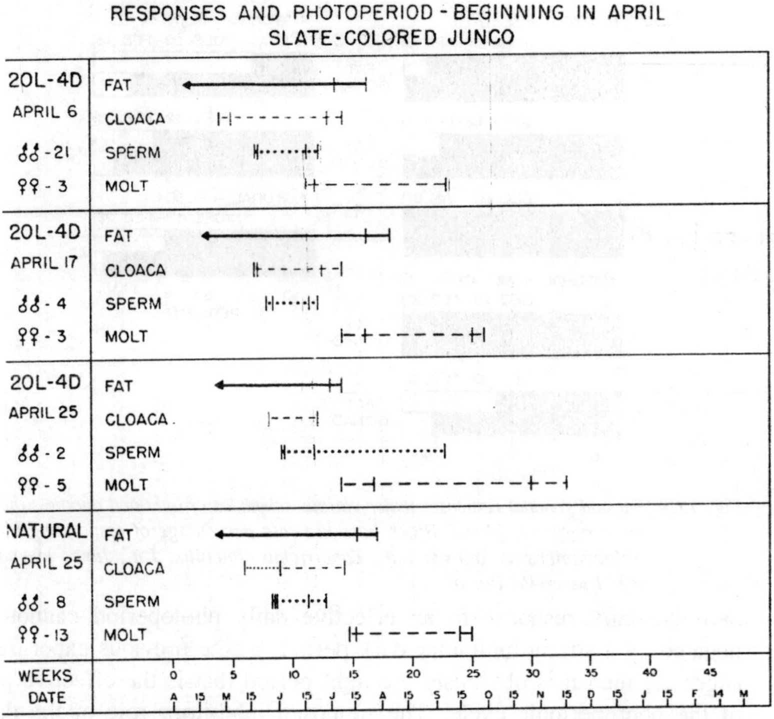

RESPONSES AND PHOTOPERIOD - BEGINNING IN APRIL
SLATE-COLORED JUNCO

20L-4D
APRIL 6
♂♂-21
♀♀-3
FAT
CLOACA
SPERM
MOLT

20L-4D
APRIL 17
♂♂-4
♀♀-3
FAT
CLOACA
SPERM
MOLT

20L-4D
APRIL 25
♂♂-2
♀♀-5
FAT
CLOACA
SPERM
MOLT

NATURAL
APRIL 25
♂♂-8
♀♀-13
FAT
CLOACA
SPERM
MOLT

WEEKS 0 5 10 15 20 25 30 35 40 45
DATE A 15 M 15 J 15 J 15 A 15 S 15 O 15 N 15 D 15 J 15 F 14 M

Fig. 13.8. Comparison of fat, reproductive, and molt responses in birds treated with 20L-4D beginning April 6, 17, and 25.

although this experiment appears to favour the long night inhibition hypothesis, it still does not rule against the "carry-over" idea, since quarter-hour interruptions of the 8*L* might not be considered long enough to allow for effective carry-over of light stimulus.

Further, we tested the following complex schedule (1*L*—2*D*) × 7— 1*L*—16*D*, an alternation of the 1*L*—2*D* stimulatory program with long nights. It proved stimulatory, and may, perhaps, be interpreted as denying the inhibitory function of long nights. However, this 38-hour cycle may not be sufficiently comparable to a 24-hour day to permit extrapolation. However, the effective light period, despite interruptions with dark periods, was 22 hours long and the dark period was 16 hours long. The proportion of light to darkness in this 38-hour cycle would have favoured a response. A similar view may be taken of the 12*L*—16*D* schedule which we found gonad-stimulating.

Experiments are now in progress employing longer dark periods, for example, 12*L*— 20*D*; 16*L*— 22*D*, 16*L*— 32*D*. If such studies show

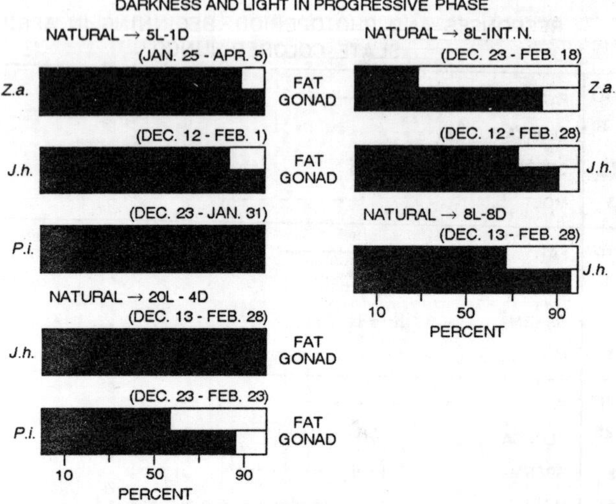

Fig. 13.9. Fat and gonadal responses under various schedules of light and darkness during the progressive phase. Black bars indicate percentage of responding birds. Abbreviations as follows: Z.a., Zonotrichia albicollis; J.h., Junco hyemalis; P.i. Passerella iliaca.

that the daily response to an effective daily photoperiod cannot be negated by a strong inhibitory dark period, as the previous experiment suggests, then it is obviously the light period that is the effective part of the photoperiodic cycle. The supposed inhibitory role of the dark period would then be merely a matter of semantics. A long night would be inhibitory only because it did not permit a longer or effective duration of light (photoperiod plus carry-over period).

Two additional schedules of light and darkness have been used: 16*L*—16*D* and 8*L*— 16*D*— 16*L*— 8*D*. The former was designed to test the inhibitory nature of a 16-hour dark period in a 32-hour cycle, the latter to test whether an effective daily response could be stored when administered on alternate days. Both schedules were effective.

It is still impossible at this point to distinguish precisely between the roles of light and darkness in the regulation of the gonadal and fat responses in the progressive phase. However, some things seem clear: (1) The total amount of darkness in a 24-hour cycle (when administered in small doses) is not the equivalent of the same amount of darkness given in a single dose. (2) Short cycles within a 24-hour period with the same proportion of light and darkness are equivalent to 24-hour cycles with regard to stimulation (5*L*—1*D* equals 20*L*—4*D*), but they are not equivalent with regard to failure to stimulate or "inhibition"

(1*L*— 2*D* and 8*L*—16*D*). (3) With different effective schedules of light and darkness within a 24-hour cycle, there is a difference in rate and duration of response. (4) Whatever the roles of light and darkness, the daily schedule of light and darkness in nature is the critical external factor. Effective daily schedules probably result in physiological "increments of response" which are eventually manifested, or at least become more readily observable and measurable, as gonadal growth, spermatogenesis, fat deposition, and increase in body weight.

Day length and the Refractory Period (Preparatory Phase)

After a period of seasonal breeding activity, the gonads regress spontaneously, sometime in July and August for most north temperate species. During this period, long days or increasing days cannot induce gonadal activity, and, hence, it has been called the refractory period. The natural termination of this period varies with the species, but it occurs usually in October or November. The refractory state can also be produced in the laboratory. Burger (1949) has reviewed the status of the problem of regulation of reproductive cycles and aptly stated that "attention has been focused too narrowly on the progressive phase of the reproductive cycle."

Juncos in nature experience a reduction in day length during the summer and fall, and it was plausible that decreasing day lengths, or short days, or long nights, could regulate the duration of the refractory period. Burger (1947) and Bissonnette showed that treatment with long days in the starling eventually induced a refractory period which could only be dissipated by treatment with short days. Miller (1948, 1951, 1954b) found that treatment with long days, begun in the fall in the migratory golden-crowned sparrow (*Zonotrichia atricapilla*), prevented the occurrence of gonadal growth at the normal time in the spring. This has been demonstrated by us for the junco and white throated sparrow. Birds refractory after long daily treatment with 20 hours of light were made responsive again to 20*L*—4*D* by six weeks of 9- or 12-hour light days. They remained refractory if kept in natural (longer) illumination during these six weeks. In related experiments, birds exposed to 20*L*—4*D*, beginning in the natural refractory period and extending for as long as six months, failed to respond, even in the normal spring breeding season.

These experiments indicated that it should be possible to induce responses several times within one year, even though juncos normally show only one period of gonadal activity and two periods of fat deposition

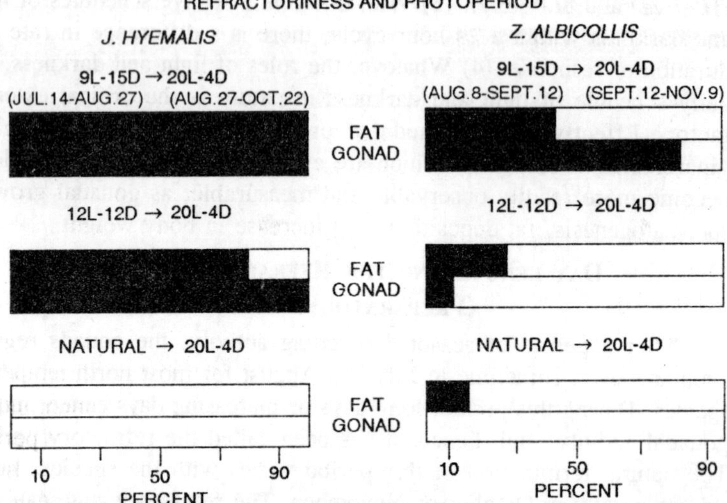

Fig. 13.10. Photoperiod in relation to completion of the preparatory phase.

annually. This was attempted by exposing birds to alternating periods of short days (*9L—15D*) and long days (*20L—4D*). Five periods of gonadal activity, five periods of fat deposition, and two molts occurred within the 369 days. Gonadal growth, fat deposition, and increase in body weight were correlated with long days. Gonadal regression, loss of fat deposits, and decrease in body weight were correlated with short days. These correlations became less distinct as the experiment progressed. Prior to this study, Rowan (1929), Miyazaki (1934), Damste (1947), and Burger (1947) had induced more than one period of gonadal activity in a year in different species of birds.

The results of this experiment confirmed the conclusion that short days in the fall regulate a reaction which enables the bird to respond to subsequent photoperiodic treatment. Without this "preparatory period" the bird does not respond. Although the bird is "refractory" to long days in the fall, it is undergoing a reaction which is regulated by day length (short days) and which is necessary for a subsequent response. Hence, it seems more appropriate to call it the preparatory period or phase. This period is then followed by the progressive phase of the cycle.

Role of Light and Darkness in the Regulation of the Refractory Period

Having established the requirement for short days during the preparatory period, the next problem was the determination of the effective part of the short days. Was it the short photoperiods or the long dark periods? A variety of schedules was used during the natural refractory period which was followed by subsequent treatment with long days (20*L*—4*D*) beginning in December or January.

The results demonstrated that it was not a short period of light *per se*, nor the total amount of separate doses of light and darkness in a given day, nor the proportion of light to darkness in a short cycle that was effective. Rather, it was a single period of darkness, 12 hours long, or longer, in a 24-hour cycle that was the effective part of a short day. A 16-hour dark period was generally more effective than a 12-hour period, which seems to be near the threshold for a minimum effective duration of the dark period.

The next question was whether a 16-hour dark period *per se* was the effective stimulus, or whether a long single dose of light was inhibitory. To answer this, birds were exposed to schedules of 16*L*—16*D* and (1*L*—2*D*) × 7—1*L*—16*D* during the preparatory period and then treated with long days (20*L*—4*D*). Both schedules were ineffective and corresponded to long days. That they were actually long days was established by the response of birds to those same schedules during the winter, as described earlier. The results indicated that a 16-hour dark period *per se* is not the effective stimulus. There is some relation

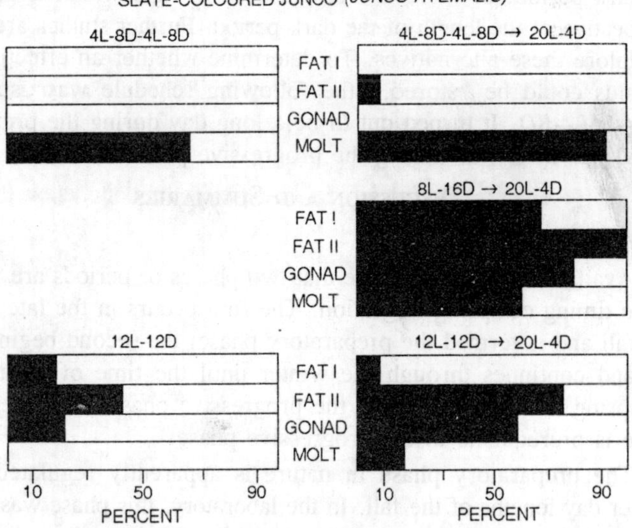

Fig. 13.11. Effectiveness of various schedules of light and darkness in inducing completion of the preparatory phase in the slate-coloured junco.

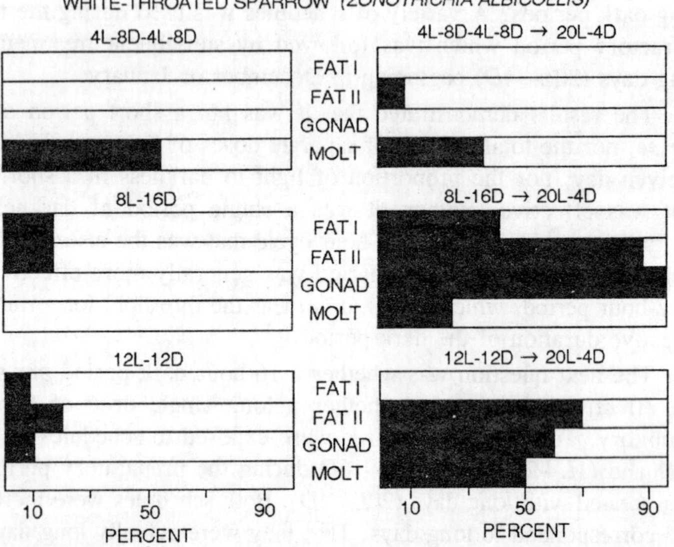

Fig. 13.12. Effectiveness of various schedules of light and darkness in inducing completion of the preparatory phase in the white-throated sparrow.

apparently between the duration of the light period and the duration of the dark period, or there is a single dose of light which is inhibitory irrespective of the length of the dark period. Further studies are needed to explore these alternatives. To determine whether an effective daily stimulus could be "stored," the following schedule was used: 8*L*—16*D*—16*L*—8*D*. It turned out to be a long day during the preparatory period just as it was during the progressive phase.

Discussion and Summaries

Day Length and Migration

Available data demonstrate that two phases or periods are involved in the timing of spring migration. The first occurs in the late summer and fall and is termed the preparatory phase; the second begins in late fall and continues through the winter until the time of migration in the spring and may be called the progressive phase. The preparatory phase is prerequisite to the progressive phase.

The preparatory phase in nature is apparently regulated by the shorter day lengths of the fall. In the laboratory, this phase was induced at different seasons of the year by subjecting birds to short days (9*L*—15*D*). Twelve hours of light per day appears to be near the threshold for the maximum day length which can act like a short day.

Long days (16*L*—8*D*) arrest the preparatory phase and the appearance of the subsequent progressive phase. Therefore, it is the short days of fall that may be assumed to regulate the migratory behaviour and breeding cycle which appear six to seven months later. Relevant data are few, but suggest that there may be some relation between the length of the dark period, the duration of treatment, and the rate at which the preparatory phase proceeds.

After the preparatory phase has been completed, the progressive phase begins. In nature, it probably begins automatically in late November and December when the days are short, but *the rate at which it proceeds* is governed by day length. The spring premigratory physiological state was induced by means of long days about 80 days ahead of its natural occurrence, in a majority of the birds. With a short day (9*L*—15*D*) the same state appeared in about 160 days, which was 40 days later than its occurrence in nature. The near-maximum rate of response occurred with a day length as short as 15.5 hours; the minimum rate occurred with a day length of 9 hours, but shorter day lengths were not tried. Nine hours is about the shortest day to which juncos are exposed in nature.

It is important to emphasize here that the progressive phase, though slowed, developed *when there was no increase in day length*. About 60% of the birds released in June after being held on 9-hour days since December undertook a migration. Increasing day lengths, whether or not they are gradual, are not necessary to induce spring migration. The role of day length, once the birds are ready to respond, is the regulation of the rate at which the response proceeds.

Once in the premigratory physiological state, the length of time birds remain in this state is a function of day length. Long days dissipate the state more quickly than short days, and under short days (12*L*—12*D* and 9*L*—15*D*) the birds may not lose their fat. A change from long days to short days will also dissipate the state rapidly.

In nature, the spring premigratory physiological state ends when the birds arrive on their breeding grounds, or shortly afterward. After the breeding season, the gonads regress, the birds molt, and subsequently, there is a physiological change, which precedes the onset of fall migration. Nothing is known about the factors which regulate this state. When the fall migration gets underway in September and October the day lengths have reached a length which is effective for the beginning of the preparatory phase of the next spring migration. And, thus, a new cycle begins.

Role of Light and Darkness in the Migratory Cycle

The available data suggest that in the preparatory phase, the effective part of the short days is the dark period. It seems likely that there is a dark-dependent response which requires a daily *uninterrupted* dark period of at least 12 hours duration. If the dark-dependent reaction goes on during short dark periods at all, it cannot summate to give an effective daily response, or a response after the lapse of many days. A long light period can negate the positive effect of a long dark period; hence, the interpretation of the role of darkness is provisional.

In the progressive phase, both the light (stimulatory) and dark (inhibitory) periods have been suggested by different groups of workers as the controlling part of the photoperiodic cycle. Our data do not permit a clear-cut choice between the two possibilities. This may be because both light and darkness play important roles in the daily photoperiodic schedule.

An interesting difference between the "light and dark" reactions of the preparatory and the progressive phases is that in the preparatory phase, where the dark period seems to be the critical factor, the effects of shorter dark periods in a 24-hour cycle do not summate to give an effective daily stimulus. In the progressive phase, where the light appears to be the most critical factor, the effects of short light periods do summate to give an effective daily stimulus when there is no inhibitory duration of darkness in a 24-hour cycle. In the preparatory phase a long light period *per day* appears to be inhibitory, whereas in the progressive phase, a long dark period *per day* appears to be inhibitory. In both phases, the data show clearly that the 24-hour cycle is highly significant. It is possible that an inherent 24-hour rhythm, or other innate rhythms in the bird, are related to the migratory and reproductive responses to photoperiodic schedules and this must be explored in future studies.

Day Length and Reproduction

The regulation of the reproductive cycle in the male also occurs in a preparatory phase and progressive phase. The role of day length in each of these phases is similar to that in migration, except for the much slower rate of gonadal response to short days.

The duration of reproductive activity and, hence, the time of regression is also regulated by day length. The shortest durations were produced by long days, the longest by 12-hour days. The long days of summer probably initiate and maintain refractoriness. Some data suggest that regression may be an innate phenomenon, whereas refractoriness may be a feature induced by the long days of summer in temperate latitudes.

The role of light and darkness in the reproductive cycle is generally similar to that in the migratory cycle.

Day Length and the Timing of Spring Migration in Equatorial and Transequatorial Migrants

The relatively constant day lengths of the equatorial region can no longer be regarded *a priori* as nonregulatory. Constant photoperiods of 12 hours have been shown experimentally to be effective, and the duration of the photoperiod, moreover, regulates the rate of response. Hence, birds wintering in the equatorial region could be responding to the relatively constant day lengths of 12 hours. The birds that cross the equator and winter in the subtropics or temperate regions of the southern hemisphere are exposed to gradually increasing day lengths after they arrive in late October or November and to gradually decreasing day lengths after December 21, which reach a length of about 12 hours on March 21. Therefore, they are exposed during their entire stay on the wintering grounds to long days which, although they increase gradually to a maximum and then decrease gradually to 12 hours, would remain at an effective photoperiodic level, judging from our experimental work with juncos.

However, the main problem in equatorial and transequatorial migrants is, perhaps, not the effect of the day lengths on the wintering grounds, but the relation between day length and the preparatory phase and the initiation of the progressive phase. Since juncos could complete the preparatory phase on 12-hour days in late summer, it seems likely that equatorial and transequatorial migrants would be exposed to a sufficient number of 12-13-hour days during the fall migration to complete their preparatory phase. It is interesting to note that in nature, juncos winter where they will experience dark periods which last longer than 12 hours; spontaneous initiation and rapid development of the progressive phase are thus assured. The regulation of the preparatory phase and of the initiation of the progressive phase is the critical problem in the regulation of spring migration in equatorial and transequatorial migrants.

Day Length and Breeding Cycles in the Tropics

It has not been possible to extrapolate directly from experimental studies of the temperate zone gonadal cycle to avian breeding cycles in the tropics. Studies of breeding cycles at all latitudes led Baker (1938) to the conclusion that "the main proximate causes of the breeding seasons of birds in nature are thought to be temperature and the length of day in the boreal and temperate zones, and rain and/or intensity of insolation near the equator." Recently, a re-examination of the problem led to the same general conclusion that day length and temperature in the higher latitudes and humidity and rainfall in the tropics are correlated with the breeding seasons. The significance of these climatic factors is believed to lie in their effect on food supply. Breeding seasons are regarded as an adaptation and apparently are timed so that the young can be reared when food supply is at a maximum. If the breeding season is, indeed, adapted to environmental conditions operating toward its close, other factors must be postulated for the initiation of the cycle. Although day length and temperature are acceptable for higher latitudes, in the tropics both "are too nearly constant to offer a possible explanation". Our experimental findings, however, suggest that day length should not be ruled out as a regulatory factor in the tropics simply because it is relatively constant. The data from two series of experiments and other observations show that the gonadal and molt cycles of some tropical and equatorial birds, African whydahs and weavers (*Steganura*, *Euplectes*, and *Vidua*), can be altered by changes in day length.

A number of other correlations point to a relationship between day length and reproduction in the tropics: (1) In many tropical species which occur on both sides of the equator the breeding periods in the northern and southern populations are correlated with the seasons and, hence, occur at opposite times of the year. Frequently this is associated with the wet or dry season, but often it is not. (2) In many groups of birds, and even in a species with wide distribution, clutch size tends to be smaller in the tropics than in the temperate latitudes. In the domestic fowl egg laying (which is not homologous to clutch size) is greatly influenced by latitude. (3) The maximum size of active testes in tropical species is only 8 to 67 times the size of minimum inactive testes, whereas in temperate species it ranges from 267 to 2096 times the minimum size.

Miller (1954a) has shown that in the uniform environment of the Magdalena Basin in Colombia (3° 12' north latitude) eight out of ten species studied showed acyclic and uncoordinated breeding; two species showed coordinated and cyclic breeding. All of the species showed "evidence of possessing the same innate mechanism basic to cyclic breeding as north temperate species. This consists of need for rest, or assumption of a refractory state, and a basic tendency to progressive recrudescence." Miller (1954a) regards tropical day length, or light stimuli, as sufficient to surpass threshold needs in the innate mechanism, and as a regulator of the duration of refractoriness, rate of recrudescence, and duration of breeding condition.

Marshall and Disney (1956) have tested the response of an equatorial bird, *Quelea quelea*, to increased photoperiods, and found some response to long days.

Keast and Marshall (1954) have made extensive studies of reproduction and breeding seasons in relation to drought and rainfall and found that the gonads may remain inactive for a succession of seasons during a prolonged drought. Desert species can respond quickly to rainfall, or its effects, and nesting may begin within a few days of heavy precipitation, irrespective of day length and light increment. Marshall and Disney (1957) on the basis of an experimental study of the induction of the breeding season in a xerophilous species concluded that an internal rhythm of reproduction exists which is modifiable by external conditions, including possibly rainfall and social stimulation. In their opinion it is unlikely that the breeding seasons of truly equatorial vertebrates are controlled by photoperiodicity, and many equatorial and other species have evolved a reproductive response to rainfall or its effects. The data enabling one to evaluate conclusions with respect to the effect of rainfall on reproduction unfortunately are still conspicuously few.

In summary, the relation between day length and the gonadal cycle in tropical species is not known, but the point that must be emphasized is that day length cannot be ruled out *a priori* as a fundamental regulator of migration and breeding cycles in the tropics simply because it is relatively constant. There is no reason to believe that all of the progressive and regressive aspects of reproduction will be regulated by day length in precisely the same manner in all species. Nor must all of these aspects be regulated by day length. Regulation of a few is all that is needed for day length to be a primary factor in the timing of breeding seasons. Other environmental factors, such as rainfall and diet, and psychic factors may prove to be important modifying factors with different degrees of effectiveness in the preparatory, progressive, and regressive, phases. Only day length has been shown so far to be a primary regulatory factor, and its relation to gametogenesis and reproductive rhythmicity in tropical species remains to be determined by extensive experimental studies.

Neuroendocrine Mechanisms in Relation to Regulation of Migration

From the excellent and ingenious experiments of Benoit and his collaborators primarily, the mechanism whereby light induces reproductive activity during the progressive phase in the duck is thought to be as follows. Receptors for light are in the retina or in the encephalic regions near the orbit, the hypothalamus and the rhinencephalon. Stimuli from receptors reach neurosecretory cells of the paraventricular and supraoptic nuclei in the hypothalamus, but the pathway from the receptors to the neurosecretory cells is not known. From the hypothalamus, Gomori-positive neurosecretory material moves down the axons to the median eminence and the capillaries of the hypothalamo-hypophyseal portal system. When the material reaches the anterior lobe of the pituitary, the pituitary secretes gonadotropins which induce development of the gonads. During dark periods, the gonadotropins are retained in the hypophysis.

With this kind of mechanism, the different rates of gonadal response to different photoperiods could be explained by different rates of synthesis and secretion of gonadotropins. But how can we explain rapid responses when only 8 hours of light in 1-hour doses are given per 24 hours and no response when the 8 hours of light are given in a schedule of 8*L*—16*D*? Farner, Mewaldt, and Irving (1953) have suggested that the response to light may involve a process which becomes active almost immediately and which continues to exert its effect after the lights go off. Therefore, the daily gonadotropic effect would be the summated gonadotropic effects of the photoperiods plus the periods which occur in the darkness and called the "carry-over' periods. With only one 8-hour period of light and one carry-over period there apparently is insufficient gonadotropic effect.

With an additional light period during the night, a greater gonadotropic effect is produced. Another way of thinking of it is that interruption of the dark period (before the effect of an 8-hour light period falls below a threshold value) produces an effect as though the light period had not ended. Obviously, very little total light is needed each day to give an effective gonadotropic stimulus, but this light must be administered in such a way that the duration of a single dark period is not 14 hours or longer in the junco. The effective stimulus appears to be the light; darkness seems to be inhibitory only in that it prevents light from acting.

After activity has been maintained for some time, the gonads regress. This regression could be induced by a feedback mechanism involving the sex hormones, the pituitary, and the hypothalamus, but attempts to inhibit the "light response" of the gonads with sex hormones have not been completely successful. Farner (1959) reports little or no gonadal suppression with testosterone or progesterone. Kobayashi (1954), on the other hand, was able to inhibit the photoperiodic gonadal response in the white-eye (*Zosterops japonica*) with either testosterone or estradiol.

More important, perhaps, in this connection are the experiments of Bailey (1950) and Lofts and Marshall (1956) with prolactin. Bailey found that prolactin inhibited the photoperiodic gonadal response in *Zonotrichia leucophrys pugetensis*, and Lofts and Marshall induced testicular involution in the house sparrow (*Passer domesticus*), the greenfinch (*Chloris chloris*), and the chaffinch (*Fringilla coelebs*). Hence, prolactin could be responsible for the cessation of gonadal activity at the end of the breeding season.

The cause of refractoriness is not known, but it seems to reside at the hypothalamo-hypophyseal, and not the gonadal level, since administration of gonadotropins to either intact or hypophysectomized birds during the refractory period induces gonadal development. With regard to the pituitary, it is known to contain gonadotropic hormones during the fall, so refractoriness may be at the hypothalamic level if it controls secretion of gonadotropins from the pituitary. Kobayashi (1957) defines the refractory period as the period during which the hypophysis is secreting enough thyrotropin to induce molt, but in our laboratory we find birds refractory many weeks after the molt has ceased.

The preparatory phase might conceivably be a period of synthesis of gonadotropins in the pituitary, the mechanism requiring long periods of darkness each day, or probably more correctly, being inhibited or arrested by stimulatory doses of light. When sufficient quantities of gonadotropins have been synthesized in the pituitary, then perhaps secretion under hypothalamic control may be initiated during the progressive phase. The long days of summer, or of experimental treatment, possibly inhibit synthesis of gonadotropins, and, hence, prevent indefinitely the occurrence of the preparatory phase. A short treatment on short days perhaps returns the pituitary to synthesis of gonadotropins which then can be secreted under the regulation of the hypothalamus and photoperiod. That the situation is probably not so simple, is indicated by Vaugien (1955b) who showed that in immature English sparrows there is a gradual increase in responsiveness during the fall.

The role of the hypothalamus as a possible regulatory center of the mechanisms involved in the production of the migratory physiological state has been under consideration for some years, but we are still a long way from understanding the mechanisms whereby it exerts this control. Nevertheless, it seems likely now that the hypothalamus is the primary target organ of the external factor, the daily durations cf light and darkness, and that the interaction between the external factor and the hypothalamus results in the occurrence of spring migration and breeding periods at the proper time of the year.

INDEX